LAWS AND OTHER WORLDS

# THE UNIVERSITY OF WESTERN ONTARIO SERIES IN PHILOSOPHY OF SCIENCE

A SERIES OF BOOKS
IN PHILOSOPHY OF SCIENCE, METHODOLOGY,
EPISTEMOLOGY, LOGIC, HISTORY OF SCIENCE,
AND RELATED FIELDS

VOLUME 31

FRED WILSON

*Dept. of Philosophy, University of Toronto, Canada*

# LAWS AND OTHER WORLDS

## *A Humean Account of Laws and Counterfactuals*

D. REIDEL PUBLISHING COMPANY

A MEMBER OF THE KLUWER  ACADEMIC PUBLISHERS GROUP

DORDRECHT / BOSTON / LANCASTER / TOKYO

Library of Congress Cataloging-in-Publication Data

Wilson, Fred,
   Laws and other worlds.


This work is cataloged under LC No. 86–14612.



CIP-data appear on separate card
ISBN 90–277–2232–3

---

Published by D. Reidel Publishing Company
P.O. Box 17, 3300 AA Dordrecht, Holland

Sold and distributed in the U.S.A. and Canada
by Kluwer Academic Publishers,
101 Philip Drive, Assinippi Park, Norwell, MA 02061, U.S.A.

In all other countries, sold and distributed
by Kluwer Academic Publishers Group,
P.O. Box 322, 3300 AH Dordrecht, Holland

Printed in The Netherlands

To Carolyn Siobhan
who, at age $3\frac{1}{2}$, said of the manuscript
"If I would have written that,
it would stand by itself"
and who may well have been right.

# TABLE OF CONTENTS

PREFACE                                                                    ix

ACKNOWLEDGEMENTS                                                           xv

CHAPTER ONE / SCIENCE AND SCIENTIFIC EXPLANATION            1
  I     The Aims of Science                                               1
  II    Ideals and Research                                             28
CHAPTER TWO / LAWS, ACCIDENTAL GENERALITIES AND
              COUNTERFACTUAL CONDITIONALS                      72
  I     Introduction: Counterfactuals and the Humean
        Account of Laws                                                72
  II    Accidental Generalities                                        99
  III   Laws as Relations among Universals: A Recent
        Defence of Necessary Connections                             111

CHAPTER THREE / POSSIBLE WORLDS: A DEFENCE OF HUME         133
  I     Introduction                                                  133
  II    The Possible Worlds Account of Counterfactuals               134
  III   How Possible Are Possible Worlds?                            139
  IV    Nomic and Causal Dependence                                  150
  V     Similarity among Possible Worlds                             155
  VI    Lewis' Regularity View of Laws                               168
  VII   Are There Counterfactual Conditionals that Involve
        No Laws?                                                      172
  VIII  The Case for Lewis' Analysis Examined                        195
        (i)    Semantics of Counterfactuals                          195
        (ii)   Causal Discourse                                      224
        (iii)  Causal Pre-emption                                    234
        (iv)   Which Belief-Contravening Assumptions Are
               Reasonable?                                           243
        (v)    Counterfactuals and Inductive Support                 280
  IX    Conclusion                                                   288

NOTES                                      291

BIBLIOGRAPHY                               315

INDEX OF NAMES                             322

INDEX OF SUBJECTS                          324

# PREFACE

Problems in the philosophy of language have the habit of not remaining just such problems. The problem of counterfactuals has been a standing problem in the philosophy of language ever since the classic discussions of Chisholm ('The Contrary-to-Fact Conditional') and Goodman ('The Problem of Counterfactual Conditionals') — though the problems had been discussed by philosophers such as Hume, J. S. Mill and Bradley, and even, in his own way, by Aristotle. There has been a tendency recently for approaches to the problem of counterfactuals to divide into two streams. In fact, the two streams are there at the origin of the contemporary discussions. One stream derives from Chisholm's approach and takes its inspiration from modal lgoic and set theory. Often enough this approach involves — as with Aristotle — an assumption of non-truth-functional connectives to define unanalyzable natural necessities, or defines such connectives and such necessities in terms of an ontology of possible worlds. The other stream derives from Goodman's approach and takes its inspiration from philosophy of science, most often an empiricist philosophy of science.

From the approach of modal logic, the empiricist philosophy of science is far too meagre in its resources to adequately analyze the logic and semantics of counterfactuals. From the approach of empiricist philosophy of science, ontologies of natural necessities and possible worlds are far too problematic to provide a *philosophically* adequate account of counterfactuals.

The present essay firmly locates itself in the second stream, arguing, on the one hand, that the other approach is philosophically inadequate and incapable of analyzing counterfactual conditionals, while arguing on the other hand that an analysis flowing from an empiricist philosophy of science can adequately solve the problem of counterfactuals.

This essay is written in the conviction that there are no necessities other than the logical and no worlds other than the actual.

It defends a Humean account of laws against the natural necessities of the rationalists. It analyzes counterfactual conditionals as condensed arguments in which Humean laws function as major premisses. It defends this analysis of counterfactual conditionals against those who argue that the logic of such conditionals can be adequately explicated only in terms of an ontology of possible worlds.

The Humean account of laws and of counterfactuals is, according to the approach deriving from Goodman, closely tied to many empiricist claims about the nature of science and of the scientific method, for example, the claim that the deductive-nomological model of explanation adequately captures the logic of scientific explanation. This essay therefore briefly sketches an empiricist philosophy of science. This is not merely descriptive of science, but descriptive of science at its best; it is a *normative* account of science. These norms are justified by reference to the cognitive interests that motivate our concern for science. These same interests define the context in which counterfactual conditionals find their major use. It is by reference to these cognitive interests that the argument is made that the Humean account of counterfactuals as conditionals sustained by laws is the analysis that one *ought* to accept.

Chapter I, "Science and Scientific Explanation," presents the empiricist account of science and scientific method. Section I, "The Aims of Science," has as its primary concern the statement and defence of the deductive-nomological model of explanation and its corollary, the thesis that explanation and prediction are symmetric. Section II, "Ideals and Research," defends certain notions of content and scope as the ideals of scientific explanation of individual facts and laws respectively. It also shows how these ideals generate research tasks, and how the latter are often guided by theories. In particular, it shows how an empiricist account of research ties in with certain views of Kuhn about the nature of science.

These two sections re-state briefly, and in a way relevant for our present purposes, claims that I have argued for in the greater detail elsewhere, in my *Explanation, Causation, and Deduction,* (Western Ontario Series in Philosophy of Science, D. Reidel, 1985) and my *Empiricism and Darwin's Science* (in preparation). The former deals largely with the deductive-nomological model of explanation, while the latter deals with theory structure and the nature of research. Scientific ideals are discussed in detail in both.

Chapter II, "Laws, Accidental Generalities, and Counterfactual Conditionals," develops the Humean account of laws and explores the connections between this analysis of laws and the analysis of counterfactual conditionals. The basic ideas are presented in Section I, "Introduction: Counterfactuals and the Humean Account of Laws." The basic proposal is the familiar one that a counterfactual conditional:

> if this were $F$ then it would be $G$

is a condensed argument

> All $F$ are $G$
> This is $F$
> _____
> This is $G$

where the generality is a law, and only implicit in the context, while the minor premiss and the conclusion are not asserted, i.e., are either denied or suspended. The case of Locke and Hume against objective necessary connections is sketched. This argument is based on an appeal to a Principle of Acquaintance. But to criticize one account of laws is not yet to have a positive position. The positive Humean account of laws is also sketched. Hume's proposal is first, that what distinguishes a law from an accidental generality is *not* something objective: in terms of truth-conditions, both are of the form

> All $F$ are $G = (x)(Fx \supset Gx)$

What distinguishes them, rather, is something *subjective*, specifically, the different assertive attitudes adopted towards them. A generality is asserted as a law (briefly: law-asserted) just in case that one is prepared to use it to predict and to support counterfactual conditionals. If one is not prepared so to use it, then it is accidental. The Humean criterion for lawfulness is thus *contextual* rather than *ontological*. Hume's proposal is, second, that a generalization is *worthy* of acceptance as a law, i.e., for purposes of explanation and prediction, and assertion of counterfactuals, just in case that the evidence testifying to its truth has been acquired in accordance with the rules of the scientific method. But ought this proposal be accepted? The Humean defence of this proposal is briefly sketched, that is, the Humean defence of the claim that it is reasonable to adopt the rules of sceince and not those of, e.g., superstition, as those we use to justify the acceptance of laws. (Some of the relevant points are treated in greater detail in my 'Hume's Defence of Causal Inference'.)

The Humean criterion for lawfulness is that those generalities that are laws can, whereas those that are accidental cannot, support the assertion of counterfactual conditionals. Section II, "Accidental Generalities," attempts to sketch what it is that accidental generaliteis lack that renders them incapable of sustaining the assertion of counterfactual conditionals. The Humean holds that, objectively considered, there is no difference between those generalities that are laws and those that are accidental generalities: both are of the form

$$\text{All } F \text{ and } G = (x)(Fx \supset Gx)$$

The criteria for a deductive-nomological explanation that are defended in Chapter I, Section I are purely objective: deduction from a true generality is necessary and sufficient for explanation — though not, of course, for the best explanation. Given these criteria, it would seem that, objectively considered, accidental generalities are as good as laws at explaining. Equally it would seem that, objectively considered, accidental generalities are as good as laws at predicting. But if laws and accidental generalities are, objectively considered, equally as good at explaining and predicting, then it would seem that they should be equally as good at supporting counterfactual conditionals. The Humean distinction between laws and accidental generalities would thus seem to collapse. But, on the other hand, this is precisely what one would expect, since, after all, the Humean holds that, objectively, there is no such distinction. On the other hand, we *do* have such a distinction, and it must be accounted for. The point to be made about prediction is that while an accidental generality, *if* true, is as good a predictor as a law, the "if" here is of the essence: we in fact often *do not know* whether the generality we treat as accidental is true, and in fact *treat it as accidental, unworthy of being relied upon as a predictor, precisely because the evidence available testifies that we ought not to accept it as reliable.* As for the support of counterfactual conditionals, what we suggest is that the generalities we accept as accidental are such that, *when the contrary-to-fact assumption is made then, in the context of accepted and law-asserted background knowledge that is not held in abeyance, it turns out that the generality cannot consistently be used to deduce the consequent of the counterfactual.*

Involved here are two sets of rules. There are, first, the rules for use in explanations and predictions. These have already been examined in Chapter I, Section II, but a further point remains which we take up in Section III, the final section of the part, which critically evaluates a

recent anti-Humean argument to the effect that the very rules of
scientific method that the Humean relies upon to distinguish science
from superstition presuppose a non-Humean account of laws in terms
of natural necessities.

But there is a second set of rules that must also be taken into
account, namely, those that govern counterfactual conditionals, and, in
particular, govern which belief-contravening assumptions are legitimate
and which are not. We delay a discussion of these rules until Chapter
III, Section VIII (iv), where it is argued, first, that if we refer to the
cognitive interests which motivate science and which, very often at
least, define the contexts in which counterfactuals are used, then certain
rules for belief-contravening assumptions are reasonable; and, second,
that these rules make sense of much of our ordinary use of counterfactuals and of our ordinary distinctions of laws and accidental
generalities. Perhaps not surprisingly, these rules turn out to presuppose the norms of scientific method for determining which generalities
are worthy of acceptance for explanation and prediction.

The discussion of these rules is delayed until Chapter III, where their
point easily becomes clear in the context of the philosophical dialectics
aimed at defending the Humean account of laws and counterfactuals by
criticizing the account of counterfactuals in terms of a "possible world"
semantics.

Chapter III, "Possible Worlds: A Defence of Hume," takes up the
wide variety of themes associated with the "possible worlds" account of
counterfactual conditionals. The basic contrast to the Humean position
on laws and explanation is brought out in Section I, "Introduction." In
Section II, "The Possible Worlds Account of Counterfactuals," the
position of David Lewis, one of the clearest and most forceful of the
possible worlds accounts, is laid out. Section III, "How Possible Are
Possible Worlds?" criticizes Lewis' realism about possible worlds,
rejecting the idea that there are unrealities that are nonetheless actual.
But it also argues that, some objections of Lewis notwithstanding, the
notion of "possible worlds" can be given a *linguistic explication* that
renders it ontologically innocuous and *philosophically unproblematic*
for the empiricist. With this explication, the possible worlds account of
counterfactuals enters the same ring as the Humean account sketched
in Chapter II. The question then becomes, which of the two accounts is
true?

Before answering this question, it is necessary to discuss Lewis' view of causation. Lewis' position on this matter is laid out in Section IV, "Nomic and Causal Dependence." It turns out that Lewis' thesis about the truth-conditions of counterfactuals and his views about laws turn on the idea of "similarity" among possible worlds. This idea is examined in Section V, "Similarity among Possible Worlds." It turns out that the idea is seriously ambiguous; contrary to Lewis, there are in fact several scales of inter-world similarity. Lewis distinguishes causal judgments from laws. The former are understood in terms of counterfactuals; laws are regularities embedded in theories. This separation of laws and counterfactuals is contrary to the Humean position. In Section VI we examine Lewis' criterion for lawfulness, and find it inadequate.

The main point remains, however. For Lewis, counterfactuals do not require laws to sustain them; for the Humean, laws are indispensable. The issue is addressed explicitly in Section VII, "Are There Counterfactual Conditionals that Involve No Laws?" The question it asks is answered negatively. Or rather, this Humean answer is defended by arguing that, in contexts defined by the cognitive interests that motivate science, one *ought not* to assert a counterfactual conditional *unless* it is sustained by a law.

Finally, in Section VIII, "The Case for Lewis' Analysis Examined," we evaluate — and find wanting — the arguments given to defend the possible world analysis of counterfactuals. Subsection (i) looks at claims that certain features of the logic and semantics of counterfactuals are better accounted for by Lewis' analysis than the Humean's. Subsection (ii) examines claims that certain features of causal discourse are better handled by Lewis' account than the Humean's. Subsection (iii) looks at an example of a complex causal process that Lewis calls "pre-emption" which he claims his analysis can handle but the Humean's cannot. Subsection (iv) discusses the problem of what Goodman has called "cotenability", and in this context attempts to sketch the norms that determine which belief-contravening assumptions are legitimate. Certain norms are defended as reasonable relative to the cognitive interests that motivate science and which define the contexts in which counterfactuals are used. These norms, we suggest, do in fact make sense of much of our counterfactual discourse. Finally, Subsection (v) examines a particularly intriguing example that Lewis claims refutes the Humean's position.

A brief "Conclusion" (Section IX) ties the results at which we have arrived once again to the empiricist philosophy of science presented in Chapter I.

# ACKNOWLEDGEMENTS

Materials from the following papers and book have been included in this book.

F. Wilson: 'Hume's Theory of Mental Activity,' in D. F. Norton, N. Capaldi, and W. Robison (eds.), McGill Hume Studies, Austin Hill Press, San Diego, 1979; copyright © 1979, Austin Hill Press; reproduced here by kind permission of the editors and publisher.

F. Wilson: 'The Lockean Revolution in the Theory of Science,' in S. Tweyman and G. Moyal (eds.), *Early Modern Philosophy: Epistemology, Metaphysics and Politics*, Caravan Press, Delmar, N.Y., 1986; copyright © 1986, Caravan Press; reproduced by kind permission of the editors and publisher.

D. Lewis: *Counterfactuals*; copyright © 1973, D. Lewis, Harvard University Press, Cambridge, Mass. and Basil Blackwell, Oxford; reproduced here by kind permission of Harvard University Press, Basil Blackwell, and David Lewis.

Editors and publishers are thanked for permission to reprint.

## CHAPTER ONE

# SCIENCE AND SCIENTIFIC EXPLANATION

## I. THE AIMS OF SCIENCE

Thomas Taylor, the English Platonist and friend of William Blake, but also of Thomas Love Peacock, was an odd sort of character. He invented a perpetual lamp, the explosion of which upon its first public demonstration was a near disaster, all but burning down the Freemasons' Tavern where the demonstration occurred. In the introduction to his translation of Proclus' commentaries on Euclid, he in passing squared the circle. He was no doubt the last serious defender of the neo-Platonic view on the explanatory power of pure mathematics. His influence was slight in his native Britain, but for some reason he was influential in the new United States. But if his positive views were even in his own time thoroughly odd, he was not without critical insights into the views he felt himself to be combatting. In particular, I think the following remark he makes on Locke in the introduction to his Proclus-translation is acute.

On Mr. Locke's system, the principles of science and sense are the same, for the energies of both originate from material forms, on which they are continually employed. Hence, science is subject to the flowing and perishable nature of particulars; and if body and its attributes were destroyed, would be nothing but a name. But on the system of Plato, they differ as much as illusion and reality; for here the vital, permanent, and lucid nature of ideas is the foundation of science; and the inert, unstable, and obscure nature of sensible objects, the source of sensation . . . . In the ancient [system], you have truth itself, and whatever participates of the brightest evidence and reality: in the modern, ignorance, and whatever belongs to obscurity and shadow. The former fills the soul with intelligible light; the latter clouds the intellectual eye of the soul, by increasing her oblivion, strengthens her corporeal bonds, and hurries her downwards into the dark labyrinths of matter.[1]

Taylor, of course, believes Locke to be utterly wrong, while, for myself, I believe Locke to be, if not utterly right, at least quite clearly on the correct side of that line that divides truth and enlightenment from error and superstition. But Taylor and I agree that what is crucial for Locke is that "science and sense are the same." And in fact what I would want

1

to argue is that Locke's metaphysics, which identifies science with sense, is one important ingredient in the complex of causal factors that led to the historical emergence, as a living fact, of an intellectual understanding of the empirical nature of science.[2] But however this latter may be, it certainly is the case that Taylor captures in an interesting way the distinction between pre-scientific or non-scientific explanations and scientific explanations of empirical phenomena.

Non-scientific explanations seek to account for what occurs in the world of sense experience in terms of an entity or entities that lie outside this world. The temporal world of sense experience, the world of flowing and perishable phenomena, is to be explained in terms of an entity that transcends such a world, an entity that is, as Taylor says, permanent, i.e., an entity that is outside of time. It is knowledge of such timeless entities that, in Russell's terms, the mystic seeks when he aims to know the world *sub specie aeterni*. For the empiricist, of course, "the principles of science and sense are the same", and the entities the mystic yearns to know are excluded from his ontology. How such exclusion might be justified, we need not go into now. Our present task is not to give *that sort* of justification of empiricist reason.[3] For what we are about, it suffices to note, with Russell, that, while empirical science eschews any appeal to the *entities* that the mystic introduces into his metaphysics, it nonetheless does not eschew the *aim* of the mystic, the aim of understanding the world *sub specie aeterni*. "Metaphysicians," Russell says, "have frequently denied altogether the reality of time. I do not wish to do this; I wish only to preserve the mental outlook which inspired the denial, the attitude which, in thought, regards the past as having the same reality as the present and the same importance as the future."[4] And he then quotes Spinoza: "In so far as the mind conceives a thing according to the dictate of reason, it will be equally affected whether the idea is that of a future, past, or present thing."[5] But in seeking to view things *sub specie aeterni* the scientist does not seek a timeless permanence that is beyond experience: the scientist seeks rather a permanence *in* experience. The scientist seeks to view the world of sensible phenomena past, present and future as indifferently exhibiting a *timeless pattern*. That is, the scientist seeks regularity, the permanent regularities of experience, and aims to understand phenomena by showing them to be instances of such regularity, by subsuming them under general laws of nature. Where the non-scientific explanation seeks to account for a set of phenomena as a manifestation of a *permanent entity beyond experience*, the scientific explanation seeks to account for those phenomena as a manifestation of a *permanent pattern in experience.*

From the viewpoint of Thomas Taylor, of course, the scientific explanation leaves everything *unexplained*, a *mystery*, when it refuses to move beyond sense experience, and the patterns in it, to entities outside that realm. In this connection, some remarks of Alexander Bain are worth quoting at length.

This word "mystery" is itself greatly misconceived. Such was the opinion of one of the ablest of biblical critics — Principal George Campbell — as to the employment of the word in religious doctrine. In Campbell's view "*μυστήριον*" means simply what we call a secret — a thing for the time concealed, but afterwards to be made known. It is the correlative term to "Revelation," which disclosed what had previously been hidden.

In another acceptation, Mystery is correlated to Explanation; it means something intelligible enough as a fact, but not accounted for, not reduced to any law, principle, or reason. The ebb and flow of the Tides, the motion of the Planets, Satellites, and Comets, were understood as facts at all times; but they were regarded as mysteries until Newton brought them under the Laws of Motion and of Gravity. Earthquakes and volcanoes are still mysterious; their explanation is not yet fully made out. The immediate derivation of muscular power and of animal heat is unknown, which renders these phenomena mysterious.

The meaning of the correlative couple — Mystery, Explanation — has been rendered precise by the march of physical science since the age of Newton. Mystery is the isolation of a fact from all others. Explanation is the discerning of agreement among facts remotely placed: it is essentially the *generalizing process*, whereby many widely scattered appearances are shown to come under one commanding principle of law. The fall of a stone, the flow of rivers, the retention of the moon in her circuit, are all expressed by the single law of Gravity. This generalizing sweep is a real advance in our knowledge, an ascent in the scale of intelligence, a step towards the centralization of the empire of science; and it is the only real meaning of Explanation. A difficulty is solved, a mystery is unriddled, according as the mysterious fact can be shown to resemble other facts. Mystery is solitariness, exception, or it may be apparent contradiction; the resolution of the mystery is found in assimilation, identity, fraternity. When all natural operations are assimilated, as far as assimilation can go, as far as likeness holds, there is an end to explanation, and to the necessity for it; there is an end to what the mind can intelligently desire; perfect vision is consummated.

But, say many persons, after resolving the fall of a stone and the sun's attraction into one force called gravity, there still remains the mystery — what is gravity? Even Newton sought to explain gravity itself. Well, if you must go farther, find some other force to *assimilate* with gravity; you will then make a new generalizing stride, and achieve a farther step of explanation. If, however, there is no other force to be assimilated, gravity is the final term of explanation, the full revelation of the mystery. There is nothing farther to be done; nothing farther to be desired. Nor have we here any reason to be dissatisfied with this position, to complain of baulked satisfaction, or of being on a lower platform than we might possibly occupy. Our intelligence is fully honoured, fully implemented, by the possession of a principle as wide in its sweep as the phenomenon itself.[6]

The intellect can, and ought to be contented with what it can achieve.[7] If the empiricist is correct — Hume's defense of empiricism is sketched briefly in Chapter II, Section I — then there is in fact nothing more to explanation than subsumption under empirical, matter-of-fact generalizations; to seek for more is unreasonable.

For the empiricist, then, explanation does not proceed from the world of sense experience to some deeper or more ultimate reality, as it does for the mystic. As Neurath once put it, "To one who holds the scientific attitude, statements are only means to predictions; all statements lie in one single plane . . . . *[Empirical science] knows no 'depth', everything is on the 'surface'.*"[8] And, he continues, this is true of all areas of experience, that of physics, but that of life, human life, as well: sociology and psychology, as much as physics, are empirical sciences. "The scientific world-view stops at nothing. Whatever is part of life, it examines. The question always is: what can we predict about this? It does not matter whether it be stellar paths, mountains, animals, men or states. Within this framework sociology is an empirical science, concerned with the behaviour of human groups."[9] The eschewing of depth, and "ultimate realities," that is, the positivist spirit, applies as equally to social as it does to physical reality. One who, like Habermas, hankers for a social science that can give more than empirical science can give must be accounted among those whose beliefs are essentially mystical. But if such beliefs are unacceptable, the aim that prompts them, the aim of understanding the changing world of experience *sub specie aeterni*, is, to repeat, not to be rejected. It is, rather, to be satisfied in a non-mystical way, by acquiring the sort of knowledge that empirical science can provide: knowledge of matter-of-fact empirical regularities.

The motive behind this aim is, often enough, disinterested, or in Veblen's term,[10] idle curiosity. Russell has remarked well on the positive aspects of this motive:

Disinterested curiosity, which is the source of almost all intellectual effort, finds with astonished delight that science can unveil secrets which might well have seemed forever undiscoverable. The desire for a larger life and wider interests, for an escape from private circumstances, and even from the whole recurring human cycle of birth and death, is fulfilled by the impersonal cosmic outlook of science as by nothing else. To all these must be added, as contributing to the happiness of the man of science, the admiration of splendid achievement, and the consciousness of inestimable utility to the human race. A life devoted to science is ... a happy life, and its happiness is derived from the very best sources that are open to dwellers on this troubled land and passionate planet.[11]

The knowledge of matter-of-fact regularities, while satisfying disinterested curiosity, is also of utility, as Russell says; that is, it also serves our pragmatic interests. This point was once put nicely by Neurath.

One may of course separate theories, but theories must be such that they can all be combined, if they are to help in predicting a given spatio-temporal process. Therefore *unified science is the stock of all connectible and indeed logically compatible laws*, i.e., the formulations of order. This unified science is the substitute for magic which also once encompassed the whole of life, although in the meantime the split into theology and technology has noticeably emerged.

The development of modern science, which in the end includes the whole of life, consolidates the close connection of theoretical and practical workers.[11a]

The knowledge at which science aims — knowledge of matter-of-fact regularities — is also knowledge of means — the means by which our various ends can be achieved. As such, the knowledge is desirable not merely for its own sake but for its pragmatic utility. The search after scientific knowledge can be disinterested but it is also, often, sought because it enables us better to control events, natural and/or social, so as to fulfill our several purposes. It is not by accident that Galileo set the scene of his *Dialogues on Two New Sciences* in the arsenal of Venice and that the theoretical discussion of projectible motion is succeeded by an attempt to apply this knowledge to calculate the trajectories of cannonballs.[11b]

The cognitive interest that motivates science is in the first instance with respect to individual facts, and is, also in the first instance *pragmatic*: we desire knowledge in order to *control*, to interfere intelligently, that is, knowledge is desired as a means that permits us to intelligently bring about various non-cognitive ends. The knowledge desired is of the *sort* that, where technically possible, permits control and intelligent interference, or, at least, potential interference, where the potentiality is that of technological potentiality. (To this extent those who describe positive science as determined by technological rationality are quite correct.) But one may also seek such knowledge for its own sake. (To this extent those who describe positive science as determined by technological rationality are quite wrong.) In that case our cognitive

interest is not pragmatic — the knowledge is not desired for its utility —
but rather is simply idle curiosity. But, of course, such idle curiosity is
nonetheless a cognitive interest, and something that must be taken into
account when one takes the "pragmatic turn" with respect to explanation. Of that, more in a moment. The present point is that the interest
— be it a pragmatic interest or idle curiousity — is in knowledge of the
*sort* that, where technically possible, permits control. The concepts of
*explanation and prediction are tied to this idea*: scientific prediction is
prediction based on knowledge of the sort that permits control, and
scientific explanation is explanation based on knowledge of the sort that
permits control.

Now, the idea of *control* is this: if we do so and so then such and
such cannot but come about: is *must* come about. And if we know that

> All *A* are *B*

then we have the desired sort of knowledge. For, if we have that
knowledge then we know that if we want *this* to be a *B*, then we can
achieve that by bringing it about that *this* is an *A*. For if all *A* are *B*,
then if *this* is *A* then it *must* be *B*. Explanations based on knowledge of
laws of this sort are often called "deductive-nomological."

In introducing cognitive interests, or, in other words, the aims of
science, we have introduced values. The motive behind an aim may be
one of two sorts.[12] The motive may, in the first place, be simply idle
curiosity. In the second place, it may be some pragmatic interest. These
two sorts of motive serve respectively to distinguish "pure science"
from "practical research" that has some "application" in mind. The
acquisition of knowledge of laws in the latter case is not an end in itself,
as in the former case, but only a means. The knowledge of laws is to be
understood as a knowledge of means to achieve some goal one has,
which goal it is depending upon one's pragmatic interests.

The relevant end need not, of course, be a direct personal or self-
interested end of the scientific investigator. It may, for example, be a
general social end, with respect to which knowledge of means is
desired.

Thus, the end which makes economic knowledge of pragmatic
interest is (often, at least) the economic health and well-being of the
community. The crucial point to be noticed, however, is this, that the
distinction between "pure" science and science which is *not* pure is not

one of subject-matter. In both cases what is aimed at is knowledge of laws. The difference, rather, is one of motive, in the one case idle curiosity, in the other some pragmatic interest.

It is clear that the motive of idle curiosity casts a wider net than that of any pragmatic interest. If one's interest is pragmatic one will stop the search for knowledge as soon as one has sufficient information about laws to enable one to achieve one's end. But if the motive is that of idle curiosity one will continue on the search for knowledge even if there are no "practical" consequences. Clearly, one can have pragmatic interests which motivate investigation without idle curiosity. The converse also holds.

The subject-matter of science is laws. The reference is, of course, not to prescriptive rules, but to natural laws, which are matter-of-observable-fact generalities,[13] in no way normative. Perhaps the most important point which follows from laws being matter-of-fact generalities is that their truth depends only on what they are about, the facts they concern, and not on who happens to believe them, or on how they happened to be discovered. That all unsupported massy objects fall depends only on massy objects and their movements, not on any person actually believing or disbelieving that this is so. In this sense, science is objective.[14] This is so even where the facts in question happen to be subjective, e.g., the facts about what values certain persons or groups of persons have. In this respect, the social sciences, which have such facts as a crucial ingredient of their subject-matter, are as objective as such sciences as physics or biology. And so, too, are the policy sciences in this sense objective: their aim is knowledge of laws, matter-of-fact generalities, the truth of which is a matter of objective fact. To be sure, the policy sciences seek such knowledge in order to apply it, but that is not what is at issue: such a point concerns their motive, and not their subject-matter, namely laws, which remains in the relevant sense objective.[14a]

Thus, if science, whatever its motive, aims at knowledge of laws, matter-of-observable-fact truths, if the latter are the subject-matter of science, *then science provides us with no values.* The body of propositions science accepts will contain no value-judgments, only matter-of-observable-fact truths. Of course, as we just noted, these matter-of-observable-fact truths may be truths about the values persons in fact have, but being about values does not transform a factual statement into a value-judgment. Furthermore, although we began our discussion by mentioning values, it is now clear that this talk of values in no way leads to the conclusion science aims to discover values.[15]

What we have done is distinguish the aims of scientists from that at which they aim.[16] The former *is* a matter of values. The latter is not. And it is only with respect to the latter, the subject-matter of science, that the empiricist is concerned to maintain the thesis that science is value-free.

Hereafter, then, we shall take as given the distinction between the aims of scientists and that at which they aim, and proceed from there. As for the aims of scientists, the idle curiosity or pragmatic interests that determine their interest in knowing matter-of-fact regularities, these we shall not here try to justify: we shall simply take them for granted. Then, in terms of these aims, we can raise such questions as: How can we best satisfy these aims? What kind of knowledge most adequately fulfills these ends? and, What are the best means for acquiring such knowledge? *In answering such questions as these we shall be able to develop and justify norms and standards for cognitive or epistemic value with respect to different kinds of scientific knowledge; and, with respect to research tools and strategies, norms and standards for their instrumental value relative to the aim of acquiring the valued sorts of scientific knowledge.* To repeat, this justification proceeds within the *given* framework established by the aim, determined either by idle curiosity or some pragmatic interests, of knowing matter-of-fact regularities. In taking that framework as given, I do not intend to say that *it* cannot be justified.[16a] The point is only that we shall not do that here. Of course, for the empiricist, the ultimate justification will derive from some ultimate value-judgment which will belong to the non-cognitive, non-scientific, part of discourse. Perhaps idle curiosity is such an ultimate value-judgment. But if it is, its non-cognitive nature does *not* preclude our making reference to it in our attempts to justify the norms and standards of science. Indeed, if science is, as the positivists claim, value-free, then the norms and standards of science, *qua* norms and standards, *cannot* be justified by cognitive or scientific considerations alone. *Qua* norms and standards they can be justified only by reference to some *value judgment,* that is, by a non-cognitive judgment. Hooker is quite wrong when he urges that ". . . positivist philosophical doctrines are normative only in the sense that they constitute the only appropriate framework from within which clear thinking can be conducted; since these doctrines do not possess a cognitively significant content there is no further problem concerning the sources and nature of their normativeness."[17] *Of course* positivists can discuss the norms and standards of science. The fact that such discussion must ultimately refer to a value judgment that is itself non-cognitive is irrelevant. Actually the point is put by Hooker in a way that partially begs the question. He refers to the non-cognitive value judgment as not being "cognitively significant." It is indeed non-cognitive but that does not mean it is without significance for the cognitive enterprise. It is this latter significance upon which we are now insisting.

We have, then, idle curiosity and pragmatic interests as the motives behind our aiming at scientific knowledge, knowledge of matter-of-fact regularities.[17a] In terms of these cognitive interests we shall argue for and justify certain norms and standards for science. To do this, we begin by asking the question, What kind of knowledge can most adequately satisfy our cognitive interests? By assumption, we can give a generic description of this knowledge: it must consist of knowledge of *matter-of-fact regularities*. But regularities may be of various forms: there are, for example, the crude generalizations of everyday experience ("water, when heated, boils"), and, in contrast, there are such mathematically formulated regularities as those that Newton discovered for the solar system. We can therefore ask, What, *specifically*, is the form that regularities must take to most adequately satisfy our cognitive interests?

A regularity is, as we said, a *permanent pattern in experience*. It is expressed in language as a *true* statement of *general fact*. And of course, as we have also said, whether or not such a statement is true or false is an *objective* matter. Now, to say that science aims at *truths* of the form

    All *F*'s are *G*

is to say something that holds almost by definition of scientists. But if we look at the requirement of truth in the light of our cognitive interests, we can discover its importance, why it *ought* to be included in the definition of science, and also, at the same time, discover something about the form of laws we *ought to prefer*, that is, ought to prefer if we want to best satisfy our cognitive interests.

The regularities science seeks to know are such as to permit *control*. That is why our pragmatic interests make such knowledge one of our aims. Control, naturally, implies prediction and the use of the generality expressing the regularity to support contrary-to-fact and subjunctive conditionals. The predictive assertion

>    This is $F$, so it is $G$

is thought of as analyzed into

(A)      All $F$'s are $G$
         This is $F$
         _____

         This is $G$

where both premises and therefore the conclusion are asserted.[18] The subjunctive assertion

>    If this were $F$, it would be $G$

is thought of as analyzed also into

>    All $F$'s are $G$
>    This is $F$
>    _____

>    This is $G$

except that in this case only the first premise is asserted while the second is either denied (in which case the conditional is contrary-to-fact) or judgment is suspended on it. The contrary-to-fact inference is of special importance when it comes to control, to intelligent interference in natural and social processes. If we can reasonably assert that if this were $F$ then it would be $G$, then we can conclude that by *making* this to be $F$ we can make it to be $G$. The task of making something $G$ then becomes the task of implementing the means, that is, bringing it about that the thing becomes $F$. In any case, the predictions and the subjunctive conditionals that are asserted on the basis of scientific knowledge of regularities turn out to receive parallel analyses. Whether this sort of analysis suffices for all subjunctive conditionals we cannot go into at this point;[19] that is the task of Chapters II and III, below. Suffice it for now to say that we here mean only to take the analysis as the one that displays *how reasoning and inference with respect to prediction and control proceeds when it is based on the sorts of regularity science aims to know.*

Where we make the scientific prediction (A), the regularity provides us with a *reason for expecting* this $F$ to be $G$. But it does something more, and this is what is vital if it is to satisfy our pragmatic interests that seek knowledge that permits control. *That the regularity obtains constitutes a reason for this F* BEING *G.* The regularity constitutes grounds not only for *expecting* an *F* to be *G* but also for that *F being G.* Of course, it is not a ground in the sense in which Thomas Taylor sought non-sensible entities as grounds for sensible facts being what they are: science eschews the search for such non-sensible grounds. The regularity is simply a general fact and in no way is some entity separate from the fact that this *F* is *G*; indeed, the latter individual fact is one of the truth-conditions for the general fact. But this fact, namely, that the obtaining of the individual fact is a logically necessary condition for the obtaining of the general fact, is the sum and substance of the point that the obtaining of the regularity or general fact constitutes grounds for the obtaining of the individual fact, grounds for the latter *being* what it is. The point can be put in another way by noting that if the general statement in (A) is *not* true, then this argument yields *no* reason for the *F* in question to be *G.* For, unless the general premiss is true, the argument is unsound, and an unsound argument tells us nothing at all about the truth-value of its conclusion. In short, *our scientific cognitive interests* MUST *remain unsatisfied if the statements of general fact we employ in our attempts to predict and control are false. Our cognitive interests* DEMAND THE TRUTH *of the generalities we employ in scientific inference.*[19a]

Reasons for expecting may be various. No doubt for some, sooth-sayers provide reasons for expecting — though whether they ever, by themselves, provide good reasons we may well wonder. But there are other reasons for expecting that are quite rational, and which are, nonetheless, not the sort that satisfy those pragmatic interests that lead us to seek knowledge of such regularities as permit control. Thus, consider the rolling of a single six-sided die under so-called chance conditions. Make the usual assumptions about the physical construction of the die, and the mechanism of its being rolled. Call these assumptions *F.* Then we know the *statistical* regularity that

(S)     Whenever *F*, then in a long sequence of throws the probability of a six is 1/6

from which it follows that under similar conditions the probability of a non-six is 5/6. Now, if the die has been tossed a number of times, then (S) gives us a good reason for expecting the next toss will be a non-six.

Of course, we are not certain that the next toss will be a non-six. But if we were to bet, then it would be more reasonable to bet on a non-six than on a six. We can in this sense tentatively expect a non-six, where the degree of tentativeness is proportioned to the probability of a non-six. However, even though (S) gives us a *reason for expecting* the next toss to be a non-six, it provides *no reason* for the next toss *being* a non-six. In giving no reason for the next toss being one way rather than the other, (S) provides us with nothing that will enable us to control the outcome of the next toss, that is, that will enable us to make the next toss *be* one way rather than the other. Indeed, (S) says nothing at all about *any* particular toss. It only says something about patterns in long sequences of tosses. Thus, as far as concerns the individual tosses, (S) says nothing that can satisfy our pragmatic interests with respect to control. Hence, if scientifically desirable regularities are those that satisfy our pragmatic interest in control, then, so far as concerns the individual tosses, (S) is *not* of the appropriate sort. On the other hand, for a situation of sort *F* and a *long sequences of tosses*, (S) provides a reason not only for expecting it to be in a certain way, viz., exhibiting the pattern of 1/6 of the tosses being sixes, 5/6 non-sixes, but it also provides a reason for the sequence to *be* that way. And it thereby provides us with information that permits control. Thus, so far as concerns certain *mass events* — the long sequences of tosses — (S) provides scientifically satisfactory information, but so far as concerns the *individual events* in the mass events — i.e., the individual tosses — then (S) does not provide scientifically satisfactory information.[20]

Scientific explanation of individual facts proceeds by subsumption of these facts under scientific laws. These laws, i.e., regularities, we prefer to be those that can best satisfy both our idle curiosity and pragmatic interests. This means that the regularities we use in scientific explanation are, we prefer, such as to permit control and prediction. It follows directly that (A) provides the fundamental model for scientific explanation. This model, which has come to be (somewhat misleadingly) called the "deductive-nomological" model was clearly formulated for the first time by John Stuart Mill,[20a] though it is implicit in the thought of both Hume and Bacon. Contemporary discussions have tended to take off from the formulation of the idea given by Hempel and Oppenheim in their classic essay 'Studies in the Logic of Explanation.'

Hempel and Oppenheim had a number of concerns in this essay, but of these the most important was a defence of the idea of a science of man. They were concerned to reply to various claims to the effect that the idea of a science of man is impossible because human behaviour as a matter of principle cannot be subsumed under laws like that in (A) which permit prediction.[21] For example, they reply to the objection that human behaviour is intentional by pointing out that intentional states can function as the antecedents in regularities of the sort that occur in (A).[22] And they reply to the argument from the uniqueness of human actions by pointing out that even the tick of a watch is unique, individuated by a uniquely specifying pattern of relations to all other events, but that such uniqueness, in both cases, does not exclude such events in respect of certain of their characteristics being subsumable under general laws.[23] Our present interest is not directly in these arguments. Besides they have been elaborated sufficiently by others.[24] The present point is the simple one of defining the context in which Hempel and Oppenheim were presenting their exposition of the "deductive-nomological" model of explanation. And in that context, what is of primary focus is the fact that where one offers a scientific explanation there, of necessity, one has offered a prediction. Their attempts to reply to the attacks on the idea of a science of man required little more than that idea. Nor did they provide much more. Certainly, they neither provided nor even claimed to provide a *complete* account of scientific explanation. This last point must, I think, be kept clearly in mind. We shall see shortly, that some critics of Hempel and Oppenheim failed to let themselves be guided by this simple hermeneutical point.[24a]

Hempel and Oppenheim begin their discussion with two simple examples. The first is this:

A mercury thermometer is rapidly immersed in hot water; there occurs a temporary drop of the mercury column, which is then followed by a swift rise. How is the phenomenon to be explained? The increase in temperature affects at first only the glass tube of the thermometer; it expands and thus provides a larger space for the mercury inside, whose surface therefore drops. As soon as by heat conduction the rise in temperature reaches the mercury, however, the latter expands, and as its coefficient of expansion is considerably larger than that of glass, a rise of the mercury level results.[25]

The second example is this:

To an observer in a row boat, that part of an oar which is under water appears to be bent upwards. The phenomenon is explained by means of general laws — mainly the law of refraction and the law that water is an optically denser medium than air — and by reference to certain antecedent conditions — especially the facts that part of the oar is in the water, part in the air, and that oar is practically a straight piece of wood.[26]

Hempel and Oppenheim point out that one can find in these paradigm cases of a scientific explanation the principle that such explanation implies prediction. The structure of their essay involves taking this idea exemplified in purely physical explanations and arguing that there is no reason to suppose it cannot be found in explanations of human behaviour also. Of less interest in the context of their exposition was the idea of control. Yet it, too, is clearly present in the examples. It is clearly the idea of Hempel and Oppenheim that a scientific explanation involves a prediction, — but not only a prediction, also knowledge that admits of control.

It is [its] potential predictive force which gives scientific explanation its importance: Only to the extent that we are able to explain empirical facts can we obtain the major objective of scientific reseach, namely not merely to record the phenomena of our experience, but to learn from them, by basing upon them theoretical generalizations which enable us to anticipate new occurrences and to control, at least to some extent, the changes in our environment.[27]

Thus for Hempel and Oppenheim, scientific explanations involve knowledge of such laws or regularities as permit of control and prediction. The relevant principle they state as follows:

. . . an explanation is not fully adequate unless its explanans [the antecedent conditions and general laws that do the explaining], if taken account of in time, could have served as a basis for predicting the phenomenon under consideration.[28]

We may call this the "Principle of Predictability", recalling, however, that the relevant sort of prediction is that which is associated with the possibility of control: the prediction or reason for expecting must also be a reason for being, so that the basic model is provided by (A) rather than (S).

The Principle of Predictability *defines* the empiricist idea of scientific explanation.[28a] It is intended that it provide necessary and sufficient conditions for an argument like (A) to constitute an explanation. Moreover, the conditions are intended to be objective conditions in the sense that whether they are fulfilled will depend only on the argument and the facts the sentences in it are about and not on, for example, who happens to accept the argument nor on, for another example, the evidence that happens to be available. Hempel and Oppenheim specifically reject the latter sort of condition being imposed on explanations.[29] But if the Principle is meant to state necessary and sufficient objective conditions for an argument like (A) to be an explanation, it is not therefore meant that it does so in full detail. To the contrary. What must be done, therefore, is more fully articulate the objective standards that are implicit in the Principle. This Hempel and Oppenheim try to do.

Hempel and Oppenheim divide the explanation into *explanans* (which does the explaining) and the *explanandum* (which is the sentence describing the phenomenon to be explained). The explanans consists of two kinds of sentence: those stating antecedent conditions and those stating general laws. Their description of the "deductive-nomological" model is this:[30]

(R₁)      The explanandum must be a logical consequence of the explanans; in other words, the explanandum must be logically deducible from the information contained in the explanans, for otherwise, the explanans would not constitute adequate grounds for the explanandum.

(R₂)      The explanans must contain general laws, and these must actually be required for the derivation of the explanandum . . . .

(R₃)      The explanans must have empirical content; i.e., it must be capable, at least in principle, of test by experiment or observation.

(R₄)      The sentences constituting the explanans must be true.

These conditions $(R_1)$—$(R_4)$ are to be thought as a more detailed expression of the Principle of Predictability.

It is worth mentioning an aspect of $(R_3)$ that has hitherto only been implicit in what we have said. In holding that the law in the explanans of (A) must have empirical content, we are ruling out "explanations" like

(D)      All bachelors are unmarried.
         This is a bachelor.
         _____

         This is unmarried.

The conclusion of this argument is not a prediction. Rather, it only restates something that already is implicit in the minor premiss. Indeed, given the meaning of 'bachelor', the minor premiss asserts nothing more than

             This is male & this is unmarried

which clearly, entails the explanandum but, equally clearly, does so without the need to introduce, as $(R_2)$ requires, general laws essential to the deduction. The general premiss in (D) is really inessential to the deduction. The linguistic rule

'Bachelor' is short for 'unmarried male'

that justifies asserting the general premiss of (D), already justifies inferring the explanandum from the minor premiss, the statement of antecedent conditions alone.

Behind $(R_3)$ is, of course, the Principle of Predictability, and it is clear that the exclusion of (D) as explanatory is justified by this Principle. For, in (D) we make no *prediction*: we do not assert *one* individual fact on the basis of *another* individual fact. Rather, we merely restate in the conclusion an individual fact that has already been asserted in the minor premiss. If we think of the major premiss of (A), viz.,

All *F* are *G*

as permitting of control, then *F* is the means and *G* is the end of the use of *F* to achieve. Here, we think of means and ends as distinct. But in the major premiss of (D),

All bachelors are unmarried

the antecedent and consequent are not distinct. Here one means, of course, *logically distinct*. One cannot bring about someone's being unmarried *by means of* bringing about their being a bachelor, for in implementing the means one has *ipso facto* achieved that end.

It follows that the Principle of Predictability requires us to draw a distinction between cases like (A) admitting of predictive power and cases like (D) which involve no prediction. The empiricism of the early positivists captures this point easily within the framework of the empiricist's language that it proposes for science. The empiricist's language has the logical framework of *Principia Mathematica*. This framework permits us to draw a sharp distinction between sentences that are analytically true or false and those that are true as a matter of fact. The latter are the synthetic statements, the former analytic. Arguments of sort (A) constitute explanations only if the generality that occurs as the major premiss is a *synthetic* or factual truth.[31]

There is one aspect of the Hempel-Oppenheim position that bears further comment. It follows directly from the Principle of Predictability that where one has a scientific explanation, there one has a scientific prediction, and that where one has a scientific prediction, there one has a scientific explanation. This has been called the *symmetry thesis* with respect to explanation and prediction.[32] Since this thesis is part and parcel of the Principle of Predictability which is itself the core idea for the empiricist of the idea of scientific explanation, any challenge to the symmetry thesis is *ipso facto* a challenge to the empiricist idea of science. And the symmetry thesis has in fact been challenged. We have been presented with examples that are supposed to be explanations which do not involve predictions and with examples that are supposed to be predictions that are not explanations.

Some of these objections are weak indeed. Thus, Scriven has challenged the "deductive-nomological" model by pointing out that any statement about the future is a prediction but that does not make "It will rain tomorrow" an explanation of the coming storm; from this he concludes the symmetry thesis is false.[33] The obvious reply to this is that Hempel and Oppenheim, and the empiricist in general, is talking about *scientific prediction*, reasoned prediction:[34] this of Scriven's objections turns on little more than a pun. Again, we often explain where we could not *in fact* have predicted. Thus, under normal conditions (e.g., no bombs are being dropped) a bridge will collapse under a certain load if and only if certain structural defects are present. A load is put on the bridge and it does collapse. We could not in fact have predicted the collapse owing to the difficulty in coming to know the relevant initial conditions, that the structural defects were present. But after the collapse we can deduce from that fact plus the law that the defects were indeed present. Then, turning this about, we can explain the collapse by deducing it from the law plus the initial condition of the presence of the structural defect. Even here, however, there is no real breakdown in the logical symmetry of explanation and prediction, for we can still say that we *could* have predicted *if* we had had knowledge of the initial conditions. The only problem here lies in the fact that it is difficult to acquire the needed knowledge of initial conditions; the knowledge can be acquired only *ex post facto*. So the logical symmetry of explanation and prediction stands. This replies to another objection of Scriven.[35] Others of Scriven's objections are more interesting, however — more interesting because they challenge, and therefore replies to them shed light upon, features of the empiricist idea of scientific explanation, requiring us to recognize that things cannot be so simply construed as the naive falsificationist and the basic "deductive-nomological" model (A) would have us suppose. But if things are not that simple, neither do they require us to give up either the empiricist Principle of Predictability or the symmetry thesis it entails.[36] Since the examples can teach us something of relevance to our purposes, we will look briefly, now, at two that challenge the idea that every scientific prediction is an explanation, and one that challenges the idea that every explanation is a prediction.

One example Scriven has used as a purported example of a lawful prediction which is not an explanation[37] (it derives originally from Bromberger)[38] involves geometrical optics. From the length of the shadow of a flagpole as an antecedent condition, one can compute the height of the flagpole, using as general premisses the law that light travels in straight lines from sources to the areas they illuminate and the law that a flagpole is a non-self-luminous opaque sort of body, and as another antecedent condition the location of the sun as a self-luminous point source.[39] Here one uses lawful knowledge to predict the height of the flagpole from the length of the shadow. But, it is claimed, the length of the shadow does not explain the height of the pole; rather, the explanatory connection goes only in the opposite direction, from the height of the pole to the length of the shadow. The reasoning behind this claim seems to be this: one can change the length of the shadow by changing the length of the pole, but one cannot change the length of the pole by changing the length of the shadow.

Now, this last point cannot be denied. On the other hand, one is not thereby forced to the claim that citing the length of the flagpole has *no* explanatory force *vis à vis* the height of the flagpole. And indeed, surely there *is* such an explanatory connection. Surely the length of the shadow explains why the height must *be* what it is. Surely, the length does constitute a *reason for being*, a reason for the height being what it is. And insofar as it does provide such a reason, surely it also can quite correctly be said to explain why the flagpole must be that height. It would seem, then, that however plausible is the inference that leads Scriven to suppose the length of the shadow does not explain the height, we must reject that inference as invalid.

The premiss is that one cannot cause the height to change by manipulating the length of the shadow, while one can cause the length of the shadow to change by manipulating the height of the flagpole. From this asymmetry is inferred an explanatory asymmetry. The latter does not follow. It can be supposed to only if one identifies the direction of explanation with the direction in which effects flow from our manipulations. But there is no reason whatsoever for making such an identification and it is in fact thoroughly confused. It is this confusion that gives Scriven's example such force as it has.

Geometrical optics, which provides the laws in the example, does not describe how things change over time. Rather, it relates certain properties at a time, in temporal cross-sections as it were. For this reason, such laws have been called laws of coexistence[40] or cross-section laws.[41] The regularities describe the geometrical relations that obtain when self-luminous bodies illuminate certain areas of non-self-luminous objects. These laws say: given such and such conditions at $t$ then so and so other things obtain at $t$. Among the things thus related are the shapes of physical objects and the areas of illumination and of non-illumination (shadows).

One the other hand, to speak of manipulation is to speak of a *process*. Manipulation in fact involves the exercise of *forces*, pushes and pulls, stresses and strains. When one changes things by manipulation, one exerts mechanical forces on them. The laws describing such forces, that is, the laws of classical dynamics, describe changes in the properties of physical bodies. They describe the interaction of physical bodies. But *not* the interaction of physical bodies and non-physical-bodies such as shadows and areas of illumination. Manipulation can therefore only change the height of the flagpole, not the length of the shadow. The latter isn't even a relevant variable in classical dynamics. But in changing the length mechanically, one changes certain conditions that are related by the laws of geometrical optics to certain *non-mechanical factors*, viz., areas of illumination and shadows. So, in changing the height of the flagpole one can thereby change the length of the shadow, while, on the other hand, one cannot change the length of the shadow by manipulating *it*; nor therefore change the height by manipulating *directly* the length of the shadow.

This makes clear, I think, why there is a manipulative asymmetry, but also why this fact does not impugn the idea that the regularities of geometrical optics that relate shadow-lengths to the dimensions of physical objects provide not just predictive but also explanatory connections. Three other points have also emerged. *First,* the regularities of science are of two kinds, those of co-existence or cross-section laws and those of succession or process laws. *Second,* scientific laws may relate physical and non-physical factors. *Third,* although any law that provides reasons for facts *being* in a certain way is important from the point of view of our pragmatic interests in control, not every such law relates things *directly* to our capacities to manipulatively interfere in the ongoing processes of the world.

Another example that Scriven has used is that of the height of a barometer indicating the coming of a storm.[42] Here we have a case where we are involved in explaining the fall of mercury columns, on the one hand, and explaining the fall of the temperature, on the other, and then, of course, what is crucial, the contrast between these.

We have the law for mercury columns in barometers that (for suitable units and calibration) height = temperature $(h = t)$. This equation is symmetrical, yet, Scriven argues, while it can be used to explain why height decreases when temperature falls, it cannot be used to explain why temperature falls when height decreases. For, we are told, in the latter case, the "explanans" can hardly be considered an explanans at all, not to say that it possesses explanatory power. This comment is, it seems to me, just wrong. Let $t = 1$. Then, by the law that $h = t$, $h = 1$. We also can deduce that if it *were* the case that $t = 9$, then it would be that $h = 9$. Moreover, given that $t = 1$, then the law explains why the height must have the value 1. Similarly, let $h = 3$; then $t = 3$. Furthermore, if it were the case that $h = 9$, then it would be that $t = 9$. Finally, given that $h = 3$, then the law *explains* why the temperature *must* have the value 3. That is, I suggest, there *is* an explanatory connection in both directions, contrary to Scriven's claim. Moreover, given the cognitive interests that determine our concept of explanation, this is as it should be. On the other hand, in neither case is the explanation *the best* we might hope for, since, clearly, both cases represent but partial descriptions of a much more complex process. I.e., both explanatory connetions are imperfect. Moreover, it is this broader context that distinguishes the two cases. In the temperature-to-height case, we have a causal relation, whereas in the height-to-temperature case we do not: the change in temperature *causes* the change in height, but not conversely. The point is, of course, that the notion of cause is intimately tied to that of *manipulation* or of *active interference*. By manipulating the weather we can change the height, but we cannot change the weather by manipulating the height. But this knowledge, about what can be manipulated to effect what changes goes beyond the simple law that $h = t$. Indeed, it is less imperfect than the latter. It is this less imperfect knowledge that justifies asserting that "mercury columns of thermometers fall *because* of temperature decrease" and justifies denying that "temperature decreases *because* mercury columns of thermometers fall". The first of these is a correct *causal* explanation, the second is an *incorrect* causal explanation. But from the fact that the latter is a bad causal explanation it does *not* follow that the law $h = t$ provides *no* explanatory connection between $h$ and $t$. The trouble with the second case is that, not only does it purport to give an explanation stronger than permitted by the explicitly stated laws, but also that the conclusion, we know from background information, to be wrong. In analysing this case, as providing a purported counter-example to the deductive-nomological model, it is necessary to take into account not only the pragmatic concept of cognitive interest but also the pragmatic concept of *presupposed background information* that so often provides the context in which "why-questions" are asked. But of course, there is nothing problematic about these pragmatic notions, and therefore no reason why the empiricist should not invoke them when he attempts to reply to criticisms such as those of Scriven.[43]

In the barometer-weather example, there is a regular, lawful, connection that permits one to infer the coming of the storm from the height of a barometer. The connection is, we might say, that of an *indicator law*. Indicator laws permit prediction. On the other hand, the height of the barometer does not *cause* the storm. Rather, the latter relation is the converse: the storm causes the height of the barometer. So, while we can predict the storm from the height of the barometer and the height of the barometer from the storm, it is the storm that explains the height of the barometer but not conversely. We thus seem to have a case where we have laws — indicator laws — that permit prediction but not explanation.

The initial reply to this challenge to the symmetry thesis is as in the previous example, to insist that the indicator law does explain why, given the height of the barometer, the atmosphere must *be* stormy. The indicator law can provide this explanatory connection precisely because it does provide a reason for the atmosphere being in such and such a way, that is, precisely because it provides a reason for being. On the other hand, the indicator law is, as Scriven's premises state, not a causal law. Scriven therefore apparently arrives at his conclusion that the example violates the symmetry thesis only because he identifies causal law with explanatory law. But, as we just argued, there is nothing to compel this identification. Scriven's example thus does not refute the symmetry thesis nor therefore the "deductive-nomological" model of the empiricist philosophy of science. It does force us, however, to recognize that not all laws are causal laws; and that not all explanations are causal explanations. (This is why the 'nomological' in the phrase "deductive-nomological model" of explanation is misleading.)

We may also draw a second moral from this example. The indicator law connecting the height of the barometer to the storm enables one to predict the latter. It moreover provides not merely a reason for expecting but also a reason for being. However, we earlier connected such reaons for being with our pragmatic interests in control. What this example makes clear is that not every law that provides reasons for being *ipso facto* is a law describing means for control. Indeed, this is part of the point of distinguishing indicator from causal laws: the latter do and the former do not describe means for control. From the point of view of our pragmatic interests, laws that provide reasons for being are better than laws, like (S), that provide only reasons for expecting. But, from the same point of view, that of the interest in control, an explanation's providing reasons for being will be only a necessary condition for that explanation *fully* satisfying our pragmatic interest. What we would really hope for is something better, a law so describing *all* details of the process as to permit not only interference but, so far as it was possible, informed and efficient interference.

What the barometer example forces us to distinguish is the *generic* category of explanations that provide reasons for being from the various *specific* instances of that sort of explanation. Generically considered, all such instances aim to satisfy our pragmatic interests in control. But specifically considered, some of these explanations will, more adequately than others, satisfy those interests. The indicator laws are at the extreme; they are the least preferable of the relevant laws. What we are admitting here is not that the indicator laws fail to explain, but only that they fail to give the *best* explanation. What Scriven's barometer example forces us to recognize is that *there are degrees with respect to scientific understanding.*[44]

There are various aspects to this idea of degrees of explanatory worth that must be explored. It is important to notice right at the beginning, however, that it is a matter of the *laws themselves* and has nothing to do with our knowledge of them. That is, it is an *objective matter*, not subjective. The difficulty with indicator laws is that, while they are true, they nonetheless fail to mention a great many of the variables that are relevant to any *complete* lawful description of the processes in question, and therefore also, of course, fail to mention how these variables interact with each other and with those that the indicator law does mention. That certain factors are or are not lawfully relevant is an objective matter. In the case of indicator laws we often do know of the existence of such omitted relevant factors. We thereby recognize the incompleteness of the explanation given by the indicator law. (Which, as we just argued, does *not* mean that the indicator law provides *no* explanation: an incomplete explanation is not a non-explanation.) And, it is clear, if we think in terms of satisfying either our idle curiosity or our pragmatic interests, then we can *more adequately satisfy them the more completely we take into account all the objectively relevant factors.*[45]

There are other forms such incompleteness may take. For example, $G_1$ and $G_2$ may be effects that are correlated because they have a common cause $F$. The correlation between $G_1$ and $G_2$ is a regularity — a lawful regularity — and therefore has *some* explanatory power. But *more complete* explanations can be offered if we also include in our law-premises reference to the common cause $F$. Again, it is clear that if we think of satisfying either our idle curiosity or our pragmatic interests, then we will rationally prefer explanations that more completely take into account all objectively relevant factors. But the example will bear further analysis.

Explanation by a common cause occurs when a piece of imperfect knowledge — some sort of correlation, perhaps — is explained by placing it in the context of broader *causal* knowledge. This, clearly, is a case of the less imperfect absorbing the more imperfect. The definition, by Salmon, Greeno, van Fraassen, and others, of $C$ being a common cause[46] of $A$ and $B$ when "$\Pr(A \cdot B/C) > \Pr(A/C) \times \Pr(B/C)$" where "Pr" is the probability function, is *decidedly misleading*. If it is a genuine case of common *cause*, then the statistics are *not irreducible*. One must distinguish *statistical laws*, which are irreducible to non-statistical laws, from *actuarial assessments*, where the statistics are used to as it were average out various causal factors which we do not know specifically but have reason to believe are there. If, and when, these latter come to be known — though coming to know them in detail may not be worth the research effort — then the imperfect statistical knowledge of the actuarial assessment will be absorbed in the less imperfect causal knowledge. In the case of the statistical laws, there is no such absorbtion into less imperfect causal knowledge — as in the cases of coin tossing and of statistical mechanics. An actuarial assessment may well lead us, though not by itself — further background knowledge is needed — to be belief that $C$ is a common cause of $A$ and $B$. Precisely because we reasonably anticipate the absorbtion of this knowledge into less imperfect, *non-statistical,* knowledge, we can assert that something being $C$ will *explain* its being $A$ and $B$, in just the way that we *cannot* claim to have explanatory connections among individual events when the laws are *irreducibly statistical.* Moreover, precisely because the knowledge *is causal* and not statistical, manipulation of a common cause is relevant to controlling other singular events (its effects), in precisely the way that there is no element of control from one singular occurrence to the next when the laws are irreducibly statistical. We explicitly introduce here the crucial cognitive interest in knowledge of the sort that permits control. If I am correct, then, as Hempel and Oppenheim also recognized, this is the crucial and most fundamental of the cognitive interests that must be invoked when attempting to determine the concept of scientific explanation.

A related case is that in which the question "Why does (person) $p$ have (disease) $D$?" is answered by citing the fact that $p$ is $C$ and the statistical generalization that 10% of $C$'s and $D$.[47] It is suggested that we would consider this answer unsatisfactory as an explanation. This suggestion is true. But the substantive question remains: *why* would we reckon it unsatisfactory? One answer to this question is that of Salmon, that this is mere stubbornness of mind.[48] The other answer makes reference to pragmatics. Salmon's point is that $p$'s being $C$ is of *some* relevance to $p$'s being $D$, and that we therefore ought *not* to reject the claim that $p$'s being $C$ explains its being $D$. The important point is that this is so *only if* the law is taken to be not irreducibly statistical. If there is no causal relationship that underlies and can explain the statistical correlation, then there is *no* explanation of $p$'s individual case. But our present example is with respect to disease, which is unlike coin tosses: for the former, unlike the latter, we not only assume but in fact have good reason to suppose that *there are* causal mechanisms that underlie the statistical correlations.[49] In particular, in the case of disease $D$ we might well have grounds for supposing that it is caused by a unique species of germ. For example, $D$ might be of a genus $\mathcal{D}$ other members of which we have discovered to be caused by unique species of germs of genus $\mathcal{G}$. So we infer that it is likely that there is a unique species of germ of genus $\mathcal{G}$ such that its presence in $C$ is necessary and sufficient for the presence of $D$:

$$(*) \qquad (\exists! f)\,[\mathcal{G}f\,\&\,(x)\,[Cx \supset (Dx \equiv fx)]]$$

This is a law, but also it constitutes the existence and uniqueness conditions for a *species* of germ. We may therefore introduce a *definite description* for these germs (cf. "the flu bug"). Let us abbreviate this definite description by

$$\bar{G}$$

It immediately follows that

$$(**) \qquad (x)\,[Cx \supset (Dx \equiv \bar{G}x)]$$

From the knowledge

$$Cp$$

and the fact

$$Dp$$

that $p$ has the disease, we can deduce the presence of the germs:

$$\bar{G}p$$

This yields the following *deductive-nomological explanation* of the fact $Dp$

$$(x)\,[Cx \supset (Dx \equiv \bar{G}x)]$$
$$Cp$$
$$\bar{G}p$$
$$\overline{\qquad\qquad\qquad\qquad}$$
$$Dp$$

However, such an explanation can be achieved only *ex post facto* since we do not know *specifically* what $\bar{G}$ is, and therefore can't decide whether it is present or not antecedently to our knowing whether its effect $D$ is present. The law (*), or, what is the same, its logical equivalent (**), is thus *not useful for prediction*. So far as prediction is concerned, the best we can do is the statistical law

10% of $C$'s are $D$.

That is, this is the best we can do until we identify $\overline{G}$. Suppose research tells us that

$$\overline{G} = G_1$$

Then we can replace (**) by

(***)     $(x)\,[Cx \supset (Dx \equiv G_1x)]$

In fact, this also replaces (*). The law (***) yields deductive-nomological explanations of course. But it can do so not only *ex past facto*, but also *predictively*. It can do this because it asserts specifically what the crucial causal factor is, rather than simply asserting that *there is* such a factor. Because it makes a determinate rather than a determinable claim, it is *less imperfect*. Moreover, it is also less imperfect than the statistical law. In fact, once (***) is available we no longer (at least in principle) have to rely on the statistical law to tell us how strongly to expect the next $C$ will be $D$, how much to bet on it; we can instead rely on (***) to *predict definitely* about each $C$ whether it will be $D$ or not.

Thus, Salmon is correct in supposing that with the 10% correlation, we do have an explanatory connection — provided, that is, that we have knowledge like (*) justifying the claim that the statistical connection is not irreducibly statistical. Except, of course, that the explanatory connection is provided by the law (*) rather than the statistical relationship itself. But Salmon is wrong to think that finding the explanation unsatisfactory is mere stubbornness. To the contrary, our cognitive interest that determines our concept of scientific explanation determines also that we prefer the less to the more imperfect. In particular, it determines us to prefer (***) to (*). And since (*) asserts that a law like (***) *does* obtain, is there to be discovered, then the law (*) that we use in the absence of anything better itself asserts that something better is there to be had. So naturally we reckon explanation based on (*) unsatisfactory, and in fact proceed (if we are scientists) to undertake research aimed at identifying the factor it asserts to exist, that is, research aimed at replacing the more imperfect with the knowledge that it asserts can be had.

The critics of the deductive-nomological model hold that we reject an explanation of $p$'s being $D$ in terms of its being $C$ as unsatisfactory because when we are asking why $p$ is $D$ we are asking a question that can't be answered by citing its being $C$. I.e., the purported explanation does not answer the question we are *really* asking. But it does not follow that the purported explanation is not an explanation. All that follows is that it is not the explanation we were asking for. The critics suggest that what we were asking for was the specific condition of *p* that accounts for the fact that, among *C*'s, he is part of the 10% that get *D*. We have argued that this is correct. For, to ask this question is to ask for an explanation in terms of the law (\*\*\*). What the critics do not do is relate this to the cognitive interests that determine our concept of explanation. We have tried to show how this can be done, by reference to the idea of knowledge being more or less imperfect. But this notion of imperfect knowledge or imperfect explanation requires further explication; we turn to this task in the next section.

Before turning to those problems, however, one point should be emphasized about knowledge of laws — anticipating in part our discussion in Chapter II of the Humean account of laws. The crucial point with respect to knowledge of laws is this: no matter-of-fact generality can be known, that is, known for certain, to be true.[50] At least, this is the main consequence of the Humean account of laws that we shall defend in Chapter II. The point is a logical one: A matter-of-fact generality makes a statement about a population, while all one ever observes is a sample. What as a matter of fact holds in a sample need not as a matter of fact hold in a population. Given this logical gap between sample and population no generality can be asserted without risk, no generality can be known for certain to be true. The scientist cannot know, in the sense of conclusively verify, the generalities he asserts. The best — that is, the best logically speaking — that he can hope for is partial confirmation. And since this is the best that *can* be achieved, one cannot *reasonably* aim for more.[51] *A scientist cannot reasonably aim to overcome the Humean limits on the assertability of matter-of-fact generalities.* When we say, then, that the scientist aims at knowledge of matter-of-fact generalities, what we must recognize is that he is aiming at *confirmed, not conclusively verified generalities.* (To say this is not to commit one to any particular theory about the nature of confirmation; indeed, it is compatible with Popper's view that a theory can be asserted only if it has survived testing.) Further, to say the scientist aims only at confirmation is to say that the knowledge at which he is aiming is *fallible knowledge*, knowledge with respect to which the scientist, in terms of the logic of the situation, must be forever open to the possibility of refutation.[52]

We must now draw a distinction that is of the utmost importance for the defence of the "deductive-nomological" model of explanation. Consider such an explanation

(A)     All *F* are *G*
        This is *F*
        _____
        This is *G*

For this argument to be an explanation, we required that the regularity appealed to really *is* a regularity, that is, that the general premiss be *true*. This is the condition that Hempel and Oppenheim labelled ($R_4$). Since truth is an objective matter, this is an *objective condition* for (A) to be acceptable as an explanation. However, we may not know whether the regularity holds or not. Even if it does, objectively speaking, hold unless evidence testifying to its truth is available to the subject who uses it or to whom it is offered, then (A) cannot be acceptable to that person as an explanation. The availability of evidence is thus a *subjective condition* for (A) to be acceptable (to a person) as an explanation. We may thus distinguish between (A) being *objectively worthy* of being (counted as) an explanation and (A) being *subjectively worthy* of so being (counted). It is clear that evidence can be better or worse, so that there are *degrees of subjective worthiness.* However, since there is the *inevitable logical* gap between sample and population, *even the best evidence must always fall short of establishing the objective worthiness of an explanation.* The best available evidence may therefore testify to the truth of the generality in (A), when in fact the generality is false. If the generality is false, then (A) is not objectively worthy of being counted as an explanation; indeed, it is *not an explanation.* Thus, the criteria of subjective worthiness may lead us to accept an argument (A) as an explanation when it is in fact not objectively worthy and therefore not an explanation.[53] What was subjectively worthy as an explanation can thus turn out never to have been an explanation at all. Given Hume's point about the logical gap between sample and population, there is, of course, nothing surprising in this: it is simply another way of stating the point that, where scientific knowledge is concerned, men are inevitably fallible.


## II. IDEALS AND RESEARCH

In the preceding section we noted that there are certain limitations on the worth of an explanation or of a prediction. Once this is recognized, there is an important issue that immediately arises. That is the issue of what criterion is used to evaluate such worth. Which is to say, what is the *proper* criterion? Another way of raising the issue is to ask what is the *ideal* of scientific explanation? Scientific explanation is in terms of laws so this question amounts to asking what science might reasonably aim at by way of knowledge of laws? The 'reasonable' here is meant to exclude two things. First, it is not reasonable to attempt to overcome the limitations of induction. So the ideal cannot be one in which such limitations are overcome. Second, it is not reasonable to attempt to achieve absolute precision of measurement. So that, too, is excluded from the ideal. In general, one excludes from the ideal anything it is not possible to attain. With this qualification in mind we can sketch an answer to our question, which, since science aims at knowledge of laws out of two motives, idle curiosity and pragmatic interest, may be rephrased as: what might science (reasonably) want to know by way of laws out of either idle curiosity or pragmatic interest? Answering this question will yield the ideal of scientific explanation and prediction. The answer, I suggest, is *process knowledge.*[1]

A system is a group of entities (bodies, fields, etc.) in an identifiable portion of space. To have process knowledge of a system one must know three things. *One.* One must know a complete set of relevant variables. That is, one must have a set of properties such that no other property within the system is causally relevant to the interaction of this set. *Two.* One must know the system is closed. That is, one must know that no property outside the system affects what goes on inside the system. The generalization of this is that of knowledge of boundary conditions, knowledge of the influences that cross the boundary from outside to inside the system. *Three.* One must know a process law for the system. A state of the system is the entities taking on a specific set of values for the relevant variables. A process law is a law such that, for any possible state of the system, given that state, then from it together with the law any future state and any past state can be deduced.[2] (Here, clearly, given knowledge of initial conditions, the given state, explanation and prediction are symmetric.)

Newton's account of the solar system provided the first approximation to such knowledge. The masses, positions and velocities of the ten objects constitute a complete set of relevant variables. Other objects are sufficiently far away that the system is in effect closed. The law of gravitation enables one to formulate a process law which permits deduction of all future and past states of the system. Such knowledge has also been available in phenomenological thermodynamics. This second example makes it clear that there is nothing specifically "mechanical" about such knowledge. And, indeed, as we have defined such knowledge there has been no limitation on what might be the relevant variables. They may be mechanical, as masses or velocities; or non-mechanical, as fields in physics are non-mechanical; or human and social. Human values, the values people in fact have, are clearly among the relevant variables for psychology and the social sciences. Thus, the notion of process knowledge is as applicable to the latter sciences as it is to the physical sciences, and specifically, the fact that the social sciences have human beings and their values as their subject-matter does not exclude the possibility that process knowledge is the ideal of explanation in these sciences as in physics.

But does process knowledge yield what one might reasonably want to know out of either idle curiosity or pragmatic interest? Well, in respect of the relevant variables, it tells one what will happen in the system, what did happen, what would happen if certain variables were to have different values, what would have had to have been the case if a variable were now to be of a different value, whether a variable can take on a certain value in the system, and if it can, what change must occur (be made to occur) if that variable is to take on that value. That, I suggest, is everything that one could reasonably want to know about the system in respect of the relevant variables out of either idle curiosity or pragmatic interest. What more could one want to know? Nothing, as far as I can see. Certainly, it is knowledge sufficient for explanation. A person understands and can explain a system, e.g., a machine or a production process, just in case he can tell us what will happen to the machine, how the action of the parts is functionally related to the action of the other parts, what would happen if the parts were changed in certain ways, and so on. Process knowledge provides just such a complete understanding. Further, process knowledge tells one everything one needs to know in order to intelligently interfere in the operations of the system and bring about (if one can) what one wants to bring about. In short, both our idle curiosity and pragmatic interests will be fully satisfied by process knowledge. I conclude that process knowledge is the ideal of science, that is, that explanation and prediction of events on the basis of process knowledge is the ideal of science.

Any knowledge of laws which falls short of this ideal we may call "imperfect knowledge."[3] To call it imperfect is not to denigrate it as knowledge. It is, rather, simply to locate it relative to the ideal. Thus, the knowledge that water when heated boils is undoubtedly knowledge, knowledge which is at least relevant to such pragmatic interests as that of making tea. But it is imperfect: it does not come up to the standards of process knowledge. In the first place, the law omits a relevant variable, atmospheric pressure. In the second place, it fails to mention the time required for the water to boil. Indeed, it fails to speak at all of the intervening states of the system. That shows how distant it is from the ideal of process. In this case, one can say, with certain reservations, that this piece of imperfect knowledge can be located within the context of a piece of process knowledge provided by phenomenological thermodynamics. Such location serves to show in just what respect the imperfect knowledge falls short of the ideal. Imperfect knowledge typically has "exceptions", a certain "looseness", perhaps "holding only under certain conditions" as does the law about heating water. What process knowledge enables one to see is the precise nature of the imperfections in the relevant imperfect knowledge. In this sense, perfect knowledge (if it is available) serves to explain the imperfect knowledge in that area.

Of course, such perfect knowledge is often not available. It remains only the ideal goal of science. The only knowledge in fact available in most areas is imperfect. This is the situation at the present time in the social sciences. This is why they must be content with the use of idealized mdoels and with statistics. With respect to models, e.g., those in economics which make use of an idealized economic man, or those embodied in Weber's "ideal types", or those which operations researchers set up of, say, a production process, — in the use of such models, it is explicitly recognized that one is ignoring (often since one is ignorant of them) certain relevant variables. Such a use of models which fit only loosely and approximately and with exceptions is a mark of the imperfection of the knowledge they embody. Statistics go beyond the use of models to a certain extent in that their use enables one to so to speak average out the effects of relevant variables one does not know. In other words, it enables the scientist to overcome some of the difficulties of not having a complete set of relevant variables. Still one has neither those variables, nor *eo ipso* a process law, and so such a use of statistics still serves to mark the area as one of imperfect knowledge.

We must not, however, think that such knowledge is located only in the social sciences. For example, in engineering the knowledge relied upon is often imperect. Often in principle the relevant process knowledge is available, as in the building of bridges, but the complexity of the phenomena is sufficient to exclude in practice any resort to such knowledge, forcing the engineer to rely upon imperfect knowledge, the "rules of thumb" which are so common in engineering texts. And there are other areas of physics, low temperature physics and high velocity aerodynamics, for example, in which all knowledge is imperfect and process knowledge is not even in principle available. The crucial point is that while the term 'inexact sciences' is often used to characterize the social sciences, this usage is decidedly misleading, for that which makes such sciences "inexact", namely, the limitation to imperfect knowledge, is present in branches of such a so-called "exact" science as physics, or engineering (which is physics, save for its motivation being one of pragmatic interest).[4]

One consequence of only imperfect knowledge being available for an area is that when a scientist comes to explain or predict he may often have to rely on his intuitive expertise, rather than explicit knowledge. Explanation and prediction still remains deductive, only it is deduction from laws which fit but loosely and have exceptions. It is because of this looseness and these exceptions that intuitive expertise is relevant. The relevant variables, the conditions of closure, the functional connections among the variables, that ignorance of which renders our knowledge imperfect, are, after all, not known. That being so, when one attempts to apply the imperfect law to explain or predict, one does not *know* whether this application is legitimate or one of the exceptional cases and therefore illegitimate, yielding false explanations and false predictions. That is, one does not know on the basis of consciously articulated knowledge of laws and initial conditions. There is, however, a substitute which is often available, a substitute which lies between our ignorance, in the sense of not being able to consciously articulate (which is, after all, the goal), on the one hand, and complete ignorance on the other. This substitute enables scientists to often make correct predictions even where they do not know how they make them, where they cannot consciously articulate the basis of their inferences. This substitute is the intuitive "know how" which learned expertise in an area provides the scientist. In medicine, most knowledge is in fact imperfect. But we train physicians in such a way that on the basis of certain unconscious cues they can intuitively apply such imperfect knowledge correctly and make successful predictions. This ability which is part of his expertise is what makes the physician's prognosis so useful in respect of his and our pragmatic interests, and so much more reliable as a basis for decision-making than pure guess-work or coin-tossing and chance. Such use of expertise is also important in such social sciences as economics and political science. It is especially important, for practical matters, when these sciences become, under the motive of some pragmatic interest, "policy" sciences. Profound social effects will follow if the ministerial advisor's expertise is relied upon by the minister and the expertise fails, the prognosis turns out to be wrong.

But at this point we must return to the ideal of science. In point of fact a scientist will often be satisfied with less than this ideal. There may be various reasons for this. *First.* It may turn out that in a certain area it is unreasonable to aspire to that ideal because we come to know that in fact the ideal does not hold in that area. For example, consider the sequence of events which consists of a long sequence of tosses of an unbiased coin. With respect to such a sequence it is known that the best we can do is acquire knowledge of statistical laws (the probability of the state heads is one-half, or tails one-half). It is known that process knowledge cannot be obtained for such a system. It is therefore unreasonable to try to obtain such knowledge in this case. *Second.* It may·be that there is no knowledge which makes process knowledge unreasonable as an ideal for an area, while there is sufficient reason, owing, e.g., to the complexity of the subject, to rule it out as unreasonable for any immediate purposes, or, indeed, for purposes in the foreseeable future. This is the state of all social science. *Third.* Both the first two reasons for rejecting process knowledge as the ideal applied whatever the motive was for the science, whether it was a "pure" science or a "policy" science. The third reason occurs only in the latter case where the motive is some pragmatic interest. Idle curiosity always wants to know as much as is reasonably possible. Not so pragmatic interest. Medicine would be satisfied with a cure for cancer. Physicians will leave it to the physiologists to work out the complete physiological process by which the cure is effected. Thus, if one's interest in an area is pragmatic, knowledge far short of the ideal may suffice to satisfy. The "policy" sciences are motivated by a pragmatic interest; they are therefore apt to be satisfied with less knowledge of an area than it is in fact possible to attain.

We noticed earlier that idle curiosity so to speak casts a wider net than any pragmatic interest. It follows that in general only idle curiosity, not pragmatic interest, will lead one to aim at the ideal, to aim at process knowledge. Pragmatic interest will lead one to be satisfied often with much less, with knowledge which is imperfect relative to the ideal. Thus, recognition of the ideal, that is, recognition of process knowledge as the ideal of explanation and prediction, presupposes the recognition that it is possible to be merely idly curious about matter-of-fact generalities. If one does not recognize the possibility of such an interest, if one recognizes only pragmatic interests as the motives for acquiring knowledge of natural laws, then one will continually rest satisfied with imperfect knowledge and simultaneously will not recognize there is a sort of knowledge in which those imperfections are overcome, with the consequence that one will not recognize that one's knowledge is in fact imperfect. Imperfection is defined only relative to the ideal; if the latter is not recognized then neither will the former be recognized. Until Galileo, matter-of-fact knowledge was not thought *worthy* of idle curiosity. For Aristotle, one had to go beyond such knowledge, to the rational intuition of natures. For the mediaevals, only religious truths were worthy of idle curiosity. For the Renaissance humanists, only the classics of literature were worthy. This accounts in part at least for their never recognizing the ideal of scientific explanation. In short, it is important to recognize idle curiosity as one motive behind the search for knowledge.

We must turn now to the idea of a scientific *method.* To adopt a scientific method is to set about actively and rationally to achieve the ends at which science aims, that is, to set about actively and rationally to acquire evidence leading to knowledge of matter-of-fact generalities, whether one's interest in such knowledge is either that of idle curiosity or of pragmatic interest. The basic idea is to weed out falsehoods, and, while weeding them out, acquire more and more supporting evidence for what remains. It is the attempt to constantly improve one's knowledge by testing it against nature. In other words, the scientific method is essentially the experimental method.

Two things are relevant to the workings of the eliminative mechanisms of the experimental method. One is, of course, the relevant data. The other is a set of relevant possible, but competing, hypotheses. The judgment, that data of *these* sorts are relevant while data of *those* sorts are not, is a matter-of-fact judgment of *lawful relevance.* So is the judgment that *these* hypotheses are possible, i.e., lawfully possible, explanations while *those* are not. In the terminology of the logicians, the former sort of judgment is known as a Principle of Limited Variety, and the latter sort as a Principle of Determinism.[5] These Principles — which, it must be emphasized, are matter-of-fact generalities, laws among laws — determine a range of relevant hypotheses. Then data within that range are collected with the aim of falsifying, and thereby eliminating, all of the possible contrary hypotheses but one, and confirming the latter. At that point, one will have an hypothesis worthy of use in explanations and predictions of the relevant data. The framework Principles constitute a body of imperfect knowledge. The aim of the research, the data gathering, is to go beyond this imperfect knowledge to something less imperfect. The framework Principles assert that these less imperfect laws are there — there to be discovered — and they delimit the range of data that must be gathered in order to discover the law that is there. Thus, imperfect knowledge can provide a theoretical framework that guides one to the discovery of less imperfect knowledge.

A scientific theory is commonly reckoned to be a body of laws that so unifies the laws in its domain that it leads to the discovery of previously unknown laws. Imperfect knowledge that functions logically as Principles of Limited Variety and of Determinism can constitute a theory in this sense. The earliest example of a theory of this sort with any scope is found in the axiom system of classical mechanics.[6] These axioms apply generically to a wide variety of specific sorts of systems. In the Law of Inertia these axioms state that neither position nor velocity but acceleration is a function of circumstances. This determines that the process laws in the specific systems all satisfy a certain generic condition. Further generic conditions are introduced in the force = mass × acceleration and action-reaction laws. Within this generic framework, different specific sorts of system have different sorts of process law. Thus, there are those systems for which Newton's Law of Gravitation describes the processes undergone, and there are those systems for which Hooke's Law describes the processes. The axioms also apply to systems of electrically charged elements (electrostatic systems). When the axioms were first formulated, the laws for electrostatic forces were unknown. But the axioms provided a theoretical framework to guide research, giving generic restrictions on the form of the law describing the electrostatic forces, and in due course Coulomb, with careful experimental work discovered the specific form of this law.

Theories are imperfect knowledge relative to the explanation of individual fact and processes. Relative to our cognitive interest in explaining individual facts, then, such laws fall short of the ideal, are not the most desirable. Yet they are of great indirect interest. For, if we desire to know laws for the explanation and prediction of individual facts, then we must be prepared to undertake research as a means to discovering such laws, and to value as means, to take a pragmatic interest in, whatever tools aid that research. But theories are just such tools: they guide the researcher in his search after laws. Hence, the researcher always has a pragmatic interest in theories.

Thus, the physician aims at a specific diagnosis of his patient's disease. To achieve that, he needs a guiding framework of relevance considerations and possible diagnostic hypotheses. These are a means to achieving that at which he aims. With these, and the data available, he then attempts to eliminate all diagnostic hypotheses but one, and to confirm that one. The process of differential diagnosis is, in terms of its logic, simply an application of the scientific method; it is of a piece with other sorts of scientific research. As for the theoretical background, in which the physician has a pragmatic interest — for otherwise the eliminative mechanisms needed to effect the differential diagnosis cannot work — this the physician in general draws from various sciences. This knowledge, in general, he himself does not discover, but is discovered by other researchers, often those who are, unlike the physician, "pure scientists". The engineer similarly often uses as a means in his problem-solving, theories drawn from "pure science." And so, often, does the operations researcher.

In the context of research there is an important distinction to be made among senses of 'acceptance': a generality may appear as a *mere* hypothesis; it is accepted in the sense that that generality is the hypothesis the scientist is going to put to the test. (Call this acceptance$_1$). But it is not yet accepted (accepted$_2$) in the sense of meeting the subjective conditions of acceptance, where this latter means one has evidence that justifies (so far as it can be justified) the use of the generality in predictions and assertions of counterfactual conditionals, or, what amounts to the same, in arguments that are subjectively worthy of acceptance as explanations. The scientist, in aiming at knowledge of laws, is aiming at the acceptance$_2$ of generalities. Acceptance$_1$ is only a stage on the way to acceptance$_2$. A scientist accepts$_1$ an hypothesis in order to test it; if it survives the test, if the research comes to confirm it, then the hypothesis ceases to be a mere hypothesis. It has come to be confirmed and therefore can be accepted$_2$ by scientists. Acceptance$_2$ involves assent to the generality, in a way that acceptance$_1$ does not. As research proceeds, and evidence acquired, those generalities that come to be confirmed move from the context of research to the context of assent or knowledge. And those generalities that are disconfirmed not only do not move to the context of assent or knowledge but disappear, too, from the context of research: being disconfirmed or even falsified, they cease to be candidates for the context of assent, they cease being able to have even the status of a mere hypothesis.[7]

Of course, every scientific generality is, in one sense, always an hypothesis: conclusive verification is not possible, the evidence remains inevitably incomplete. That is the essence of Hume's argument: there is always a logical gap between total population and observed sample. It follows that to accept$_2$ a generality, to assent to it as knowledge, does not imply absolute certainty; nor does it imply acceptance$_2$ is infallible. What is accepted$_2$ as true today, on the basis of presently available (but not conclusive) evidence, may be refuted tomorrow, and will therefore tomorrow not be accepted$_2$ as true. There remains nonetheless a distinction between accepting$_2$ an hypothesis as true on the basis of confirming evidence, on the one hand, and, on the other, accepting$_1$ an hypothesis as a *mere* hypothesis, prior to such evidence being available. If one takes an hypothesis as a not implausible candidate for truth and then sets out to find evidence to confirm or disconfirm it or, more generally, evidence that testifies for or against its truth, then one is using it *only* in the context of research, not in the context of assent or knowledge. That is, one is then accepting$_1$ the generality only as a *mere* hypothesis. Every scientific generality may be an hypothesis, but not every generality is a *mere* hypothesis, and, however much one indulges the rhetoric that science is always conjectural, one cannot obliterate this distinction.

Before going on to try to get a better focus on the method of science two further points are perhaps worth making.

*One.* We must, I think, distinguish what we may call the "mere artisan" from the "applied scientist" such as the engineer or the physician or the operations researcher. For both the interest is in matter-of-fact generalities, and for both the interest is pragmatic, the *laws yielding means* to certain ends. The difference lies in the attitude taken by each towards improving his knowledge. The "mere artisan", the traditional craftsman, more likely than not will simply not attempt to improve his knowledge: he will simply accept the traditional rules of his craft. Even if he does try to improve his knowledge, it will be by a process of simple trial and error, rather than systematic experiment, one of systematically putting the question to nature. The physician or the engineer or the operation researcher, in contrast, does proceed by systematic observation, inference, and hypothesis-testing, and often even by systematic experimentation. He uses the method of experimentation to arrive at the knowledge of matter-of-fact generalities on the basis of which he formulates his procedural rules to achieve whatever end it is he has in mind. The experiments are *not* done for their own sake. The interest of the scientist, be he a "pure" scientist or an "applied" scientist like the engineer, in any experiment is not idle curiosity. Interest in the experiment is always pragmatic. For, the experiment is always a means to another end, namely, the acquisition of evidence and ultimately of knowledge of matter-of-fact generalities. The interest in *that* end may be idle curiosity, but then the experiment is simply a means to achieving an end aimed at out of idle curiosity: it is not itself aimed at out of idle curiosity. The experimenting scientist is in this respect always an engineer, whether he is aiming at knowledge out of pragmatic interest (as is the engineer) or out of idle curiosity (as is the "pure" scientist).

*Two.* It is true, of course, that direct experiment is but the simplest way of putting the question to nature. Far more sophisticated techniques have been developed. Thus, we have the use of statistics which makes possible systematic investigation of matter-of-fact connections where experiment is not possible, either owing to the complexity of the phenomena, or owing to moral reservations about performing the relevant experiments. It is not accidental that statistics have found their greatest utility in the area of the social sciences. Because of the development of these techniques more widely applicable than that of direct experimentation, we might more generally speak of "research", rather than experimentation. But, for all the sophistication, the basic logic of the inferences remains that of the experiment, the logic of eliminative induction.[7a] That idea therefore suffices for our purposes.

One thing must be made clear: the scientific method is no recipe for discovering fruitful laws and theories. No such recipe exists. Good science, like good art, requires the exercise of creative imagination for its growth and development (and for its understanding, also!). The scientific method is a means for evaluating evidence, and for evaluating theories. Its use occurs in the context of justification, not that of discovery. Having said this, one must balance with the remark that of course such rules of justification also provide guide-lines for the researcher. Roughly, the rules of justification generate the policy rule that one should avoid research that one can reasonably foretell will produce only results which violate the rules of justification, of the scientific method. But such policy rules can at best succeed in eliminating bad research. They cannot guarantee good research.

But now let us return, for a more detailed look, to the notion of theories that can guide research.

The basic idea, here, is that laws have certain features of logical form in common. This shared form can be abstracted and used to speak generically about the systems each law talks about specifically. The *generic form* is *abstracted* from the *specific* laws. The theory will thus make use of generic concepts to speak indirectly about the same systems the specific laws speak about. The specific laws will in effect be instantiations of the generic laws of the theory. Such a theory clearly unifies the laws. We may refer to such unification as *unification by abstraction*.[8] The nature of such unification can become clearer if we look at a few examples.

Perhaps the most significant example, in historical terms at least, of an abstractive theory is to be found in classical (particle) mechanics.[9] The axioms of this theory — very roughly, Newton's three laws — apply to a wide range of different species of closed two-body systems, e.g., sun-comet systems, systems of two objects connected by a spring, and so on. We may say these various species belong to the genus of (closed) mechanical two-body systems. The axiom called the Law of Inertia states[10] that

(1)     For each species $s$ of mechanical system, there is a unique pair of functions $f_x$, $f_y$, satisfying certain conditions, such that for any pair of objects $x$, $y$ of species $s$, and for any moment $t$, the accelerations $a_x$ and $a_y$ of $x$ and $y$ are functionally related by $f_x$ and $f_y$ respectively to the circumstances obtaining in the system at that moment.

$f_x$ and $f_y$ satisfy the conditions of being continuous differentiable functions, and the accelerations are accelerations along continuous non-intersecting orbits. The other axioms place further restrictions on $f_x$

and $f_y$. (These are called the "force functions" or even just "forces".) In particular, these other axioms state what the "relevant circumstances" are. The second law introduces a constant, called "mass", and asserts that the product $m_x a_x$ is a measure of the effect of the *other* object $y$ on $x$, or, as it is usually written,

$$f_x = m_x a_x$$

The third law states that what is relevant about $y$ to the behaviour of $x$ is precisely the mass and acceleration of $y$. The rule is

$$f_x = m_x a_x = -m_y a_y = -f_y$$

(1) and the further axioms are, we must note, *generic*. They assert that for any system there are functions $f_x$, $f_y$ which satisfy the indicated conditions but that is not the same as saying that a certain determinate force function holds for a determinate sort of system. Technically, once one has the accelerations and the *determinate* law about how the accelerations depend on what circumstances, then one has the process law for the system, the laws for how the motions change over time. For, the laws concerning the accelerations form a set of second-order differential equations which, when integrated, yield the process law describing how the system develops over time from instant to instant. But (1) and the other axioms do not give such process laws; they only place generic conditions upon the forms such laws must take. In particular, they assert that masses and (relative) accelerations are what are causally related in mechanical systems, or, since acceleration measures rate of change of velocity which measures rate of change of distance, they assert that masses, (relative) velocities, and (relative) positions form a complete set of relevant variables for mechanical systems. And in addition certain common generic conditions are imposed upon the different specific functions relating these variables in various different specific sorts of mechanical system.

It was, of course, this theory of mechanics that Hertz was talking about when he described it as "embracing all the natural motions" but also as "including very many motions which are not natural."[10a] Hertz's point is that all actual process laws for mechanical systems are described by the axioms of classical mechanics — e.g., when the force function is as the inverse square of the distance. But so are others never found in nature — e.g., when the force function is as the inverse cube of the distance. Obviously, what is crucial here is the generic nature of (1):

it is this which permits the theory to describe generically a wide variety of specific forms of motion, or specific forms of process laws, only some of which are actually found in nature.

But if the generic feature, the abstraction from specific forms, permits (1) to describe many different specific sorts of system, then, on the other hand, such descriptions must, obviously, be generic and not specific, and predictions based on (1) can only be determinable, not determinate and specific. This is easily enough seen in the case of (1), but applies equally to any generic theory having the same general sort of structure as (1).

For simplicity, let us state (1) more briefly as

(2)     For each species of mechanical system there is a (unique) force function.

Now, mathematically speaking, to know *the* force function for a system is equivalent to knowing a process law for it. Thus, what (1) and (2) state is that for any mechanical system there is a process law to be discovered. But note that this axiom yields only the prediction that *there is* a force function; it does not say *specifically* what this unique function is. In a sun-comet system the force function is gravitational. In a system of two objects joined by a spring the force function will be that of Hooke's Law. Contrast (2) to the specific law

(3)     For each sun-comet system, the force function for each object is a function of the inverse square of the distance from the other object.

(3) is more specific than (2), yielding a determinate prediction about the force function for such systems rather than a determinable prediction. (3) says specifically what the process law for sun-comet systems is. (3) is not entailed by (2). The latter does entail, given sun-comet systems are mechanical systems, that

(4)     For all sun-comet systems there is a force function.

(3) follows from (4) given the identificatory hypothesis

(5)     An inverse square of distance function = the force function for sun-comet systems.

For any mechanical system *s*, (1) predicts there is a force function. Finding such a function means discovering the truth of an identificatory hypothesis of the sort (5). Furthermore, finding such a function will confirm (1).[11] On the other hand, failure to find such a function would not be sufficient to require us to abandon the Law of Inertia. These features are, of course, a matter of the way the quantifiers are arranged in (1). Below we shall lay out more formally these quantificational structures, using a non-mathematical example. It is these quantificational structures which are crucial to unification by abstraction, and the simpler non-mathematical example we shall soon develop will, it is hoped, make these features more perspicuous.

Another example of an abstractive theoretical law is the axiom of thermodynamics which asserts there are equations of state:[12]

(6)     For each kind $k$ of chemical substance there is a unique function $f$ such that for any sample of $k$ and for all time, the deformation variables for that sample are functionally related by $f$ to the non-deformation variable for that sample.

For gases, the deformation variables are pressure and volume, the non-deformation variable is temperature.

Still other examples of such abstractive theoretical laws can be found in the Laws of Constant and of Multiple Proportions in Chemistry. One might also mention the van der Waals' Equation in gas theory.[13] Indeed, examples are legion.

The crucial logical feature of abstractive theories is that they are *generic. There are mixed quantifications in them and the quantifiers range over both particular systems and species of systems.* This permits us to make *determinable but not determinate* predictions about particular systems. The determinable prediction amounts to a prediction that there is a law for the particular system insofar as it is a species within the genus the theory speaks about. The quantificational structure of the abstractive theory is such that finding the law will confirm the theory, while failing to find it will *not* falsify the theory. The axioms of the abstractive theory abstract a form common to all species by introducing existential quantifiers. *Theories become generic by ceasing to be falsifiable.* But a statement of existential or mixed quantification does not thereby cease to be empirical. "There are dogs" is a good empirical statement, non-Metaphysical; for all that, it is not falsifiable. What Kuhn has done is direct our attention to the mixed quantificational features of theories, the fact that they are both empirical and not falsifiable.

It will, I think, be useful if we examine in more detail the logical structure of abstractive theories, if only because it has been too often neglected in the literature. But instead of an example like classical mechanics, which involves numerical laws, a simpler example will be better for these purposes.[14] So consider the following.

We have a number of kinds of disease, i.e., a number of species of "symptoms", say $G_1$, $G_2$, .... . Each of these is of the genus $\mathscr{G}$. About these there is the hypothesis that each is caused by some distinct but unique species of germ. The task for the researcher is to isolate for each disease $G_i$ the unique species within the genus $\mathscr{F}$ of germ which is the cause of that disease. $\mathscr{F}$ has the various species $F_1, F_2, \ldots$ . The scientist has the working hypothesis

(7)     For each specific disease $g$ of genus $\mathscr{G}$ there is a unique species $f$ of germs of genus $\mathscr{F}$ such that, for any human $x$, the presence of $f$ in $x$ is necessary and sufficient for the presence of $g$ in $x$.

Suppose he is concerned with disease $G_1$. He knows

(8)     $G_1$ is $\mathscr{G}$

From (7) and (8) he deduces that

(9)     There is a unique species $f$ of germs of genus $\mathscr{F}$ such that, for any human $x$, the presence of $f$ in $x$ is necessary and sufficient for the presence of $G_1$ in $x$

(9) tells the researcher that bacteriological causes of $G_1$ exist, and tells him, generically, of the sort they are. Assuming that

(10)     $F_1$ is $\mathscr{F}$
$F_2$ is $\mathscr{F}$
. . . . . .

what (9) tells the researcher is that he must consider all and only the hypotheses

(11—1) For any human $x$, the presence of $F_1$ in $x$ is necessary and sufficient for the presence of $G_1$ in $x$.

(11—2) For any human $x$, the presence of $F_2$ in $x$ is necessary and sufficient for the presence of $G_1$ in $x$.
. . . .   . . . . . . . . . . . . . . . . . . . . . .

(9) asserts that exactly one of these hypotheses is true. Research then proceeds to eliminate the hypotheses that are false, and to isolate that unique species he knows to exist. If the scientist accepts$_2$ (7) and (8), then he can accept$_2$ (9). In his concern about the cause of $G_1$ in humans, this generality (9) that he accepts$_2$ guides his reserach: together with (10) it leads him to accept$_1$ one of the hypotheses (11) as one worthy of research. Indeed, the generality (9) that he accepts$_2$ tells him that exactly one of (11) is true. His research is aimed at discovering just which one it is. The generality he accepts$_2$ tells him *there is* an answer to the question he is asking, "What causes $G_1$ in humans?", and it tells him what *sort* of answer it is. The scientist's task is then simply to find the answer he already knows (on the basis of what he accepts$_2$) is there to be found.

The hypotheses (11) are all specific, and they are used to make determinate predictions about humans. They quantify only over individuals. (9), in contrast involves quantification over species of germ, as well as over individuals. The quantification over species means that (9) makes only a generic claim, not a specific one. Moreover, the quantification over species is existential. Given (10), the generality (9) expresses what each of (11) have in common. The relation of each of (11) to (9) is, given (10), that of existential instantiation of (9); each of the former entails the latter as "Fido is a dog" entails "there is a dog", while, of course, (9) entails none of (11). Insofar as (9) universally quantifies over individuals it cannot be conclusively verified. And insofar as it quantifies existentially over species it cannot be conclusively falsified either. (9) predicts that *there is* a species of genus $\mathscr{F}$ that causes $G_1$ in humans. Failure to find such an $\mathscr{F}$ does not falsify (9). *In this sense*, failure to verify the prediction of the theory does not require us to give up the theory. *Predictive failure is not sufficient for giving up an abstractive theory.* The discovery that several hypotheses of (11) were false would not require us to abandon the theory (9): rather that rejecting (9) as false, we could continue to accept$_2$ it, and conclude instead that *we have not looked sufficiently hard for the $\mathscr{F}$ it says is there as the cause of $G_1$.*

Another way of putting some of these points, one which is especially illuminating, is the following. The generic theory (9) asserts the existence and uniqueness of the cause of $G_1$ in humans. We can therefore form a *definite description* that refers to this species of germs:

(12)      *the* species $f$ of germs such that $f$ is $\mathscr{F}$ and for any human $x$, the presence of $f$ in $x$ is necessary and sufficient for the presence of $G_1$ in $x$.

(9) guarantees the success of this definite description, just as

($\alpha$)      $E!(\iota x)(\varphi x)$

guarantees the success of the definite description

($\beta$)      $(\iota x)(\varphi x)$

Upon Russell's analysis' of definite descriptions ($\alpha$) is the same as

($\alpha^*$)      $(\exists x)[\varphi x \,\&\, (y)(\varphi y \supset y = x)]$

Given $(\alpha) = (\alpha^*)$, we can use ($\beta$) as if it were a singular term. In particular, it turns out to be true that

($\gamma$)      $\varphi(\iota x)(\varphi x)$

Upon Russell's analysis, this is the same as

($\gamma^*$)      $(\exists x)[\varphi x \,\&\, (y)(\varphi y \supset y = x) \,\&\, \varphi x]$

which is, of course, logically equivalent to $(\alpha^*) = (\alpha)$. But, in general, $(\alpha) = (\alpha^*)$ is synthetic. Therefore, ($\gamma$) is also synthetic.

Similarly, if we know

There is exactly one $x$ such that $x$ is $R$ to $a$

then we can successfully form the definite description

The $x$ that is $R$ to $a$

and, parallel to ($\gamma$), we can assert that

The $x$ that is $R$ to $a$ is $R$ to $a$

which will turn out to be a synthetic statement.

Suppose, now, we abbreviate the definite description (12) to something more manageable; let us abbreviate it, say, to "$\gamma_1$-germs". Then we can assert

(13)      For any human $x$, the presence of $\gamma_1$-germs in $x$ is necessary and sufficient for the presence of $G_1$ in $x$.

This is parallel to $(\gamma)$, and, just as $(\gamma)$ asserts what $(\alpha)$ asserts, so (13) asserts what the existence and uniqueness condition (9) asserts. *The latter is synthetic; therefore, so also is (13) synthetic.*[15]

We can use laws of the sort (13) to provide explanations of why individual humans are $G_1$. Suppose we discover that

(14)     Joe is $G_1$

Knowing

(15)     Joe is human

we can deduce via (13) that

(16)     $\gamma_1$-germs are present in Joe.

With this information we can provide the following deductive-nomological explanation of Joe having disease $G_1$:

(17)     For any human $x$, the presence of $\gamma_1$-germs in $x$ is necessary
          and sufficient for the presence of $G_1$ in $x$.
          Joe is human
          $\gamma_1$-germs are present in Joe
          _____

          Therefore, Joe is $G_1$

Explanations of the sort (17) do not violate the thesis of the symmetry of explanation and prediction that the deductive-nomological model requires. That thesis is simply that the explanatory argument must be such that we *could* have known the premises to be true prior to knowing the conclusion to be true. (17) satisfies this. Only, the premise (16) involves, once the definite description is expanded, an assertion of the sort

There are $f$'s of genus $\mathscr{F}$ in $x$

In the absence of any other information we are unable to assert this without our knowing some specific $\mathscr{F}$ is true of $x$; just as, we cannot assert "there are dogs" without our knowing some indivdiual fact to the effect that Fido is a dog. Well, suppose we know such a fact; suppose we know that

$F_1$ is the species $f$ of germs such that $f$ is $\mathscr{F}$ and for any
human $x$, the presence of $f$ in $x$ is necessary and sufficient
for the presence of $G_1$ in $x$.

or, what is the same, that

(18)    $F_1 = \gamma_1$-germs

Then, from

(19)    $F_1$'s are present in Joe

we can deduce that

$\gamma_1$-germs are present in Joe

and use (17) predictively. The symmetry of explanation and prediction therefore holds for (17). However, we must note what (18) commits us to: it commits us to the first of the hypotheses

(11—1) For any human $x$, the presence of $F_1$ in $x$ is necessary and sufficient for the presence of $G_1$ in $x$.

With this we can form the deductive-nomological explanation of Joe being $G_1$:

(20)    For any human $x$, the presence of $F_1$ in $x$ is necessary and sufficient for the presence of $G_1$ in $x$
Joe is human
$F_1$'s are present in Joe
_____
Therefore, Joe is $G_1$

*The explanation (17) uses a law more imperfect than, one with less content than, the law that appears in (20).* We thus see that if we know the third premises of (17) to be true antecedently to knowing its conclusion to be true, then we will always have a better, more contentful, explanation of the same fact. Whenever (17) could be used predictively, it would not be so used, for we would have available a better argument and it is the latter we would use. So, (17) would be used to explain only in those epistemic contexts in which the only way the third premises could be known to be true would be by an *ex post facto* inference from a knowledge of the occurrence of the fact to be explained. Still, that does not mean (17) is not an explanation, only that it is imperfect.[16]

An important point about the imperfect explanation (17) is that the premiss (16)

$\gamma_1$-germs are present in Joe.

already asserts what is contained in the law premiss (7). For, if we expand (16) according to the rule (13), then what (16) asserts is that

(21)    There is a unique species $f$ of germs of genus such that, for any human $x$, the presence of $f$ in $x$ is necessary and sufficient for the presence of $G_1$ in $x$, and, moreover, $f$ is present in Joe.

which, of course, entails (7). In a sense, then, the explanation (17) of $G_1$'s being in Joe says nothing more than what is already contained in the statement of initial conditions. Indeed, we can go further. We can say that the law-premiss of (17) is superfluous. For, given the just-noted expansion of the premiss about $\gamma_1$-germs, it is clear that (17) without the law-premiss, that is,

(22)    Joe is human
       $\gamma_1$-germs are present in Joe
       ―――――――――――――――――
       Therefore, Joe is $G_1$

is valid. However, *appearances notwithstanding, (22) is still a "deductive-nomological" explanation.* It appears not to be, because there is apparently no law-premiss. What must be recognized, however, is that the second premiss of (22) already implies the truth of a law ― for, as its expansion (21) makes clear, the truth of a generality is among the truth-conditions for this second premiss ― and moreover, this generality that is among the truth conditions for the premiss is essential for the deduction of the conclusion of (22).

The *ex post facto* explanations, like (17), have an air of uninformativeness about them. What is true is that when they are employed then we are *not* in a position to predict the facts they are to explain. Moreover, it follows that they cannot be taken as representing cases in which, by means of a prediction, a law has been confirmed. But for all this *air* of uninformativeness, it must be emphasized that they are *not* uninformative ― at least, not uninformative in the sense of being analytic or tautologous[17] ― and that they *are* perfectly good deductive-nomological explanations, though, to be sure, of an imperfect sort. Such explanations, for all their "uninformativeness," do regularly appear in science.[18] But, as we have seen, that would not imply they were somehow vacuous, somehow analytic. For the premises of the imperfect explanation (17) are all synthetic and jointly they entail the conclusion. To be sure, among the premises are generalities that involve mixed quantification. These generalities will be neither conclusively verifiable nor conclusively falsifiable. But, Popper notwithstanding, the non-falsifiable is not non-scientific. Or at least, it is an unreasonable stipulative definition so to define 'science' that no statement involving an existential quantifier is among the statements of law assertable by science; unreasonable, because it would exclude as non-scientific such laws as (1), the Law of Inertia.[19]

Scientists must rely upon explanations like (17) in cases where explanations like (20) are not available: where the less imperfect is not available one must use the more imperfect laws one has available. They rely upon explanations of the sort (17) in those cases where research has not decided among the various hypotheses (11). What the hypotheses (11) amount to is the set of *identificatory hypotheses*.

$$(23) \qquad F_1 = \gamma_1\text{-germs}$$
$$F_2 = \gamma_1\text{-germs}$$
$$\cdot \ \cdot \ \cdot \ \cdot \ \cdot \ \cdot \ \cdot \ \cdot \ \cdot \ \cdot$$

If we accept the existential hypothesis (9) that

$\gamma_1$-germs exist

and we know (10) then we can form the identificatory hypotheses (23). Now, even if it turns out to be false that $F_1$ is necessary and sufficient for $G_1$ in humans, and therefore that the first of the identificatory hypotheses (23) is false, it does not follow that the existential hypothesis is false. For, some other $\mathscr{F}$ species may be the $\gamma_1$-germs even if $F_1$'s are not. Similarly, the existential hypothesis

phlogiston exists

is not falsified if we happen to falsify the identificatory hypothesis

hydrogen is phlogiston

by showing hydrogen does not have the properties essential to any substance which is to be phlogiston. Or, the Law of Inertia (2)

For any species of mechanical system there is a (unique) force function

which is an existential hypothesis, is not falsified if the identificatory hypothesis (3)

> For each sun-comet system, the force function for each
> object is a function of the inverse square of the distance
> from the other object

is falsified; for the force function might be a function of the inverse
2.0015 of the distance. The existential hypotheses, that $\gamma_1$-germs exist,
that phlogiston exists, that the Law of Inertia holds, are all laws which
involve generic concepts and mixed quantification. None can be conclusively falsified by a single counterexample. In the face of a false
identificatory hypothesis we could as easily conclude that we have not
searched hard enough for the facts which the existential hypothesis
asserts to obtain.

*It is precisely this point that must be emphasized. Kuhn argues
theories cannot be falsified; it follows they are not of the simplistic form
"$(x)(fx \supset gx)$" exemplified by the hypotheses (11); but it does not
further follow from this that we must give up the empiricist account of
theories; all we need do is recognize that the non-falsifiability derives
from the mixed quantificational structure of theories.* The identificatory
hypothesis

$$F_1 = \gamma_1\text{-germs}$$

identifies $F_1$ germs with an instance of a type of genus in falls under;

> hydrogen is phlogiston

also identifies hydrogen with a species falling under a certain genus; and
the inverse square of distance falls within the genus "force function for
sun-comet systems". *Recognizing the generic and mixed quantificational
structure of abstractive theories, we can offer an empiricist account of
the features of science to which Kuhn directs our attention.*

We have said that existential hypotheses of the sort (9) can guide
research when they are accepted$_2$ by the scientist. Why should they be
accepted$_2$ prior to any research, however? We indicated, above, one
good reason: They can be deduced from a background theory of the
sort (7) which is antecedently accepted$_2$. But why accept$_2$ such
theories?

We can begin, at least, to see part of what is involved. Suppose that
of the hypotheses (11), the first turns out to be the only one the
evidence confirms. What the others deductively imply about the data
turns out to be false; only the first, through its deductive consequences

being bourne out in the data, is confirmed. But it entails (9), so the latter is confirmed also. However, given (8), (9) is a deductive consequence of (7). So the latter is confirmed also. Suppose the same holds for $G_2$ and $G_3$: we find unique $\mathscr{F}$'s that cause these diseases. We now have data that tend to justify an accepting$_2$ (7). We turn to $G_4$. $G_4$ is $\mathscr{G}$. From this an (7) we deduce and therefore are justified in accepting$_2$ the existential hypotheses

(24)     There is a unique species $f$ of the genus $\mathscr{F}$ such that, for any human $x$, the presence of $f$ in $x$ is necessary and sufficient for the presence of $G_4$ in $x$.

On this basis we form the definite description "$\gamma_4$-germs" to refer to the germs causing $G_4$ and we could accept$_1$ one of the following identificatory hypotheses

(25)     $F_6 = \gamma_4$-germs
$F_7 = \gamma_4$-germs
$. \ . \ . \ . \ . \ . \ . \ . \ . \ .$

as the basis for our research.

What happens when research thus confirms a specific law in a new area? Prior to the research, we had only imperfect explanations, those like (17), which can be obtained from the theory itself. After the research, these imperfect explanations can be replaced by less imperfect ones like (20). That does not, of course, mean that (17) ceases to be an explanation; all that follows is that when we have better, more contentful, explanations of individual facts we prefer to use these to the less contentful. And why do we prefer them? Well, as was argued above, they better satisfy our cognitive interests. So, as research proceeds, imperfect explanations of individual facts that are supplied by the generic theory come to be replaced by less imperfect explanations of those facts. The theory cannot provide the *best* explanations of *individual facts.* On the other hand, the best explanations of individual facts do not use laws with unifying power. The specific laws cannot function as theories, precisely because they are what makes them good at explaining individual facts: specific. The theory, in abstracting from specific cases, lacks the capacity to offer the best explanations of individual facts, but in being in this way generic has just the logical feature that enables it to describe the underlying unity in the laws in several specific areas, the underlying generic structure shared by the specific laws in the several areas. As we said earlier, there is tension being the goal of unity and the goal of content.

The patterns we have discovered in our examination of the simple germ theory of disease can all be found in the more complicated case of classical mechanics. In particular, one can find both examples of *ex post facto* explanations, and examples of where such explanations were replaced by less imperfect explanations. Consider, for example, pairs of electrostatically charged small objects that, when brought into proximity with each other, attract or repel. Since accelerations are involved, the Law of Inertia immediatey implies that *there are* (unique) force functions that describe this interaction. These force functions are said to describe *electrostatic forces.* Given the uniqueness implied by the Law of Inertia, one could speak unproblematically of *the* electrostatic force acting on a charged object, the definite description being justified by the theory, i.e., the Law of Inertia. One also needs the assumption that no other object besides the electrostatically charged object is relevant to the acceleration of the other charged object if one is to speak of the *electrostatic* force, but there is inductive support for the existence of effectively closed systems that is sufficiently strong to justify so speaking. No doubt one should also mention that one needs to distinguish *two* electrostatic forces, a repulsive force between like-charged objects and an attractive force between oppositely charged objects. With these qualifications, the notion of electrostatic forces is fully justified by appeal to the Law of Inertia. One can use this concept to give explanations of the motions of electrostatically charged particulars. Such explanations will, however, be imperfect and *ex post facto* explanations, exactly analogous to our explanation (17) of the presence of a disease in terms of *the* germs that cause it: they will cease being imperfect and *ex post facto* only when the force the explanation mentions is actually identified. We would need an identificatory hypothesis like (18). In the case of the electrostatic forces the job would be to specify the force law the Law of Inertia says is there, that is, the law that describes specifically the electrostatic interactions. It was, of course, Coulomb who first described this law that is now named after him. Once that was available the previous imperfect explanations could be replaced by less imperfect explanations, as we suggested (17) would be replaced by (20). Indeed, in Coulomb's case, what we were provided with were not merely less imperfect but actually process laws.

The actual history of the emergence of Coulomb's law is, naturally,

complicated, and involves a series of steps in which qualitative limitations were placed on the functions that could describe the electrostatic
forces. As these became available, the purely *ex post facto* explanations
in terms of "*the* electrostatic force" could be replaced by explanations
in terms of "the (qualitatively described) force" that is identical to the
electrostatic force. Such partial identifications would yield less imperfect explanations and which, indeed, as we just said, were not subject to
the limitation with respect to predictions that infects *ex post facto*
explanations. It was the final, full identification of Coulomb that
removed *all* imperfections, and made available the process knowledge
the Law of Inertia asserted was there to be discovered. It is, of course,
this complicated pattern of partial followed in stages by less partial
followed by full identification that is more characteristic of much of
scientific research than the rather abrupt identification of the germs
suggested by our simple model theory. But the latter is not wholly out
of place, either.[20]

One other example from classical physics will suffice. Classical
mechanics and Newton's Law of Gravity permit the prediction with
great accuracy of the planetary orbits. In these there is one serious
discrepancy between prediction and observation. This concerns the
perihelon (point of closest approach to the Sun) of the planet Mercury.
If Mercury were the only planet, its orbit would (by Kepler's first law)
be a perfect ellipse, and the perihelion would not shift with time. But
the perturbational effects of the other planets cause it to precess, not
always occuring at the same angular position. Because of Mercury's
high velocity and eccentric orbit, the perihelion can be determined
accurately by observation. It turns out that the difference between what
is observed and what is predicted by Newtonian theory on the basis of
the effects of known planets is about 43 seconds of arc per century.
Though small, this is about 100 times the probable observational error,
and so represents a serious anomaly.

This motion of the perihelion of Mercury is, of course, an acceler-
ated motion. The Law of Inertia immediately entails the existence of a
force function that describes this motion. When the Law of Gravity is
added to classical mechanics, and the usual assumptions made about
the solar system, one can further specify that this force that the Law of
Inertia asserts to exist must be gravitational. There are, however, no
gravitational forces without masses. And, in fact, given the structure of
theory, it is possible to deduce fairly precisely specifically what mass is

needed in order to account for the anomaly in the orbit of Mercury and fairly precisely where that mass must be located. So the abstractive theory of classical mechanics and Newton's Law of Gravity permits us to infer that

(26)     There exists a definite mass at a definite place the gravitational force of which is necessary and sufficient for the anomalous motion of Mercury.

Since the theory is acceptable$_2$, so is this existential hypothesis. Furthermore, since (26) refers to a definite mass at a definite place, it justifies the introduction of the definite description

(27)     The definite mass located at so and so the gravitational force of which is necessary and sufficient for the anomalous motion of Mercury.

Since (27) is successful, that is, more accurately, since its success is acceptable$_2$, we can, as we saw in our discussion of $\gamma_1$-germs, cite it in imperfect *ex post facto* explanations of the otherwise-unaccounted-for advance of the perihelion of Mercury.

Now, it turns out that, given the close proximity of Mercury to the sun, there are (at least) two possible identificatory hypotheses. Here the model is, of course, in our simple germ case, the set of hypotheses (23), though the logical structures of the alternative hypotheses is much more complicated. Both of these identificatory hypotheses will be acceptable$_1$. It is then up to research to decide which of the two acceptable$_1$ hypotheses is acceptable$_2$. It may turn out, of course, that neither is acceptable$_2$. If that is so, then that will raise problems for the researcher, to find, and, he would hope, confirm yet another specific hypothesis that is acceptable$_1$, i.e., that can account for the anomolous motion of Mercury. But failure to find or confirm such an hypothesis does not necessarily falsify the theory.

What we are here noticing is, for our purposes, a crucial logical feature of theories. The point will bear emphasis.

All theories place constraints upon what we may reasonably expect. They state we may expect such and such and if such and such is not observed then they are to be rejected as false. But there are also theories which are not straight-forwardly falsifiable. *These are permissive* with respect to what we may reasonably expect to observe. *The crucial logical feature which makes such permissiveness possible is that the theory contain generic laws involving mixed quantification.* In other words, the theories must be abstractive. It is evident that such permissiveness may come in degrees. A theory asserting a disease is caused by germs is less permissive than a theory asserting the same disease is caused by micro-organisms. Neither will be falsifiable, but the more generic will place fewer restrictions upon what we may reasonably expect: in that sense, it will be more permissive. *What makes such permissiveness possible is the existence of laws involving generic concepts and mixed quantification.* In general, a theory which unifies by abstraction will have several abstracted generic laws for its axioms. These axioms will constitute a theory in the deductive axiomatic sense. They will unify by abstraction the laws, either imperfect or process, which are instantiations of the axioms. The abstract generic characteristics which are mentioned in the axioms will constitute the form which the theory says process laws must take in the systems to which the theory applies. Given such a theory, for example mechanics, for which the Law of Inertia is an abstracted generic law, one will be able to deduce, for the relevant sorts of systems, that a force function exists for those systems. One has thereby predicted the existence of a law. Laws can be thus predicted because the axioms of the theory will involve mixed quantification: they will assert *there is* a law of such and such a generic sort. The theory will say that a law of such and such a kind *is there to be discovered.*[21] In other words, the theory guarantees that if the researcher is sufficiently ingenious then he will in fact discover the law, which, in the example of classical mechanics, would mean that he would discover the force function for the (sort of) system in question. Thus, assuming this theory as background, what research tests is not so much the theory but the ingenuity of the researcher. Of course, the theory is confirmed each time the researcher finds the appropriate law (e.g., force function in the case of classical mechanics). But his failure to find the law does not falsify the theory: the permissiveness secured by the generic concepts and the existential quantifier serves to protect it against such falsification.

The first acceptable$_1$ hypothesis with respect to the anomalous motion of Mercury consisted in hypothesizing the existence of a new planet inside the orbit of Mercury, and quite close to the sun. This hypothesis was to the effect that

(28)     There is a unique planet such that it is (identical with) the
         definite mass located at so and so the gravitational force of
         which is necessary and sufficient for the anomalous motion
         of Mercury.

*If this hypothesis is true*, then it justifies as successful the definite
description

(29)     The planet of such and such definite mass located at so and
         so the gravitational force of which is necessary and sufficient
         for the anomalous motion of Mercury.

This planet, the existence of which was thus hypothesized, was tentatively named "Vulcan", though it was, of course, not really a case of
*naming*, since according to the rule for introducing names into the
empiricist's language, i.e., the language of science, one cannot *name*
what one has not observed. Rather, 'Vulcan' functioned simply as an
abbreviation for the definite description (29). The problem is whether
(28) really is true. The hypothesis (26) was, of course, acceptable$_2$,
which meant that there was indeed good reason to suppose the definite
description (27) was successful. But (28) was not acceptable$_2$: it was
only acceptable$_1$. What is the same as saying the hypothesis that (29)
was successful, i.e., the hypothesis that Vulcan exists, was only accept-
able$_1$. Thus, while (27) could be used in acceptable$_2$ *ex post facto*
explanations of the motion of the perihelion of Mercury, (29) could not
be used in acceptable$_2$ explanations.

    The best that could be done by way of explanation using (29) is, on
the one hand, the "deductive-nomological" argument having (28) as a
premiss and the event to be explained as its conclusion, and, on the
other hand, the explicit recognition that this is *not* being offered as an
acceptable$_2$ explanation but as nothing more than an acceptable$_1$
*hypothetical or tentative* explanation. If (28) *were* to become accept-
able$_2$, then the argument *would* be worthy of being taken not merely
hypothetically but as acceptable$_2$. (28) could become acceptable$_2$ in
either one or both of two ways. One might obtain grounds for supposing that (28) was the *only* acceptable$_1$ hypothesis that was compatible
with (26). Since (26) is taken to be acceptable$_2$, and since it asserts the
existence of something which the additional grounds show could be
none other than that of which (28) asserts the existence, then it would
follow that (28) was acceptable$_2$. It would follow, too, that the claim

that (29) was a successful definite description was also acceptable$_2$. In this case, one would not actually have identified the referent of this definite description. One would therefore use the definite description in explanations, which would be *ex post facto*, and, of course, imperfect. The other way (28) could become acceptable$_2$ is by actually identifying the referent of (29), that is, by actually finding the planet (28) asserts to exist. In that case, one could, but would not, use (28) in *ex post facto* explanations — could, because it would have been shown to be successful, but would not, because by actually discovering the planet the imperfect *ex post facto* explanation would be replaceable by a less imperfect explanation which would be advanced instead.

It turns out, however, that neither of these things happened. What did happen was that, after very careful observation, it was decided that Vulcan did not exist. Instead of acceptable$_1$ explanations being replaced by acceptable$_2$ explanations, and instead of *ex post facto* explanations being replaced by less imperfect explanations, all talk of Vulcan was simply dropped, and the existential hypothesis that it existed abandoned, and, indeed, rejected as false. This raises an interesting issue. The law that the planet that causes the advance exists is mixed quantificational. Since it involves the existential quantifier, it cannot, strictly speaking, be falsified by observations. What, then, led to its abandonment? Clearly, careful observation is the reason it was abandoned. But how does this lead one to reject such an hypothesis? After all, *failure to observe* does not require rejection: that is, surely, the whole point of our emphasis upon the role of the existential quantifier. We have suggested both that existential hypotheses, e.g., like (9) or concerning Vulcan, *can* be falsified *and* that they are unfalsifiable: it is therefore incumbent upon us to show how this tension can be resolved and, more specifically, to show how, in spite of their logical form, research can, at times, lead to the rejection of existential hypotheses as false. For our purposes, it suffices to make only the more essential remarks, describe only the central logical point.[22]

Consider the existential hypothesis

(30)    There is a watch in this drawer.

And suppose we open the drawer, examine the objects in it, say two, and discover neither is a watch. (30) would then, surely, be falsified. Indeed so; for the falsity of (30) follows from the truth of

(31)  | *a* is in this drawer and *a* is not a watch
      | *b* is in this drawer and *b* is not a watch
      | *a* and *b* are the only objects in this drawer

The crucial clause is the final, which, in symbols, is

(32)    $(x)(x$ is an object in this drawer $\equiv x = a \lor x = b)$

This itself is an inductive *generalization*. How is it known? It would seem that we come to believe generalizations like (32) because we deduce them from such generalizations as

(33)    For all *y*, if *y* is a finite volume of space and if *y* is examined in such and such a way, then for any physical object *x*, *x* is in $y \equiv x$ is observed.

Since

(34)    This drawer is a finite volume of space

and assuming that

(35)    This drawer was examined in such and such a way

and provided that *a* and *b* were the only objects observed:

(36)    For all *x*, (*x* was observed if and only if either $x = a$ or $x = b$)

then we can conclude that (32) obtains. And once we have (32) we can conclude that the existential hypothesis (30) is false. Another pattern by which the falsity of (30) might be concluded is this: We examine the drawer and observe that it is empty: this yields

(37)    $\sim(\exists x)(x$ was observed)

or

        $(x)\sim(x$ was observed)

Assuming (34) and (35), we can deduce from (37) and (33) that

(38)    For any physical object *x*, $\sim(x$ is in this drawer)

which contradicts, and therefore falsifies the existential hypothesis (38). What is crucial for inferences such as these to the falsity of existential hypotheses are *laws* of the sort (33). Logically speaking, (30) remains unfalsifiable. But in the context of other knowledge, we can often deduce that it is false. *Existential hypotheses, in the context of other knowledge, can, in this sense, be falsified.* An example perhaps more appropriate for our purpose is

(39)      There is a planet at such and such a point in space

There is a law corresponding to (33) that enables us to deduce that if no object is observed at that place then (39) is false. This law is, of course, the theory of the telescope. What matters for present purposes is that *often the theoretical structure enables us to infer from the failure to observe something to the falsity of some existential hypothesis.*[23] *Equally, however, for many existential hypotheses there is no theory which enables us to deduce the falsity of that hypothesis from a failure to observe entities of the sort it asserts to exist.* Thus, such hypotheses cannot, even in the contextual sense, be falsified. Consider, for example, our disease case once again. We have disease $G_1$ and the existential hypothesis (9)

> There is a unique species $f$ of germs of genus $\mathscr{F}$ such that, for any human $x$, the presence of $f$ in $x$ is necessary and sufficient for the presence of $G_1$ in $x$.

We know the germ has to be *in* the person if it is to cause that person to have disease $G_1$. We can therefore delimit the relevant $\mathscr{F}$'s in the following way. (9) permits the introduction of the definite description "$\gamma_1$-germs" and thereby the assertion of (13):

> For any human $x$, the presence of $\gamma_1$-germs in $x$ is necessary and sufficient for the presence of $G_1$ in $x$.

We now find a person $a$ who has $G_1$. From (13) we can deduce that

$\gamma_1$-germs are present in $a$

This asserts that the species of germ that causes $G_1$ is present in $a$. We know, therefore, we have only to examine $a$ in order to discover the relevant species. But alas, no method of observation guarantees that by following it we can discover and exhaust all species of germ on a person. With telescopes, we can train them upon the relevant spot, look, and if nothing is seen, conclude nothing is there. For germs there is no such neat way of proceeding. Which is to say that, for germs there is no theory of the instrument, no theory of observation, that could enable us to conclude from

> We have not observed germs of genus in person $a$

that

> It is not the case that there are germs of genus $\mathscr{F}$ in person $a$

The relevant points are three:

(1) Logically speaking, no existential hypothesis asserting "there are $\phi$'s" is falsifiable; that is, failure to find a $\phi$ does not *entail* we must reject the hypothesis as false.

(2) Some existential hypotheses, in the context of theories, are contextually falsifiable in the sense that failure to find a $\phi$ together *with the theory* entails we must reject the hypothesis as false.

(3) Some theories involve existential hypotheses and even in the theoretical context these hypotheses are not falsifiable.

The deepest question is, of course, when does one *decide* that the theory of the instrument is to be taken as more strongly supported than the theory itself, so that the former is to be relied upon as constituting *rational grounds* for rejecting the latter. More generally, when one has two theories that are contrary to each other, when does support for one come to be considered strong enough to justify rejecting the other? We shall address this issue below when we discuss Kuhn's notion of "revolutionary science."[23a]

The points that we have been making about research have also been emphasized by N. R. Hanson and Thomas Kuhn. It will pay to look at these in a bit of detail since the positions of these thinkers have often been taken to be at odds with the Humean empiricism or positivism that we are defending.

Hanson has been taken by F. Suppe as one of those who has most effectively challenged the "positivist" or Humean position. Hanson's central thesis is, according to Suppe, the thesis that "the positivists and other adherents to the Received View were wrong in insisting that the domain of philosophy of science is limited to context of justification and that the context of discovery is the domain of psychology and history."[24] Hanson has indeed adumbrated a "logic of discovery."[25] This, in a phrase, consists of the reasons for accepting hypotheses as worthy of test. Since the testing, or the surviving of tests is what justifies an hypothesis, the reasons of which Hanson spaks are somehow not part of the logic of justification; therefore, Hanson suggests, they must be part of the context of discovery. Since the latter has usually been understood by positivists in terms of the psychology and sociology of discovery, Hanson's way of putting the matter has all the air of paradox. Certainly, it functions somehow as a criticism of the usual positivist way of putting things. But we should not, I think, be too hasty. At least, we should not hastily conclude, as Suppe does, that Hanson has offered criticism of any aspect of positivism at least so far as concerns the logic of theories and how they can guide research. In fact, as we shall now argue, Hanson has succeeded in describing, in his "logic of discovery" so-called, some important features of rational decision-making in the on-going process of scientific research. It will turn out that these points are more or less those that Kuhn makes in terms of paradigms and the role of paradigms in guiding research. And as we shall see, both Hanson's ideas and Kuhn's are explicable in terms of the concept of abstractive theories just adumbrated. But perhaps equally significant from our point of view will be the discovery of what Hanson takes to be the "Received View" that he is criticizing. As will appear, what is being attacked is the naïve falsificationist view of the logical empiricists, rather that any view that is essential to the empiricism of the logical positivists or of Hume.

Hanson begins by distinguishing[26]

(1)    reasons for accepting an hypothesis *H*,

from (2)    reasons for suggesting *H* in the first place.

This, though, is not sharp enough. Obviously, reasons of the first kind can be reasons of the second kind. What Hanson wants are reasons of the second that do not amount to those reasons that constitute tests. And there indeed are, Hanson suggests — correctly, I think, — reasons which do not justify accepting an hypothesis *H* as true but which make *H* a *plausible type of conjecture*, that is, one worthy of further investigation.[27] This leads Hanson to sharpen his distinguishing (1) from (2) to distinguishing

(1′)    reasons for accepting a particular, minutely specified hypothesis *H*

from (2′)    reasons for suggesting that, whatever specific claim the successful *H* will make, it will, nonetheless, be of *one kind* rather than another.[28]

He argues that reasons of the second kind are legitimately called "reasons", that such "reasons" are not *merely* a matter of psychology or sociology or whatever (though, of course, they are that: good reasoning is still reasoning and a process of thought). His argument does succeed. He cites in detail the researches of Kepler which clearly establish that Kepler had *reasons*, after he had established the elliptical nature of Mars' orbit, for supposing Jupiter's orbit was elliptical. These reasons include Jupiter being, like Mars, a typical planet, and Mars having (as Kepler had already established) an elliptical orbit.[29] But the argument remains curiously without force. Its problem is that it proceeds only in terms an example. Hanson speaks of hypotheses being of *kinds*; he speaks of "*H*-as-illustrative-of-a-type-of-hypothesis," and contrasts it to "*H*-as-empirically-established"; and he draws the conclusion that

... reasoning from observations of *A*'s as *B*'s to the proposal "All *A*'s are *B*'s" is different in type from reasoning analogically from the fact that *C*'s are *D*'s to the proposal "The hypothesis relating *A*'s and *B*'s will be of the same type as that relating *C*'s and *D*'s". (Here it is the *way* *C*'s are *D*'s which seems analogous to the way *A*'s are *B*'s).[30]

— all without providing any more content for the notion of *type* or *kind* of hypothesis than our intuition can garner from the Kepler example. Hanson never makes it clear just how he expects us to generalize from the concrete example he uses to state his case.[31]

It is, I believe, just the notion of *kind* or *type* of hypothesis that we have spent so much time trying to become clearer about that is relevant to Hanson's "logic of discovery". Again revert to our disease example. We may contrast

(40)    For any human $x$, the presence of $F_1$ in $x$ is necessary and sufficient for the presence of $G_1$ in $x$

stating that germs of this species $F_1$ cause this disease $G_1$ with the generic hypothesis (9)

There is a unique species $f$ of germs of genus $\mathscr{F}$ such that, for any human $x$, the presence of $f$ is necessary and sufficient for the presence of $G_1$ in $x$

I would suggest that (40) is what Hanson has in mind when he speaks of a "minutely specified" hypothesis. And (9) asserts that an hypothesis of a certain *kind* or *type* exists. So I would suggst that Hanson's idea of asserting that an hypothesis of a certain type or kind exists is captured by our notion of generic hypothesis involving mixed quantification. What reason might we have for supposing an hypothesis like (9) is true? Well, as we saw before, $F$ may be of a certain type itself; we may have (8)

$$G_1 \text{ is } \mathcal{G}$$

and we may know the more general law (7)

> For each specific disease $g$ of genus $\mathcal{G}$ there is a unique species $f$ of germs of genus $\mathcal{F}$ such that, for any human $x$, the presence of $f$ in $x$ is necessary and sufficient for the presence of $g$ in $x$.

In Hanson's own Kepler example, the hypothesis corresponding to (7) — and which we now call "Kepler's First Law" — goes roughly like this:

(41)   For any planet, there is a formula coordinating all future positions to the present position, and this formula is that of an ellipse.

'Planet' corresponds to '$\mathcal{G}$', and 'elliptiform' corresponds to '$\mathcal{F}$': they are all generic concepts. (7) could be confirmed by diseases other than $G_1$; then using (8) we can deduce (9). Similarly, Kepler determines the specific formula for Mars:

(42)   For any position of Mars, the subsequent positions are ordered according to the (specific) formula $F$.

This corresponds to specific hypotheses like (40). Knowing

(43)   $F$ is elliptiform

we can deduce that

(44)   There is a formula which is elliptiform and which describes the orbit of Mars

This asserts an hypothesis of a certain *kind* or *type* obtains; it corresponds to (9). The confirmation of (44) and the fact that

(45)   Mars is a planet

constitute data confirming (41). The confirmed (41) now provides Kepler with a reason for supposing that another analogue of (9), namely,

(46)    There is a formula which is elliptiform and which describes the orbit of Jupiter.

is true.

If we understand Hanson's idea of *kind* or *type* of hypothesis in this way then we can see the difference between

(1′)    reasons for accepting a specific $H$

and (2′)    reasons for asserting $H$ is of a certain kind.

The former sort of reasons are those of direct confirmation: the discovery of none but positive instances provides one with reasons for accepting specific $H$'s. The reasons of the other type are those of indirect confirmation: they are constituted by direct confirmation of hypotheses of the same type; the generalization to other cases of the same type; the recognition that the present case falls within the scope of that generalization; and the deduction that in the present case the law will be of the same generic sort as in the previous cases. The logic in case (2′) is more complex. It involves not just positive instances but also a reliance upon the consequence and converse consequence conditions of confirmation. Upon this explication of Hanson's view it does follow that

... Kepler's analogical reasons for proposing that $H'$ would be of a certain type were good reasons. But, logically, they would not then have been good reasons for asserting the truth of a specific value for $H'$ — something which could be done only years later.[32]

and, more generally, that

... analogical and symmetry arguments could never *by themselves* establish particular $H$'s. They can only make it plausible to suggest that $H$ (when discovered) will be of a certain type.[33]

Apropos these analogical and symmetry arguments Hanson remarks that

However, inductive arguments can, by themselves, establish particular hypotheses. So they must differ from arguments of analogical or symmetrical sort.[34]

Hanson never lays out the logic of analogical or symmetrical support. This is because he never clarifies the notion of *type* or *kind* of hypothesis. Once this is done the logical differences become apparent. In particular, it becomes clear the logic is just part of the logic of confirmation, legitimately part of the context of justification rather than the context of discovery. Or rather, it is so once one recognizes that mixed-quantificational laws — i.e., abstractive theories — like Kepler's First Law (41) are parts of science. From this point of view, Hanson is doing no more than we are when we developed our account of theories above. Nor, as we shall see, if we are correct in our interpretation of Kuhn, is Hanson's view seriously different from the latter's.

From Hanson's discussion two things become clear. *First*, he takes the "Received View" to concentrate on reasons for accepting "particular, minutely specified hypotheses." And *second*, he takes the "Received View" to ignore the reasons for suggesting that an hypothesis is a plausible type of conjecture. Now, what the latter amounts to, we now see, is the claim that the "Received View" ignores hypotheses of the abstractive sort, the hypotheses that can provide reasons for believing a successful hypothesis will have to be of a certain *kind*. Furthermore, it follows that what Hanson means by "particular, minutely specified hypotheses" are hypotheses that are *not* of the mixed-quantificational generic abstractive sort. Rather, they must be specific hypotheses of the naïve falsificationist form. Thus, Hanson is criticizing naïve Popperian views rather than anything that is specifically Humean or empiricist or positivist. Indeed the views of these latter are, in their essentials, untouched by Hanson's remarks; their position is not the "Received View" that Hanson attacks!

Let us now turn to Kuhn. In his *The Structure of Scientific Revolutions*, he has a plausible picture of *normal* scientific research that goes roughly like this. Most research is guided by a theory or paradigm. This theory applies to a variety of specific areas.[35] Past successes in guiding research testify to its utility.[36] When applied to a new, as yet unexamined, specific area, it asserts that *there is* a law of a certain sort there to be discovered, while not, on the other hand, asserting *specifically* what that law is.[37] The research task, or puzzle, is to find the law that the theory asserts to be there. Failure to find the law does not falsify the theory or paradigm, it testifies more to the lack of skill on the part of the researcher.[38] If the theory that is used to guide the research leads to the confirmation of a law of the sort it asserts to hold and to the elimination of its competitors, then the theory is in turn confirmed as worthy of acceptance for purposes of explanation and prediction,[39] but failure of a theory already successful in guiding research does not require its rejection: failure most often shows only that more work is needed.[40] Still, if the theory fails often enough, then since it is a tool for the researcher, then he will begin to think about the possibility of re-tooling,[41] looking for a new research-guide. For the "applied scientist" this often means no more than retreating to a more generic level of theory, taking account of a broader range of possible hypotheses and of a range of data previously judged to be irrelevant or of minor significance. This is generally true, too, in research in the "pure sciences". But in the latter especially it sometimes happens that the researcher comes to suspect even his most generic level of theory. At that point he has no broader theory to guide his research. Normal research practice has ceased, and the scientist must resort to other, non-theory, guides for research.[42] Kuhn has described such periods in the history of science as "revolutionary science."[43]

We have said enough already to recognize that it is easy enough to interpret Kuhn's account of normal research in terms of the account of theories with which we are by now familiar.

Let us suppose that we have a number of causal laws something like this:

(a)      For any $x$, $x$ is $F_1$ only if $x$ is $G_1$.
          For any $x$, $x$ is $F_2$ only if $x$ is $G_2$.

Let us suppose further, that the $F_i$ are of a common genus $\mathscr{F}$:

$$F_1 \text{ is } \mathscr{F}, F_2 \text{ is } \mathscr{F}$$

and the $G_i$ are of a common genus $\mathscr{G}$:

$$G_1 \text{ is } \mathscr{G}, G_2 \text{ is } \mathscr{G}$$

From the laws (a) we are in a position to generalize to a *law about laws*.[44]

(b)      For any species $f$ of genus $\mathscr{F}$, there is a species $g$ of genus $\mathscr{G}$ such that, for any $x$, $x$ is $f$ only if $x$ is $g$.

The schema (b) provides a model for the *logical structure* of a Kuhnian paradigm.

Consider a researcher approaching a specific sort system $F_n$ that has not yet been investigated. Since

$$F_n \text{ is } \mathscr{F}$$

the theory (b) applies, and we can assert of such systems the law that

(c)          There is a species $g$ of genus $\mathscr{G}$ such that, for any $x$, $x$ is $F_n$ only if $x$ is $g$.

This asserts, as Kuhn says, that *there is* a law for systems $F_n$ without, however, asserting specifically what this law is. It is the task of the researcher to find this law. Now, if he does *not* find such a law, does not isolate such a $\mathscr{G}$, then *that* does *not falsify* (c), nor, therefore, the paradigm (b). This is due to the mixed-quantificational structure of (c) and (b): by virtue of the particular (existential) quantifier appearing in (c) and (b), those laws cannot be falsified by observational data that fail to testify to the existence of what is asserted to be there.[45]

The law (c) poses a puzzle to the researcher. The latter aims to discover precise laws, laws which make determinate predictions.[46] But (c) yields only *determinable* predictions. Let us say that, if $\mathscr{H}$ is a genus such that $H_1$, $H_2$, ... are species under it then the predicate '$x$ is $\mathscr{H}$-ed' — cf. '$x$ is coloured' — is defined as short for

there is an $h$ such that $x$ is $h$ and $h$ is $\mathscr{H}$

Now consider a particular $F_n$, say $a$. From

$$a \text{ is } F_n$$

we can use (c) to deduce *and therefore predict*

$$a \text{ is } \mathscr{G}\text{-ed}$$

But the scientist also hopes to discover a law which yields a more specific or *determinate* prediction. Moreover, (c) asserts that *there is* a law which *can* yield such predictions. So, the task is to find that law.

(c) delimits a *range* of hypotheses that are *worthy* of the scientist's consideration. These are, let us say,

(d)          For any $x$, $x$ is $F_n$ only if $x$ is $G_n$
             For any $x$, $x$ is $F_n$ only if $x$ is $G_n{}^1$
             For any $x$, $x$ is $F_n$ only if $x$ is $G_n{}^{11}$

where each of the $G$'s is $\mathscr{G}$. In contrast, if $K$ is not $\mathscr{G}$, then

(e)         For any $x$, $x$ is $F_n$ only if $x$ is $K$

is *not* among the hypotheses the scientist must consider; (c) deems it unworthy of his attention. The task is to find data that confirm one and eliminate the others of the set (d), that is, to discover (so far as one can, given the Humean limits) which of (d) is not only plausible — as (e) is not — but also true. Moreover, the scientist proceeds in the knowledge that, if he is skillful enough, he will succeed: the paradigm (b) assures him of that![47]

If we assume that the hypotheses (d) are *contrary* to each other, then data that confirm one will thereby eliminate the others.[48] Suppose our researcher examines the particular system $a$, where

(x)         $a$ is $F_n$

and discovers that

(y)         $a$ is $G_n$

This is what is entailed, and therefore *predicted by*

($d_1$)       For any $x$, $x$ is $F_n$ only if $x$ is $G_n$

The latter is therefore *confirmed*, and the other hypotheses are *eliminated* as false. The researcher has now discovered the specific law that he desired to know and which the paradigm asserted was there to be discovered. But we must also notice that these same data *also* confirm both (c), the law predicted by the paradigm (b), and the paradigm itself. For, since $G_n$ is $\mathscr{G}$, if (y) is true, then so is

($y^1$)       $a$ is $\mathscr{G}$-ed.

which is what (c) predicts of $a$. Moreover, if (x) is true, then so is

($x^1$)       $a$ is $\mathscr{F}$-ed.

and given this the paradigm (c) entails, and therefore *predicts* ($y^1$). Moreover, of course, when the laws (a) are used to successfully predict, *those* data, too, count for the same reasons as data confirming the paradigm (b) which is generalized from the laws (a). Thus, the data testifying to the truth of the specific laws (a) and ($d_1$) *also* testify to the truth of the paradigm. Thus, while the paradigm — unlike the specific laws (d) — cannot be falsified by particular data, it nonetheless *can be confirmed* by such data.

Another way of putting the same point is equally illuminating. Since $(d_1)$ has (c) as its existential generalization, the former entails the latter. So the data that confirm (d) also testify to the truth of (c). But (c) is a prediction of the theory (b). Hence, the latter is also confirmed.[49] In Lakatos' terminology, in predicting (c) which goes beyond the laws (a) that led to its proposal, the theory (b) is *theoretically progressive*; when (c) is confirmed by the discovery of (d), (b) is shown to be *empirically progressive*.[50] For Kuhn, as for Lakatos, such empirical progressiveness is the test for a theory being acceptable for purposes of explanation and prediction, that is, acceptable as true — with, however, the Humean proviso that since these are laws (cf. the universal quantifier) no set of observational data can *conclusively* testify to their truth.

Note also that the data which support (b) *prior* to the discovery of $(d_1)$ also tend to support $(d_1)$. For those data justify using (b) to predict, and therefore support the prediction (c). But the data eliminate all possible hypotheses save $(d_1)$. Hence, (c) could hold only if $(d_1)$ were true. Hence, when one takes the paradigm (b) and the data that eliminate all hypotheses but $(d_1)$, then (b) predicts $(d_1)$. Thus, the data that justify using (b) to predict also support $(d_1)$. In short, the data supporting the acceptance of the laws (a) also support the acceptance of $(d_1)$. Conversely, of course, once $(d_1)$ is discovered, then the data that confirm it also tend to support the acceptance of the laws (a). Thus, paradigms set up interlocking patterns of inductive support.

It would thus seem that by construing Kuhnian paradigms as having the logical form (b), we can account for much of what Kuhn says about the nature of the process of research in science. Certainly, theories that have in fact guided research do, as we have noted, have this structure.

Let us now turn briefly to "revolutionary science."

For Kuhn, normal research is puzzle-solving. A paradigm is therefore a tool: it poses puzzles and guides one to their solutions. However, sometimes it ceases to do this. Sometimes, when scientists repeatedly attack a puzzle, they fail to find solutions, even though the paradigm says they exist. Such failures do not falsify the paradigm, but they call into question its utility as a tool. If such failure is repeated, the puzzle becomes an anomaly,[51] a crisis begins to appear,[52] and some, at least, begin to cast around for a new tool:[53] when the efficiency of a paradigm as a guide in normal research declines beyond a certain point, then scientists begin to look for alternative theories to guide them.[54] Now, by hypothesis, such a search for a new theory will not itself be paradigm-guided. It will therefore not be normal science. Rather, Kuhn suggests, it can reasonably be called "revolutionary science".[55]

The problem is that when the paradigm is given up, then one has no accepted *theory* to guide one's search. But given the infinite range of possibilities, no search can proceed unless *some* guide is available. How can one select from the infinite range of possibilities which hypothesis to accept, and put to the test? In part one must be guided by the idea that the new paradigm must account for the successes of the old one.[56] Even then, however, an infinite range of contrary possibilities confronts one. Kuhn argues that in such situations, while no individual scientist can confront the totality of possibilities, the community as such can. It does this by letting scientists in periods of revolutionary science be guided by metaphysical and/or aesthetic considerations that vary subjectively from scientist to scientist.[57] *A new candidate for paradigm is a theory that has never been projected. The subjective values provide one with reasons in periods of revolutionary science to take up a theory never previously tested:*[58] "If they [those who initially adopt an untested candidate for a new paradigm] had not quickly taken it up for highly individual reasons, the new candidate for paradigm might never have been sufficiently developed to attract the allegiance of the scientific community as a whole."[59] N. R. Hanson also points to the role of such subjective values in the "logic of discovery." As we saw, Hanson, like Kuhn, argues that research can be guided by laws — that is, laws about laws — such as Kepler's first law, by which Kepler inferred the *form* of the law describing the orbit of Jupiter.[60] But Hanson also allows another kind of factor to play a role:

Other kinds of reasons which make it plausible to propose that an [hypothesis] *H*, once discovered, will be of a certain type, might include ... the detection of a formal symmetry in sets of equations or arguments. At important junctures Clerk Maxwell and Einstein detected such structural symmetries. This allowed them to argue, before getting their final answers, that those answers would be of a clearly decipherable type.[61]

The important point is that this sort of reason is provided not by a tested law but by an appeal to aesthetics or perhaps metaphysics. Hanson does not, however, where Kuhn does, provide an account when it is *permissible* to involve such subjective values in making one's scientific judgments. Note that this is *not* to introduce irrationalism into science.[62] Rather, the introduction of such things as aesthetic value is but a *means* to an end[63] — where the end is, of course, the discovery of a new paradigm.[64] In fact, the older paradigm might turn out to solve the puzzle after all. To allow for this, the community permits some to tolerate a lower rate of efficiency than do others; the former will continue to use the old paradigm while the latter search for a new.[65] Thus, there is no *precise* point at which a paradigm is to be given up; that, too, varies subjectively from individual to individual. But that, too, is part of the *rational research strategy* of revolutionary science.[66] In periods of crisis the rule is: *let theories proliferate;*[67] *let scientists be guided in their search by their subjective values;*[68] *and let them accept for purposes of explanation and prediction those theories that satisfy those subjective values.*[69] In this way the community *as a whole* is able to deal with the vast range of possibilities it must investigate. Thus, this rule is adopted because having its members conform to it in periods of crisis is a *means* for *efficiently achieving* the *end* shared by all members of the community, namely, that of discovering a new paradigm capable of successfully guiding research in the normal way. Thus the resort to subjectively variable criteria of choice is a *rational response to crisis* in the scientific community.

The problem, raised above,[70] when does accepting one theory constitute rational grounds for rejecting a contrary theory, is now solved.

Suppose we have a theory $T_1$ which entails that *there is* a law that explains certain facts $f$. Failure to discover such a law does not falsify $T_1$. But if there is a contrary theory $T_2$ which also asserts that there is a law that explains $f$, and one discovers a law of the sort that it predicts, then $T_2$ is confirmed, or is empirically progressive, where $T_1$ is not. This renders $T_2$ acceptable₂ and that is reason to reject $T_1$. But if one has $T_1$ and not $T_2$, and $T_1$ is not falsified, when should one start to search for a contrary and in *that* sense reject $T_1$? Answer: when the puzzle-solving success-rate of $T_1$ falls below a (subjectively variable) acceptable level.[71] And in the search for an alternative to $T_1$, why accept₂ from among all the possibilities $T_2$ as one's research-guiding theory, prior to $T_2$'s confirmation by the discovery of the law explaining $f$? Answer: accept₂ $T_2$ prior to its demonstrated superiority because it satisfies one's (subjectively variable) aesthetic, etc., criteria.[72]

I conclude that the empiricist account of laws and theories that we are defending can accommodate the major Kuhnian insights into the research process.

## CHAPTER TWO

# LAWS, ACCIDENTAL GENERALITIES, AND COUNTERFACTUAL CONDITIONALS

## I. INTRODUCTION: COUNTERFACTUALS AND THE HUMEAN ACCOUNT OF LAWS

Philosophers have recognized for some time that counterfactual conditionals like

(1)     If Sam had come to Eve's party, he would have enjoyed himself

present problems. Of these, two are the most important. *First.* A counterfactual conditional cannot be translated in the usual way by means of the horseshoe of material implication. What, then, is its logical form? *Second.* A counterfactual seems to describe not a fact but an unreality. Thus, in (1) a situation, namely, Sam's coming to the party, which, it is admitted, has not happened to Sam, that is, is unreal, is further determined by another unreality, namely Sam's enjoying himself.

These two problems confront the empiricist or Humean with the task of proposing an analysis of counterfactuals that is, in the first place, logically adequate, but in the second place, since what is true is actual, does not commit one to claiming that counterfactuals are truths about unrealities, that is, claiming that there are unrealities which are also actual. The world we experience through our senses is *the* real world, not a world of unrealities. In Hume's terms, our impressions are of the real, not the unreal. No empiricist, no Humean, could hold that counterfactuals really are what they purport to be, truths about unrealities.

The problem with material implications is clear enough, but hardly surprising. The suggestion that (1) be analyzed as

(2)     Sam came to Eve's party $\supset$ Sam enjoyed himself

clearly will not do. For

$$p \supset q$$

is true if and only if

either (i)     '$p$' is true and '$q$' is true

or      (ii)    '$p$' is false and '$q$' is true

or      (iii)   '$p$' is false and '$q$' is false

and otherwise false, i.e., false just in case that '$p$' is true and '$q$' is false. In that case, since the antecedent of (2) is false, (2) will be true. But for the same reason

(3)       Sam came to Eve's party $\supset$ Sam did not enjoy himself

will also be true. If (2) analyzes (1), then (3) analyzes

(4)       If Sam had come to Eve's party, he would not have enjoyed himself.

But, while both (2) and (3) can be accepted, it is clearly not possible to accept both (1) and (4): there is something incompatible in (1) and (4), where there is nothing similar in (2) and (3). Hence, counterfactuals cannot be represented in their logic by material conditionals.

Since (1) and (4) are somehow incompatible, there must be something about them that makes them so, something that is not present in the assertion of a mere material conditional. This "something more" can be approached once we recognize that while the situations described by counterfactuals are unreal, they are *not wholly unreal.* In the case of (1), we know Sam and Eve well enough that we can say

(5)       Eve's parties are always of the sort that when Sam attends them, then he enjoys himself.

While the *situations* described by (1) are unreal, the *connection* (5) between them is not: *the connection really obtains,* even if what are connected does not. And, of course, if (5) means that the situations described in (1) *are* connected, then it must also mean that the situations described in (4) are *not* connected, for a person cannot both enjoy and not enjoy the same party.

It is to be expected that material implications are not as such commonly to be found in ordinary discourse. The material conditional

(6)       The moon is made of green cheese $\supset$ De Gaulle is tall

is true, since the antecedent is false, but is clearly "odd", or, at least, "odd" if rendered into the ordinary idiom as

(6*)        If the moon is made of green cheese then De Gaulle is tall

The latter is sufficiently "odd" that some have argued that (6) cannot be
an adequate translation of (6*), concluding that one needs a further sort
of logic — "informal" logic — to complement formal logic. This conclusion does not, of course, follow. It follows only if the "oddity" cannot be
accounted for save by assuming there is a sort of logic beyond formal
logic. Those who, like empiricists, wish all necessity to be understood in
terms of the necessity of formal logic, can avoid the conclusion by
offering an alternative explanation of the "oddity". Nor is it difficult to
find such a alternative.

The major use of a conditional

>        If $p$ then $q$

is to permit people to draw inferences from it by *modus ponens*

>        If $p$ then $q$
>        $p$
>        $\overline{\hspace{3cm}}$
>        so, $q$

and *modus tollens*

>        If $p$ then $q$
>        not $q$
>        $\overline{\hspace{3cm}}$
>        so, not $p$

It is easy to see that if a material conditional

(7)        $p \supset q$

is accepted as true only on the basis of knowing the truth-values of its
components, i.e., only on the basis of either knowing that $p$ is false or
knowing that $q$ is true (or both), then the conditional could not be used
to draw conclusions by means of either *modus ponens* or *modus tollens*.
For, *if an argument is to be useful in bringing an audience to rationally
accept the conclusion,* then the following conditions must be met:[1]

(i)        the argument must be (formally) valid

(ii)        the argument must have true premises

(iii)        the audience must know that the premises are true

(iv)        the audience must not know that the conclusion is true

Even if (i) and (ii) are fulfilled, an argument cannot rationally persuade unless the first epistemic condition (iii) also is fulfilled; while the argument is not needed, of no use, if the second epistemic condition (iv) is not fulfilled. Now, if we know $p \supset q$ to be true because we know $p$ is false, then *modus ponens* will be useless because condition (ii) is not fulfilled, while *modus tollens* will be useless because condition (iv) is not fulfilled. Similarly, if we know $p \supset q$ to be true because we know $q$ is true, then *modus ponens* will be useless because condition (iv) is not fulfilled, while *modus tollens* will be useless because condition (ii) is not fulfilled. It follows that a material conditional $p \supset q$ will be useless for purposes of inferences by means of either *modus ponens* or *modus tollens* if the *only* grounds for accepting it as true are a knowledge of the truth-values of the components.

A conditional like

(8)     If De Gaulle is a bachelor then he is unmarried

which is not at all "odd", is accepted as true not merely because we know the truth-values of the components but because we know a *connection* between antecedent and consequent. This is the connection established by the fact that

(9)     All bachelors are unmarried

is true by definition and entails (8). Similarly, a conditional like

(10)     If De Gaulle steps out the window then he falls

which is also not at all "odd", is accepted as true because we know a connection between antecedent and consequent, namely, the connection established by the known law of nature

(11)     Unsupported heavy objects fall

which entails (10) — given that De Gaulle is a heavy object.

In contrast, a conditional of the "odd" variety, like (6*), is such that *there is no connection* between antecedent and consequent. The only grounds for accepting them as true could be a knowledge of the truth-values of the components. Thus, the "odd" conditionals cannot be used for purposes of inference by *modus ponens* or *modus tollens*, whereas those that do not seem "odd", like (8) and (10), *can* be used to draw such inferences because a known connection justifies their acceptance independently of a knowledge of the truth-values of their components. But this, surely, is a sufficiently major difference to account for why conditionals like (6*) seem "odd" while those like (8) and (10) do not seem "odd": conditionals for which there is no connection between antecedent and consequent are "odd" just because they are inferentially useless. The "oddity" of the "odd" conditionals is thus accounted for without the need to introduce some sort of logic beyond formal logic; their oddity is not due to their being material conditionals, but rather due to there being no connection linking antecedent and consequent.

One cannot say this straight off, however, for there are *some* uses for the "odd" conditionals. Consider examples of the "monkey's uncle" sort:

    (a)      If Plato was an empiricist, then I'm a monkey's uncle.

Clearly there is no connection between antecedent and consequent; (a) is known to be true, but only because we know both antecedent and consequent to be false. But if we know the consequent to be false:

    (b)       ~ (I'm a monkey's uncle)

then, by *modus tollens*, we can infer

    (c)       ~ (Plato was an empiricist)

However, we of course *know already* that this is true. We have used *modus tollens* but *not* so as to *lead us to knowing the conclusion is true*. I.e., the second epistemic condition is not fulfilled. The point of using (a) is thus not to enable us to infer new knowledge. It is, rather, to *emphasize more strongly* the conclusion the *modus tollens* entails from (a). We already are prepared to affirm (c); asserting (a) is a rhetorical way of re-emphasizing that affirmation.

Thus, the absence of connection does not entail non-utility — even "odd" conditionals can have a use — but IF *the conditional is to be used to lead us inferentially to the truth of the conclusion,* THEN *the conditional is useful only if it is affirmed in the context of a law that establishes a connection between antecedent and consequent.*

If the connection is linguistic, as with the conditional (8) and the definitional truth (9), then, clearly, the gain in knowledge that the inference to the conclusion generates is not a matter of *predicting* certain things. In contrast, the conditional (10)

           If De Gaulle steps out the window, then he falls

is a *predictive conditional.* That is, it is a conditional that can be used in a *modus ponens* — or a *modus tollens* — to *predict* something about the subject of its antecedent. And this is to say that *for a predictive conditional there is a law connecting antecedent and consequent such that the constitutive and epistemic conditions (i)—(iv) can simultaneously be fulfilled.* In general, then, if a conditional is used in a context determined by *a cognitive interest in prediction,* then there must be present in the context a law that connects antecedent and consequent.

Not every conditional is used predictively, however. Consider the *factual conditional*

(12)     Since Armbruster stepped out the window, she fell

Here the conditions (i)—(iv) are *not* fulfilled. The grammatical form clearly implies that the first epistemic condition is fulfilled but the second is not: asserting (12) gives it to be understood that one knows already that the consequent is true. What the form suggests is not prediction but *explanation*: (12) provides an explanation of *why* Armbruster fell. This explanation is constituted by the *law* that connects the antecedent and consequent. Here, *a cognitive interest in explanation* determines that, for a conditional asserted in a context determined by that interest, there must be a law that connects antecedent and consequent.

We may say, then, that if a conditional is used predictively or used to explain, then there must be in the context a law that connects antecedent and consequent. There is, of course, a link between these two cases of explanation and prediction. As we saw in Chapter I, Sec. I, there is a symmetry between these: every prediction is, if it is successful, an explanation, and every explanation could have been used as a prediction had the initial conditions being known prior to the explanation.

Now, the general idea that constitutes our idea of the laws that science uses for explanation and prediction is that of *control,* or intelligent interference. Thus, our point is this: the acceptance of any non-"odd" conditional in a context determined by a cognitive interest in control or intelligent interference implicitly presupposes in the context the acceptance as a law of a generalization such that this generalization and the antecedent of the conditional jointly entail the consequent. If

(12)     If $a$ is $F$ then $a$ is $G$

is an accepted ordinary conditional, then there is an accepted generalization

(13)      All $F$ are $G$

such that

(14)      All $F$ are $G$
          $Fa$
          _____
          so, $Ga$

is valid. Thus, to give another example, if we accept the factual conditional

(15)      Since Sam came to Eve's party, he enjoyed himself

then back of it we have a generalization like (5) to justify its acceptance. Similarly, if we accept the predictive conditional

(16)      If Sam comes to Eve's party, then he will enjoy himself

then back of it we will have a generalization like (5) to justify its acceptance. These forms have the same conditional element, the same asserted connection over and above the material conditional; they differ only in what they say or suggest or imply or give to be understood about the antecedent and consequent: in (15) both the event mentioned by the antecedent and that mentioned by the consequent have already occurred, while in (16) they have yet to occur.

And now we can note the parallel with the counterfactual conditional (1)

> If Sam had come to Eve's party, he would have enjoyed himself

This shares with the factual conditional (15) and the predictive conditional (16) the same conditional element and the same asserted connection over and above the material conditional. It differs only in what it says or suggests or implies or gives to be understood about the antecedent and consequent: it says or suggests or hints that the antecedent and consequent are false.

In each case we have, implicitly, something like the following *argument*:

(A)      ($P_1$) Eve's parties are always of the sort that when Sam attends them he enjoys himself [(5)].

          ($P_2$) Sam comes to Eve's party
          _____
          (C)   Sam enjoys himself

In each, the antecedent appears as a minor premiss and the consequent as the conclusion. The major premiss is a generalization that effects a connection between antecedent and consequent. In each, this major premiss is law-asserted, accepted as true. The difference lies with respect to the propositional attitude adopted towards the minor premiss and, therefore, the conclusion. In the factual conditional, the minor premiss is accepted as true. In the predictive conditional, the minor premiss is accepted as true, but as mentioning an event that has not yet occurred. In the counterfactual conditional, the minor premiss is rejected as false, or, at least, not accepted as true.[2]

Yet, to assert either a factual, or a predictive, or a counterfactual conditional is not to give a complete inference like (A). In asserting such a conditional we do *not* give all the premisses and intermediate steps; in particular, we omit the law that effects the connection between antecedent and consequent. These premisses are only implicit in the context of assertion. Let us say that the law premiss that connects antecedent and consequent *sustains* the assertion of the conditional.

A non-material conditional thus seem to be an incomplete or condensed argument in which there are one or more implicit premisses, including the law premiss that enables non-material conditionals to function as premisses in *modus ponens* and *modus tollens* inferences in everyday contexts: the law-premiss sustains the assertion of the conditional in the sense of justifying that assertion independently of a knowledge of the truth-values of the components. Our consideration of why certain conditionals are "odd" where others are not led to the conclusion that associated with the latter are grounds that justify their assertion independently of a knowledge of the truth-values of the components. A sustaining law provides just such grounds. It is therefore reasonable to expect non-"odd" conditionals, the non-material conditionals of ordinary discourse, to have implicitly connected with their assertion a law that sustains that assertion by deductively linking antecedent and consequent, providing a *general guarantee* that there could be *no case* in which an event of the sort mentioned by the antecedent could obtain and an event of the sort mentioned by the consequent could fail to obtain.

*It is plausible, then, to construe many non-material conditionals, in particular, factual conditionals, predictive conditionals, and especially counterfactual or subjunctive conditionals, as condensed arguments.*

Now, as we shall see in Chapter III below, there are some who disagree with this, and propose to use the language of counterfactuals in

a way that involves no connections. On this view, a counterfactual is to
be asserted even if there is no connection between antecedent and
consequent provided that the corresponding material conditional is
true, roughly, in a possible world most similar to the actual world.
Provided that this involves no ontologically odd entities, e.g., unrealities
that are actual — and it can, as we shall argue, be so construed — then
no empiricist could object: people can use language as they wish, and
there is no *Academie anglaise* to enforce conformity to the rule that
counterfactual conditionals be asserted only in a context in which one is
prepared to assert a law that effects a connection between the antecedent and the consequent. However, we shall also argue that *where
certain purposes, cognitive interests, are to be served by the use of
counterfactual conditionals then it is* appropriate, given those ends, *to
assert counterfactuals, as well as factual and predictive conditionals, only
where their assertion can be sustained by a law connecting antecedent
and consequent.* The relevant cognitive interest is in *control and
intelligent interference in a natural or social process.* In any context
defined by this cognitive interest, we shall argue, a counterfactual
conditional *ought to be asserted* only if a law is available in the context
that sustains its assertion. This means, of course, that the alternative
proposed, in terms of possible worlds similar to ours, ought, in such
contexts, to be rejected.[3]

It is, of course, not accidental that argument (A) is just that sort of
argument that constitutes an *explanation* of the events mentioned in the
minor premiss and the conclusion. At least, that is the substance of the
deductive-nomological model of explanation, the defence of which we
sketched in Chapter I, above. Indeed, as we pointed out in that context,
an explanatory argument is also a possible prediction: this is the thesis
of the symmetry between explanation and prediction that is entailed by
the deductive-nomological model of explanation. We have suggested
that a factual conditional is a condensed argument; that argument, when
made explicit, constitutes an *explanation* of the events the conditional
describes. Similarly, if a predictive conditional is a condensed argument, then when that argument is made explicit it becomes a *reasoned
scientific prediction.* Our suggested analysis of non-mateial conditionals,
including counterfactuals, as condensed arguments, is thus of a piece
with the account of science that we have developed.

Nor should this surprise us, for as we also argued, the cognitive
interests that motivate our concern for scientific explanations have as

their determining notion that of *knowledge that permits control.* And if the suggestion we made above, and are to defend in Chapter III, is correct, then counterfactuals used in a context determined by a cognitive interest in control *ought to be,* when made fully explicit, arguments of the same general from as (A). Since the determining cognitive interest is the same, it is not accidental that our account of counterfactuals is of a piece with our account of scientific explanation.

We shall return in Chapter III to the theme of why we ought to accept an account of counterfactual conditionals in which they are condensed arguments sustained by general laws. But there are some other themes that must be dealt with first.

Upon the proposed account of counterfactuals, the assertion of the latter is sustained by *laws.* Now, the usual empiricist account of laws is that of Hume. Upon this account, a statement of law

> All *F* are *G*

is analyzed as a true matter-of-fact generalization

> (x)        $(x)(Fx \supset Gx)$

But consider the true generalization

> All the coins in my pocket are copper

It is evident that this will *not* sustain the assertion of the counterfactual

> If this coin [which is silver] were in my pocket, it would be copper

It thus seems that generalizations cannot sustain the assertion of counterfactual conditionals. And since the Humean analyzes laws as generalizations, it seems that the Humean is unable to account for how laws, in contrast to *mere* generalizations, can sustain counterfactuals. In fact his account of laws seems to preclude the Humean from developing any account of counterfactuals along the lines we have been suggesting. To put it another way, the account of counterfactuals as collapsed arguments is plausible only to the extent that one can draw a distinction between laws, which sustain counterfactuals, and generalizations, which do not; and since the Humean insists that, ontologically or objectively, there is no difference between laws and (mere) generalizations, he is precluded from adopting the collapsed-agreement analysis of counterfactuals.

The Humean answer to this consists in arguing that, while there is *no objective* difference between laws and more generalizations — objectively, they are both of the form (x) — it does not follow that there is no difference between them. Even if objectively or ontologically there is no difference between *propter hoc* and *post hoc*, it does not follow that there is no difference. Indeed, as commonsense insists, and as Hume recognized,[4] there is a distinction to be drawn.[4a]

The rationalist insists that the distinction between laws and mere generalizations consists in there being an *objective necessary connection* that links the properties mentioned in the law. Rationalists from Aristotle to Descartes and beyond held that this connection is not presented in sense experience;[5] one knows it, rather, by a rational intuition. This intuition is arrived at either by abstraction (Aristotle) or it is a matter of innate ideas (Descartes); but in either case the intuition provides insight into objective necessary connections that constitute a level of reality not given to us in sense experience. Locke is the first to shatter this world-view on which reality comes in two levels, a lower one known to sense and a higher one known to reason.[6] *Locke's world, the world of the empiricist, unlike that of the rationalists, is all at one level, the level of sense.* And with the elimination of the higher level, the need to introduce rational intuition is gone. No longer are there two ways of knowing, sense and intuition: all knowing now begins and ends in sense. For a rationalist like Descartes, man's reason is his spiritual element: it is his point of contact with God. For such a one, man is semi-divine, split between a heavenly rational part and a sensible corporeal part. Locke attacks this idea of man, insisting that man is thoroughly of *this* world. The rational powers that supposedly lift man out of this world are otherwise to be accounted for. Reason is as natural as sensation. There is nothing about science and knowing that marks a break in man between a sensible level and some higher level. Any claim that reason can do more than begin and end in sensible knowledge is dismissed. Locke, the first of the great philosophers to dismiss the inhuman view of man (that he is partly not of this world in which he lives and breathes), was the first to adopt the truly humane Enlightenment view of man as human and no more than human.

It is Locke's position that "If we can find out how far the Understanding can extend its view, how far it has Faculties to attain Certainty, and in what Cases it can only judge and guess, we may learn to content ourselves with what is attainable by us in this State."[7] His purpose is "... to inquire into the Original, Certainty and extent of human Knowledge, together with the Grounds and Degrees of Belief, Opinion and Assent";[8] the method he proposed to use, we all know, was the "Historical, plain Method"[9] (where 'historical' meant experimental or observational, i.e., the method of empirical science). The result was an attack on the metaphysics that from Plato and Aristotle had supported the view of man as a being divided from himself, a being with two ways of knowing. The "real essences" that were the core of that metaphysics Locke declared unknowable. Since the method of empirical science relies not on rational intuition for knowing particulars but on sense observation alone, Locke declares "real essences" not a proper object of science. For Aristotle, for Descartes, for the rationalist, method moves from sense to intuition and concentrates on the latter, valuing the former only as a trigger for the latter. Method, for such a philosopher, is not unitary. For Locke, on the other hand, method *is* unitary. He succeeds in making it unitary by distinguishing empirical science from metaphysics: the latter is a "science" of essences, or natures, or what-have-you lying, as Aristotle, Descartes and other rationalists would have it, beyond and above the world of sense. For Locke that world is unavailable, and any problems with somehow understanding it are not those of science.

Now, I think that one must recognize that in adopting this unitary method, that of empirical science, and on its terms declaring the realm of essences, natures, and real definitions unknowable, Locke's basic decision is to opt out of the game of "science" where knowledge of these trans-empirical entities is supposed to be fundamental to success. Indeed, one could say with some justice that there is a sense in which the *Essay's* defence of empirical science as the sole route to knowledge *is* one long refusal to engage in the game played by Aristotle, Descartes, and Taylor. But if that is all that it comes to, then one can legitimately wonder whether Locke's opponents have been *refuted*, whether Locke has *established* that science is a unity and this unified science is empirical science.

Aaron, in his *John Locke*, has observed judiciously that

As a matter of logic we may not, strictly speaking, be in a position to deny the possibility of discovering a solution to the most abstruse problems, but practically we frequently find ourselves in a position, as the result of repeated failure, in which we feel able to say it is most unlikely that this problem will ever be solved by us.[10]

Perhaps in some sense of 'can,' we *can* come to know real essences, to grasp real definitions, to formulate a methodology of rational intuition. On the other hand, especially in view of repeated failures of "realist" approaches to foundations and method in science, it is hardly incumbent upon Locke to provide an impossibility proof of what is highly unlikely.

The problem with metaphysical explanations is that they turn out *in fact* to add nothing to what the "historical, plain method" reveals of the world as we experience it. His point is that of Molière:[11] pointing to the essence of opium as involving a dormative virtue far from establishing a necessary connection does in fact nothing more than re-state in other terms the matter-of-fact generalization that, whenever it is taken in appropriate quantities opium puts one to sleep.[12] The *Essay* makes this point systematically. One can find in experience none of the entities, notions, natures nor necessary connections that the Aristotelians assert are there, known, familiar as objects in every day experience, and centrally ingredient in the apprehensions we have of things.[13] Locke argues on the basis of the "historical, plain method" that neither simple apprehension nor judgment nor discourse provide any grounds for the claim that we know the real essences of material objects. Our ideas of substances, which in the Aristotelian tradition consist of parts necessarily tied together in a real definition and are therefore in that tradition called "simple apprehensions," are disclosed by the "historical, plain method" not to be a group of ideas tied of their own necessity into a unity but no more than a collection of independent simple idea of sense:

. . . our specifick Ideas of *Substances* are nothing but *a Collection of a certain number of simple* Ideas, *considered as united in one thing.* These *Ideas* of Substances, though they are commonly called simple Apprehensions, and the Names of them simple Terms; yet in effect, are complex and compounded . . . of various Properties, which all terminate in sensible simple *Ideas,* all united in one common subject. (Book II).[14]

Upon the Aristotelian view notion or nature or essence provides the ontological ground of the unity; but, given the view that this very notion is in the mind when we know the substance, it is also this notion that must provide the epistemological ground for our knowing that certain sense properties are regularly connected in experience and others are not. If the Aristotelians were correct, then according to Locke, these simple notions must always be present in our judging. Yet the "historical, plain method" shows that *they are in fact never there when we judge*, and that our ideas of substances are complexes of simple ideas *in no way* bound together by ontological or logical necessity.

Locke returns to the same point later, in Book III;[15] and again in Book IV. His discussion in Book IV is particularly revelatory of his attitude towards necessary connections that the Aristotelians believed bind simple ideas together into resulting real definitions. It is evident, Locke says, that we do not know the necessary connections required for an Aristotelian understanding of why parts of things cohere.[16] But even if we knew why the parts cohere, we still would not know everything necessary for a grasp of the *notion* or essence of the thing. For the notion must account for all the causal activities of the substance of which it is the notion, insofar as these activities are not merely occasional. Now, the *regular* activities of external substances include the production of the ideas of the secondary qualities, that is, the production of the simple ideas red, sweet, and so on. For these activities to be knowable scientifically, in the Aristotelian sense, regularities revealed by sense about such activities must be demonstrable by syllogisms grounded in natures. But for that to be possible there must be necessary connections between red, sweet, etc., and the natures of the substances that cause these qualities to appear. These necessary connections must be both ontological, in the entities themselves, and epistemological, giving us, when in the mind, scientific knowledge of those entities. But, Locke argues, we grasp no such connections:

'Tis evident that the bulk, figure, and motion of several Bodies about us, produce in us several Sensations, as of Colours, Sounds, Tastes, Smells, Pleasure and Pain, etc. These mechanical Affections of Bodies, having no affinity at all with those *Ideas*, they produce in us, (there being no conceivable connexion between any impulse of any sort of Body, and any perception of a Colour, or Smell, which we find in our Minds) we can have no distinct knowledge of such Operations beyond our Experience; and can reason no otherwise about them, than as effects produced by the appointment of an infinitely Wise Agent, which perfectly surpases our Comprehensions . . . .[17]

Locke's appeal is to a Principal of Acquaintance.[18] This is not to say that Locke was systematic in his development of this Principle: he did not give up a substantialist account of mind largely incompatible with the Principle;[19] and his nominalism and nominalistic account of relations was largely incompatible with the Principle.[20] Nonetheless, we can see in Locke the first systematic use of this Principle. He did at least deploy the Principle sufficiently systematically and sufficiently deeply to remove forever from rationalism whatever plausibility it has previously had.[21]

For the tradition, certainty of inferences from samples to populations is achieved through a grasp of the essences or natures of things. Causal activity of a thing in accordance with its essence guarantees that always, if again in similar circumstances, it would again behave in similar ways.[22] Hume, like Locke, argues vigorously against this view,[23] adopting arguments from Malebranche[24] and, indeed, almost quoting Malebranche verbatim,[25] as Jones[26] points out. Malebranche argued that there are no objective necessary connections other than those of God's causal activity: "les causes *naturelles* ne sont point de véritable causes . . . . Il n'y a donc que Dieu qui soit véritable cause, & qui ait véritablement la puissance de mouvoir les corps".[27] Hume refers to this "Cartesian" doctrine,[28] and argues that it is untenable:[29] if, as these philosophers hold, there are no objective necessary connections among bodies because we have no impression of such a connection, then neither do we have any idea of it — ideas being derived from impressions — and therefore cannot have an idea of God that includes within it the idea of causal power or activity.[30] "We never therefore have any idea of power".[31] Like Locke, Hume rejects objective necessary connections on grounds of a Principle of Acquaintance.

As does Locke, Hume argues that properties are separable, that there is nothing in one property that ties its exemplification necessarily to another, different property. Contrary to the rationalists, "There can be no *demonstrative* arguments to prove, *that those instances, of which we have had no experience, resemble those, of which we have had experience*". For, "we can at least conceive a change in the course of nature; which sufficiently proves, that such a change is not absolutely impossible".[32]

. . . as all the distinct ideas are separable from each other, and as the ideas of cause and effect are evidently distinct, 'twill be easy for us to conceive any object to be non-existent this moment, and existent the next, without conjoining to it the distinct idea of a cause or productive principle. The separation, therefore, of the idea of a cause from that of a beginning of existence, is plainly possible for the imagination; and consequently the actual separation of these objects is so far possible, that it implies no contradiction nor absurdity; and is therefore incapable of being refuted by any reasoning from mere ideas; without which 'tis impossible to demonstrate the necessity of a cause.[33]

The ideas are separable, because "all ideas are copy'd from impressions"[34] — this is Hume's appeal to a Principle of Acquaintance (PA) — and "objects have no discoverable connexion together . . .",[35] or, as he puts it elsewhere,

Suppose two objects to be presented to us, of which the one is the cause and the other the effect; 'tis plain, that from the simple consideration of one, or both these objects we never shall perceive the tie, by which they are united, or be able certainly to pronounce, that there is a connexion betwixt them. 'Tis not, therefore, from any one instance, that we arrive at the idea of cause and effect, of a necessary connexion of power, of force, of energy, and of efficacy. Did we never see any but particular conjunctions of objects, entirely different from each other, we shou'd never be able to form any such ideas.[36]

or, again, as he puts the same appeal to PA in the *Enquiry*,

The mind can never possibly find the effect in the supposed cause, by the most accurate scrutiny and examination. For the effect is totally different from the cause, and consequently can never be discovered in it. Motion in the second Billiard-ball is a quite distinct event from motion in the first; nor is there anything in the one to suggest the smallest hint of the other. A stone or a piece of metal raised into the air, and left without any support, immediately falls: but to consider the matter *a priori*, is there anything we discover in this situation which can beget the idea of a downward, rather than an upward, or any other motion, in the stone or metal?[37]

For the empiricist, this appeal to PA remains sound:[38] there are no objective necessary connections.

But that means that ontologically and objectively there is no difference between a law and a mere generalization. What, then, is the difference, which we know commonsensically to exist, between *propter hoc* and *post hoc*? Or, since there is no objective difference between laws and generalizations, *what is it about* some *generalizations that makes them laws*? That is, as Hume himself put it, "What is our idea of necessity, when we say that two objects are necessarily connected together"?[39] What he finds is that when two sorts are causally connected, upon the appearance of an object of the one sort "the mind is *determin'd* by custom to consider its usual attendant" and that it is this "determination, which affords me the idea of necessity".[40] This yields Hume's second definition of "cause" as "An object precedent and contiguous to another, and so united with it in the imagination, that the idea of one determines the mind to form the idea of the other, and the impression of the one to form a more lively idea of the other."[41] If a propositional content of the form "$(x)(Fx \supset Gx)$" is such that *as a matter of psychological fact* it is used to sustain assertions of subjunctive conditionals ("the impression of the one [determines the mind] to form lively idea of the other") *then* the assertion of that proposition is the assertion of a causal generality. A generality is lawlike just in case it is in fact used to predict and to sustain subjunctive conditionals; otherwise it is a statement of "mere regularity". The connection between lawlike generalities and subjunctive conditionals was noted above. Chisholm once put the point very neatly:

> Can the relevant difference between law and nonlaw statements be described in familiar terminology without reference to counterfactuals, without use of modal terms such as "causal necessity", "necessary condition", "physical possibility", and the like, and without use of metaphysical terms such as "real connections between matters of fact"? I believe that no one has shown the relevant difference *can* be so described.[42]

Hume agrees that the law/nonlaw distinction cannot be made without reference to counterfactuals, and without reference to a notion of "causal necessity". But he denies the need to introduce non-empirical "real connections". The latter would introduce an *objective* difference between lawful and nonlawful regularities, and such a difference is denied by the Humean. Rather than put "causal necessity" into the objective facts, the Humean puts it on the subjective side, and identifies it with a preparedness to assert counterfactuals. Far from denying the connection between laws and counterfactuals the Humean uses the connection to *define* lawlikeness. Or, more accurately, *he takes the preparedness to use a generality to sustain the assertion of subjunctive conditionals as the defining characteristic of a lawlike or "causally necessary" generality.* Thus, a generality is causal or not depending upon its psychological context.[43] Lawlikeness is a matter of the propositional attitude which obtains with respect to the generality: a generality is lawlike only if it is not merely believed or asserted, but law-asserted, that is, asserted with a preparedness to take risks with it, to predict and to assert subjunctive conditionals.[43a]

But this, surely, is circular! We may justifiably assert a counterfactual only if it is sustained by a law, so that the justified assertability of counterfactuals is defined by reference to the notion of law; while the notion of law is defined with reference to the assertion of counter-factuals. The reply is that this, of course, is correct, but that so far as *justification* is concerned there is no vicious circularity.

A generalization L is lawlike if and only if it is used to sustain the assertion of counterfactuals, that is, if and only if it is law-asserted. But it is a *further* issue whether the law-assertion of L is *justified*. If the

Humean can distinguish evidence on which the law-assertion of a generalization is justified from evidence on which it is not, then he can go on to hold that a counterfactual conditional can be *justifiably* asserted only if it is sustained by a generalization L for which the law-assertion is justified by the appropriate evidence. Provided that the rules for the justification of law-assertions are not defined in a way that makes use of the idea of counterfactuals, then, so far as justification is concerned, there is no circularity.

What are these evidential rules? To begin with, we know what is *not* involved. *Given* Hume's argument that, objectively considered, all causal assertions are assertions of constant conjunction,[44] *then* it is not reasonable to demand an objective justifictaion for the adoption of the law-assertion attitude toward some generalities rather than others.[45]

However, from the fact that no objective justification is possible, it does not follow that all adoptions of the law-assertion attitude are equally justified. Though the demand for objective justification is unreasonable, not all the law-assertive attitudes are therefore equally reasonable. Hume clearly recognizes this point and explicitly draws our attention to cases where the law-assertion attitude holds but where it is also not justified. His discussions of credulity,[46] of the often adverse effects of education,[47] of the role of imagination,[48] of unphilosophical probability (to which a whole chapter is devoted),[49] all make evident that the *Treatise* draws a distinction between these cases where the attitude is unjustifiably held and cases where it is justifiably held, where its adoption is in accordance with the "Rules by which to judge of causes and effects," which appear in their own chapter with that very title.[50]

An assertive attitude is objectively justified if and only if the proposition in question is true. But this holds equally for laws and for "mere regularities"; provided that the latter are true, they may justifiably be asserted. So truth, while sufficient to justify the attitude of assertion or mere assertion, is only necessary to *justify* the attitude of law-assertion. Thus, if a generality is false, one is objectively unjustified in holding toward it the law-assertion attitude. This will be so even if the other necessary conditions of justification (whatever they may be) are all fulfilled. And it will be so even if we have *all possible reason* to believe that that necessary condition of truth is fulfilled.

A generality is a statement about a total population. Normally all one ever observes is a sample. Between sample and population there is a logical gap.[51] The gap is such that properties regularly associated or constantly conjoined in the observed sample may not be constantly conjoined in the population. Hume's argument based on PA establishes the existence of this logical gap, which he at one point[52] expresses as the principles *"there is nothing in any object, consider'd in itself, which can afford us a reason for drawing a conclusion beyond it"* and *"that even after the observation of the frequent or constant conjunction of objects, we have no reason to draw any inference concerning any object beyond those of which we have had experience"* (that is, reason drawn from those objects "consider'd in themselves" rather than "never any sort of reason," for, after all, Hume does go on to give us the "Rules by which to judge of causes" where he sketches the conditions under which one *can* reasonably infer from a sample to a population). That properties are constantly conjoined in a sample is a necessary condition for their being constantly conjoined in the population. A necessary condition for justifiably making a law-assertion is that the generality asserted be true. Given the *logical* gap between sample and population, it is not *possible* to know whether this necessary condition is fulfilled simply by observing that the regularity obtains in the sample. The *best* we can do is to know that a necessary part of this necessary condition obtains, namely, that the regularity holds in a sample. That is the *best* we can do, short of omniscience. And since it is the best we *can* have, we must make do with it. *If we observe that a regularity holds in a sample, we thereby have every objective reason it is (at that point)*[53] *possible to have to justify one in believing that the regularity holds in the population.*[54] Subjectively, the only and best objective evidence a regularity obtains overall is that we have observed it to obtain among the facts we already know. So, subjectively, we may be justified in asserting a generality when objectively the assertion is not justified.[55] Still, if we have done the best we can do, if we assert only when we are subjectively justified, then we cannot be blamed for not having done more, even where we are objectively unjustified. Fallibility is not a vice.[56]

We have still not distinguished lawful from "mere" regularities. The remarks just made, drawing attention to our fallibility, apply equally to both sorts of regularity. An observed constant conjunction is a necessary condition for our being subjectively justified in adopting the law-assertion attitude toward a generality, but it is not a sufficient subjective condition. It is sufficient only if the acquisition of this evidence has proceeded in accordance with the "rules by which to judge of causes." Hume notes carefully that there are many situations in which the mind is confronted with a set of contrary hypotheses, each of which initially fits the data, fulfills the necessary condition for being subjectively justified.[57] Since these are contraries, the law-assertive attitude toward any particular one cannot be justified initially. But now suppose more data is sought, in accordance with the "rules by which to judge of causes." Suppose further that these data ultimately eliminate all but one of the contraries. In that case the adoption of the law-assertive attitude toward that hypothesis will be (subjectively) justified, that is, justified to the extent that it is reasonable to seek such justification. On the other hand, other principles of selection are possible.

Hume cites the principle that we choose as worthy of law-assertion the hypotheses we want to be true, quoting Cardinal de Retz on the principle that the wish is the father of the belief, "*that there are many things, in which the world wishes to be deceiv'd.*"[58] Where the world wishes to be deceived it can avoid trying to gather together the *evidence* relevant to reasonably deciding among the possible hypotheses.[59] In the chapter "Of unphilosophical probability"[60] a number of such unreasonable principles are mentioned. Those who desire a *reasonable decision* among contrary hypotheses must go out and actively collect additional observational evidence that will permit a decision to be made. The data that are given are often acquired only with great difficulty. As Hume puts it, directly after stating the "rules by which to judge of causes":

There is no phaenomenon in nature, but what is compounded and modify'd by so many different circumstances, that in order to arrive at the decisive point, we must carefully separate whatever is superfluous, and enquire by new experiments, if every particular circumstance of the first experiment was essential to it. These new experiments are liable to a discussion of the same kind; so that the utmost constancy is requir'd to make us persevere in our enquiry, and the utmost sagacity to choose the right way among so many that present themselves.[61]

One makes a reasonable decision among alternative possible hypotheses when one has *actively* sought out such data as would permit one logically to make such a decision, eliminating or falsifying hypotheses until exactly one is rendered subjectively worthy of law-assertion. *The choice among hypotheses may be made* EITHER by collecting relevant observational evidence according to the "rules by which to judge of causes" OR by some other principle. If on some basis other than such obervational evidence, then the resulting law-assertion is unjustified.

Hume's account of mind, his account in terms of dispositions and propensities, is an account in terms of laws. These *laws* are themselves subject to a Humean analysis, of course. As Hume remarks, "The same experienc'd union has the same effect on the mind,[62] whether the united objects be motives, volitions and actions; or figure and motion."[63] What these laws of learning describe are (among other things) the conditions under which law-assertive attitudes are acquired. There are a number of sufficient conditions for acquiring this attitude. Frequent repetition by itself is one.[64] So is frequent repetition under the further condition that the "rules by which to judge of causes" have been conformed to. So is the wish that is father to the belief.[65] Let us call the acquisition of a law-assertive attitude *successful* just in case the predictions it yields turn out to be true. *If we are guided by the aim of discovering the truth, so far as we can, then we shall be aiming to acquire successful law-assertive attitudes.* What we discover are regularities to the effect that certain conditions bringing about law-assertive attitudes are more likely than others to yield successful law-assertive attitudes. These regularities are inductions about making inductions. The infirmities attaching to all inductive inference thus attach to these, and doubly so, if you wish. Yet we can and have tested them. Thus, of the three conditions I have mentioned, the third is much more likely than not to yield *unsuccessful* law-assertive attitudes. *This is a lawful fact we discover about ourselves, the world, and our interactions with it.* We discover that if we are serious about the truth, then we should put aside the pieties of religion (it is a "blameable" method of reasoning to condemn a doctrine because it is dangerous to religion)[66] and the platitudes put into us by education ("an artificial and not a natural cause" of belief)[67] and rely instead upon the "rules by which to judge of causes," the norms of "experimental philosophy."[68]

Previously I asked why it is reasonable to adopt Hume's "rules by which to judge of causes" as the conditions of subjective justification for adopting the law-assertive attitude, rather than some other rules. We now have our answer to that question. *Given that our end is to come to believe general truths, then what we discover is that, as a matter of lawful fact, conformity to these rules is the best or most efficient means for achieving that end, so far as we are able.* What we must do, then, is so discipline ourselves that we adopt the law-assertive attitude under only these conditions, not the others. Such self-discipline is, of course, difficult: "The utmost constancy is requir'd to make us persevere in our enquiry, and the utmost sagacity to choose the right way among so many that present themselves."[69] But such discipline is possible. And if our end is that of truth, then it is a discipline we *ought* to practice. We earlier emphasized that causal inference involved an element of conative impulse, of mental activity. What we are now seeing is that, like all appetites and impulses, these, too, can be disciplined. As Butler indicated, particular impulses are not to be denied, are not to be reduced to self-love,[70] though they are to be disciplined by the principle of self-love[71] and also by other higher principles.[72] So, also, according to Hume, are our mental impulses to be disciplined to higher ends: "Who indeed does not feel an accession of alarcrity in his pursuits of knowledge and ability of every kind, when he considers, that besides the advantage, which immediately result from these acquisitions, they also give him a new lustre in the eyes of mankind, and are universally attended with esteem and approbation?"[73]

Curiosity moves us to seek knowledge of matter-of-fact generalities. The habits of causal inference, fallible though they may be, are the rational means to employ in trying to satisfy this interest. In order to justify the "rules by which to judge of causes and effects" Hume presents an elaborate argument in Bk. I, Part III of the *Treatise* to the effect that among all the possibilities — e.g., induction by simple enumeration, choosing to accept judgments according to whether we want them to be true, or judging causes according to the rule that if A is like B then A causes B — the rules of science alone can, within the fallibilist limits that the world imposes, lead to judgments satisfying the motive of curiosity.[74]

Now, this is not the place to enter into a long discussion of Hume's argument. For our purposes, the relevant point is that in fact Hume's "rules by which to judge of causes" yield a picture of science and of scientific research which is, in subtance, the picture of science that was sketched in Chapter I.[75] What we can say, then, is that for the Humean, for the empiricist, a generalization will *justifiably* be reckoned a law, that is, will justifiably be law-asserted just in case the evidence in its support has been collected in accordance with the research procedures sketched in Chapter I, Section II. Even so, however, there are problems that remain, which we shall try to bring out in the next section, below. Nonetheless, the solution we shall propose — it will be sketched in

Chapter III — will still be Humean in the sense that it is *contextualist rather than ontological* and will involve essential reference to the *empiricist account of scientific method* that was given in Chapter I.

The contextualist feature is crucial to the Humean. For, it affords him an answer to many criticisms of "regularity" accounts of causation which, like Hume's, insist that there are no ontological and objective differences between laws and mere generalizations. Thus, to take one example, Kneale[76] has criticized a view of Popper's that is somewhat akin to the Humean account of natural laws or causes. Popper[77] once proposed to analyze contrary-to-fact conditionals in terms of universal material implications of the form "$(x)(Fx \supset Gx)$", where '$F$' and '$G$' are not lists of proper names but rather unrestricted descriptive predicates. Given the connection between laws and subjunctive conditionals, it follows that Popper is proposing to transcribe natural laws into universal implications. Kneale argued against such transcriptions by means of two examples. The first turns on the logical point that "$(x)(Fx \supset Gx)$" is logically equivalent to the statement of non-existence of things which are both $F$ and not-$G$: $\sim(\exists x)(Fx \ \& \ \sim Gx)$.[78] Kneale lets '$Fx$' be '$x$ is a chain reaction of plutonium' and '$Gx$' be '$x$ is outside a strong steel shell containing heavy hydrogen'. He then considers the statement that "there never has been and never will be a chain reaction of plutonium within a strong steel shell containing heavy hydrogen", which upon Popper's account, will be the statement of law that "No chain reaction of plutonium occurs within steel shells containing heavy hydrogen." Now, it is at least conceivable that the statement of non-existence is true, but in order to accept that statement we do not need to accept what Popper's account would entail, that there is a law of nature excluding such a combination of events.[79] This is clear since the non-existence of such combinations can clearly be accounted for, on the one hand, in terms of the improbability of such a combination occurring without human planning, and on the other, in terms of the prevalence of the belief that such an event would have disastrous consequences. The Humean response here is that not every statement of non-existence is equivalent to a law of nature, and that, in this case in particular, the generality "No chain reactions etc." is not a law of nature. The Humean can so reply where Popper cannot because he distinguishes, as Popper does not, between universal material implications as such and universal material implications that are laws of nature. The Humean will claim that "No chain reactions etc." is not a law because one has not adopted with respect to it the law-assertive attitude.[80]

In his second example Kneale supposes a musician, lying on his death-bed, composes in his imagination an intricate tune he is too feeble to write down or even speak. In his last moment the thought comes to him that "No human being has ever heard or ever will hear this tune", meaning by 'this tune' a certain complex pattern of sounds which could be described in general terms using unrestricted descriptive predicates. Here it is clear the musician means to assert no law of nature. So not every universal material implication is a law of nature. This tells against Popper. But it does not touch the Humean, who will agree with Kneale that the musician's assertion was not an assertion of law. In fact, the Humean will argue, a law was not asserted simply because the musician's attitude was just that of assertion, rather than that of law-assertion.

No doubt, however, Kneale would not take the Humean to be really siding with him as against Popper. For, so far as the objective facts are concerned, Popper and the Humean agree: there is nothing more to laws than regularities, universal matter-of-fact implications. Kneale would not agree with Hume that lawlikeness is a subjective matter, not objective, a feature of the logical form of causal propositions. Probably he would disagree for the same reason Aristotelians and rationalists tend to disagree with a similar thesis in ethics. The thesis in ethics that value is not a matter of propositional content but rather of the psychological attitude is often called "emotivism". So Hume's account of causation may therefore perhaps not unreasonably be characterized as an emotivist account of causation. And, as with those who defend an emotivist theory in ethics, the immediate question raised in this: does not emotivism, when it denies the existence of a standard, entail that whether the attitude is adopted is not something that permits of justification? Or, in other words, does not Hume's account reduce reasoning to irrationalism?

Now, the reply to this, in Hume's case, as in ethics, is to challenge the presupposition of the objector's question, and ask what might one *reasonably* mean by "justification" in this context? In ethics, if the emotivist's arguments that objectivism is false are accepted, then it is no serious objection to his account that he leaves no room for an objective justification of value judgments. *Of course* he has left no room; he has just finished arguing that such justification is not possible. And if it is not possible, it is not reasonable to insist upon it. Whatever justification amounts to in ethics the one thing that cannot *reasonably* be demanded is objective justification. And as we have argued in the case of Hume's account of causation the same sort of response is called for. *Given* that, objectively considered, all causal assertions are assertions of constant conjunction, *then* it is not reasonable to demand an objective justification for the adoption of the law-assertion attitude towards some generalities, rather than others. What Hume must do is give an alternative account of what justifies adopting the law-assertion attitude. This, of course, he does, as we saw, and does in a way that makes science and scientific method the judge of laws. The *Treatise* draws the distinction between these cases, where the attitude is unjustifiably held, and cases where it is justifiably held, by arguing that the latter obtains when the adoption of the attitude conforms to the "Rules by which to judge of causes and effects," which, as we argued, are in fact the rules of scientific method.

However, the point remains that both laws, or law-asserted generalities, and accidental generalities are *objectively* indistinguishable. Since *objectively* they both have the form of matter-of-fact regularities, then *there is nothing peculiar in the facts asserted themselves* that could justify adopting the law-assertion attitude rather than the attitude of asserting a generalization as an accidental generality. Indeed, the very question, "What is the difference between a law and an accidental generality?" becomes decidedly misleading. It suggests we must look for some special virtue of laws, over and above their matter-of-fact universality that distinguishes them from the other sort of matter-of-fact generalizations called accidental generalities. But if the Humean is correct, there is no such special virtue, and to seek one is unreasonable. It follows that it is *not* unreasonable to draw the distinction subjectively; that is the only way that remains for it to be drawn.

Moreover, it must be emphasized once again, to say this, that the distinction is one of subjective attitude, is *not* to fall into an irrationalism that collapses all differences between science and superstition. The rules of scientific method serve to provide a criterion that can reasonably and effectively distinguish science and superstition. Or at least, so the Humean argues.

It is not possible here to go into all the relevant issus on the justification of the rules of the scientific method. One point, however, is of special importance for the Humean. Some have recently argued that *the rules of the scientific method cannot be successful unless we presuppose there are non-Humean objective necessary connections.* Clearly, the Humean analysis of laws cannot be considered secure until one has disposed of this argument that the very rules that the Humean relies upon to make his case presuppose a non-Humean or rationalist account of laws. We confront this case in Section III, below, and argue that it fails.

Before we turn to that discussion, however, some themes concerning counterfactuals should be dealt with. Just above we pointed out the misleading nature of the question, "What is the difference between a law and an accidental generality?" Equally misleading is the question, "Why do laws justifiably sustain counterfactuals while accidental generalities do not?" Again it suggests some virtue of lawful generalizations over and above their universality that enables them to sustain counterfactual conditionals, those mysterious statements that manage to assert truths about worlds beyond the actual. Since counterfactuals, on the Humean account, are condensed arguments, there is no question of their being truths about mysterious other worlds, about things that are both unreal and actual. *The mystery has disappeared.*

However, if a matter-of-fact generalization

(*)    All $F$ are $G = (x)(Fx \supset Gx)$

is — objectively — true, then it can be used to predict that *if*

$x$ is $F$

then

$x$ is $G$

It can therefore sustain the assertion of a predictive conditional. Similarly it can sustain the assertion of a factual conditional. Moreover, given the symmetry thesis entailed by the deductive-nomological model, it follows that such a generalization also explains — though we must hasten to add, the explanation may be imperfect indeed.[81] There seems no reason then why an objectively true matter-of-fact generalization like (*) should not sustain the assertion of counterfactual conditionals. But accidental generalities are of the form (*). Hence, *so long as accidental generalities are in fact true then they should be capable of sustaining the assertion of counterfactual conditionals.*[82] This, however, conflicts with our previous conclusion, again definitory of the Humean position, that accidental generalities are precisely those that do not sustain the assertion counterfactuals.

It would seem, then, that the Humean is committed to the view that accidental generalities both can, and do not, sustain the assertion of counterfactuals. Resolving this apparent conflict is the task of the next section, which attempts to answer the question, "Why do some matter-of-fact generalizations, particularly those called 'accidental generalities', fail to sustain the assertion of counterfactual conditionals?" Given that objectively both laws and accidental generalities are of the form (\*), the problem is not to find some extra virtue in laws, but to find what it is that accidental generalities lack.

The answer that we shall propose is the Humean one of *context*. Often it is simply a matter of evidence: the data are such that it seems very likely that the accidental generality lacks truth. But if an accidental generality is a generality for which there is reason to suspect that it might be false, then it *does not*, after all, *so far as we can tell*, effect a connection between two sorts of events. It cannot, after all, be used to explain or make predictions, or sustain the assertion of factual and predictive conditionals, nor, finally, sustain the assertion of counter-factual conditionals. Often enough, however, it is not so much the direct context of evidence for truth that is relevant as the context of well-supported background knowledge. The pattern characteristic of acci-dental generalities is this: when the contrary-to-fact assumption is made then, in the context of the background knowledge which is not held in abeyance, it turns out that the generality *cannot consistently* be used to deduce the consequent; that is, in the context of accepted and law-asserted background knowledge, an accidental generality cannot, save at the cost of inconsistency, sustain the assertion of a counterfactual conditional. Thus, the problem with accidental generalities is nothing objective — if true, they can, be taken by themselves, be used to predict, explain and sustain the assertion of counterfactual conditionals; rather the problem is one of epistemic context: the context of evidence and background knowledge is such that they cannot, or cannot con-sistently, sustain the assertion of counterfactuals. What accidental generalities lack, then, we shall propose, and what laws have, is not something objetive but a certain epistemic context.

It is evident, then, that if this answer with respect to what accidental generalities lack is correct, then the Humean is, after all, not committed in any way that is paradoxical the position that accidental generalities both can, and do not, sustain the assertion of counterfactuals. The apparent conflict is resolved.

Let us turn, then, to the task of seeing more precisely why accidental generalities cannot be used to sustain the assertion of counterfactuals. As we might suspect, the issues here cannot be divorced from consideration concerning explanations that were raised in Chapter I, Section I.

## II. ACCIDENTAL GENERALITIES

Kneale has stated the case against the regularity account of causation in terms of the supposed incapacity of mere matter-of-fact regularities to support counterfactual conditionals.

... laws of nature are normally expressed in the timeless present, and are assumed to be concerned not only with actual instances of some kind, but with anything that might have satisfied a certain description. If on the strength of our records [of a complete enumeration of all dodos that ever existed] we suggest that there is a law of nature that all dodos have a white feather in their tails, we say in effect that, if there had been any dodos other than those mentioned in our records, they too would have had a white feather in their tails. But an unfulfilled hypothetical proposition of this kind cannot be derived from a proposition which is concerned only with the actual. A contingent universal proposition can always be expressed in the form 'There are in fact no $\alpha$ things which are not $\beta$', and from such a proposition it is impossible to deduce that if something which was not in fact $\alpha$ had been it would also have been $\beta$.[1]

Now, whatever force this has against the simple regularity view, it has no force against the Humean, who agrees that there is a difference between an assertion of law and an assertion of an accidental generality, and that this difference consists in the former being used to sustain the assertion of counterfactuals. The Humean moreover also agrees that an assertion of law cannot be "derived from" a *mere list* or *enumeration* of data, even if the list be complete and the assertion of its conjunction be in fact equivalent to the assertion (*not*: law-assertion) of the corresponding matter-of-fact universal. At least, the Humean agrees with this point in the following way: (1) the data that rationally justify the law-assertion of a generality must be acquired in accordance with the rules of the scientific method ("the rules by which to judge of causes and effects"); (2) a *mere list* of data, while perhaps establishing a generality is true, and therefore worthy of assertion, is not, as a mere list, acquired in accordance with the rules of scientific method, and therefore cannot justify the *law*-assertion of the generality. In this sense a statement of law cannot, given the rationality of the rules of the scientific method, be derived from any statement of mere matter-of-fact regularity.

Nonetheless, the Humean insists, this does *not* entail that, in terms of truth-conditions, there is any objective difference between laws and accidental or mere matter-of-fact generalities. Kneale's conclusion, that there is such an objective difference, from his premises, that laws do, and mere generalities do not, support counterfactuals, is a simple *non sequitur* so far as the Humean position is concerned.[1a]

There is no doubt, however, that Kneale would likely not be satisfied with this answer. The basic point would be that, if objectively there is no difference between lawful and accidental generalities, then there seems to be no good reasons for supposing that they cannot both be used to make the same inferences, be they predictive, explanatory, or counterfactual. If an accidental, or mere matter-of-fact generality is in fact *true*, then it *can* be used to predict, and therefore, given the symmetry thesis with respect to explanation and prediction, it can, surely, be used to explain; and if it can be used to explain and predict, why can it not be used to sustain counterfactuals?[1b] The suggestion would be that the Humean has no reason to deny that it can be so used. In spite of his claims, then, he does not show that there is a difference between laws and accidental generalities: according to the Humean, objectively there is no difference, and this means that he must treat them all of a piece, all as laws or all as accidental generalities.

Consider a standard example of an accidental generality, the regularity that Chisholm observed to hold about a certain park bench on the Boston Common:[2]

(I)        Only Irishmen sit on this bench.

The difference between a law and an accidental generality like (I) is that the former can and the latter cannot be used to predict and to assert counterfactual conditionals. Given the variety of persons who appear on the Boston Common it would be unwise to rely upon (I) to predict that the next person who sits on the bench will be an Irishman, and one would be unwise to assert on the basis of (I) that if that man (who is a Lett) were to sit on the bench then he would be an Irishman. On the other hand, it is the mark of a law that it directly justifies such inferences. This being so, the conclusion would seem to be that an argument (A)

> All $F$'s are $G$
> This is $F$
> _____
> This is $G$

with an accidental generality like (I) as its major premise is *not* a scientific explanation. Arguments like (A) constitute scientific explanations only if the major premise is a law. The issue this introduces into our considerations derives from the fact that, in Chapter I, Section I, we took the Principle of Predictability, where this is (partially) elaborated by means of $(R_1)$–$(R_4)$,[2a] to state a *sufficient condition* for an argument to be a scientific explanation. (I) raises the question whether this is, after all, sufficient.

If we accept the Humean account of laws that we have been defending, then there is no way to construe accidental generalities as logically different from those generalities that are laws. For, at bottom, both kinds of generality will be construed as having the same logical form, namely:

$$(x)\,(Fx \supset Gx)$$

Of course, if one was an Aristotelian or a rationalist and had unanalyzable modalities and non-logical necessities of various sorts one could begin to characterize the difference logically, but such modalities are not part of the empiricist's language. And without such a logical difference it seems difficult to exclude from being objectively explanatory those arguments which have as their major premises accidental generalities but which meet such objective conditions, like $(R_1)$–$(R_4)$ of Chapter I, Section I, as are imposed by the Principle of Predictability. If those arguments meet all the objective conditions, then on what basis *can* one exclude them from being explanatory? and, equally, on what basis *can* one exclude them from being used to sustain the assertion of counterfactual conditionals?

This poses the problem for the Humean, to state precisely what it is that so-called accidental generalities lack — it is not something objective! — that renders them precarious, and incapable of use in predictions or of sustaining the assertion of counterfactual conditionals.

Now, I think if one reflects more closely upon the situation, it is perhaps not so difficult as it first appears. Leave aside the issue of counterfactual conditionals for a moment, and consider only the case of prediction. What I think should be insisted upon is that *if* (I) *is* true, then it *can* be us used to predict. Objectively speaking, *any* true generality can be used to predict. As the length of the shadow provides *a reason for* the height of the flagpole *being* what it is,[3] so also, *if* (I) is true, then provides *a reason for* a person on the park bench *being* an Irishman. Insofar as our cognitive interests are directed at generalities that can be used to predict, then any true generality will satisfy that interest. That being so, there seems no reason to assert that such arguments are *non*-explanations, or, what is the same, to assert that they do not have *some* explanatory content. It would seem, then, that we need not object to the objective criteria including in the class of explanations arguments with so-called accidental generalities as their major premisses.

On the other hand, as we know, to say an argument has some explanatory content is not to say it has the highest degree of explanatory content. This was made evident to us by the existence of indicator laws of the sort that appeared in Scriven's barometer example, and by the example of laws describing the correlation of several effects dependent upon a common cause. Even if we suppose (I) is true, we will still want to say that it does not hold *unconditionally.* There are many further relevant variables — e.g., the intentions of the persons of other national origins that walk by the bench — such that, if these had been different then the generality would have turned out not to be true. This latter counterfactual assertion is based on our *lawful* knowledge of how these other variables interact. Because we have this background knowledge of some of the factors that would appear in a *more complete* explanation than (I) alone can provide, we recognize that any explanation in terms of (I) alone will have a low degree of explanatory power. But, to repeat, that does not mean it has no explanatory power. In general, I think that what one wants to say about so-called accidental generalities, is that they do provide explanations as required by the "deductive-nomological" model as formulated in the empiricist's language of the early positivists, but that the generalities, even if true, are always very much conditioned generalities so that their degree of explanatory power will always be relatively low.[3a]

There is another aspect of the issue, however, that must also be

brought in. I have emphasized: "an accidental generality *even if* it is true
. . . ." This is deliberate. With respect to most accidental generalities,
there is genuine doubt as to whether they are even true. The evidence
available may well testify to their truth, but does so only very weakly.
Arguments involving them can therefore claim only a low degree of
subjective worthiness as explanations. Explanations in terms of laws, in
contrast, are generally of a high degree of subjective worthiness. Thus,
not only would any explanation based on an accidental generality have
a low degree of explanatory power, it would in addition be much less
subjectively worthy than any explanation based on a lawful generality.
This low degree of subjective worthiness amounts to a weak belief in
the truth of a law and therefore in its capacity to yield scientific
predictions. Accidental generalities are therefore at most rarely to be
used in making predictions — not because they could not be so used *if*
they were true, but because we are terribly unsure about whether they
*are* true. Failure to be used predictively and failure to appear in
subjectively acceptable explanations are therefore the hallmark of
accidental generalities. On the other hand, this point, which turns on
certain features of the subjective conditions of acceptability that are
characteristic of accidental generalities, does not mean that the
Principle of Predictability does not give sufficient conditions — that is,
sufficient objective conditions — for arguments to be acceptable as
scientific explanations.

The evidential difficulty with so-called accidental generalities lies in
the fact that we recognize them to be conditioned generalities. We
recognize that they have held in the observed sample only because
certain conditions obtain, and that they will continue to hold in the
population only if those conditions also continue to obtain. Most often
in the case of accidental generalities, at best we have no evidence to
suppose those conditions will continue to obtain, and often good
grounds for supposing they won't. Thus, the generality (I) about the
park bench on which only Irishmen sat holds only because

(CI)     No non-Irishman decided to sit on that bench

also holds. But, given the usual vagaries and impulsiveness of human
intentions and decisions with respect to such things as to where to sit in
a park, it is unlikely indeed that this condition will continue to hold in
the population. And if it does not continue then the accidental
generality which has held up to now will cease to hold. That is, it will be

falsified. If it is, then the objective condition, that the major or general premiss of the explanatory argument be true, will not be satisfied. Accidental generalities are evidentially weak not because their known positive instances do not testify to their truth but because background knowledge testifies to the unlikelihood of the generalities continuing to hold in the total population.

It is in this context that one must put the issue of the support that accidental generalities can give for the assertion of counterfactual conditionals. Laws, of course, support such assertions, and this is what makes them valuable from the viewpoint of our pragmatic interests. It is these counterfactual assertions that appear when we engage in practical reasoning about hypothetical consequences of possible alternative actions when we consider how best to interfere in natural and social processes. Insofar as scientific explanations are supposed to involve laws knowledge of which can satisfy pragmatic interests in control, it would seem accidental generalities do not meet this condition, *even if they are in fact true*. For, even if only Irishmen ever sit on this bench, we could still not reasonably assert of this man (who is a Lett) that if he were to sit on this bench he would be an Irishman. It would seem that the objective conditions laid down by the Principle of Predictability are as we suspected, not sufficient to pick out the arguments that can count as explanatory, that is, as satisfying not only our idle curiosity but also our pragmatic interests in control.

This last conclusion would, however, I think, be hasty. The counterfactual assertion

> If this man (who is a Lett) were to sit on this bench then he would be an Irishman

is to be supported by the accidental generality (I). However, if the antecedent of the conditional were to be true, then a Lett would have decided to sit on the bench in question. In that case the generality (CI) would be false. But the truth of (CI) is a condition for the truth of (I): if (CI) is false, so is (I). Hence, if the antecedent of the conditional were true, then (I) would not be true, and we could not conclude the Lett who sat on the bench was an Irishman. The difficulty with accidental generalities supporting the assertion of counterfactual conditionals is this. One begins by supposing the generality can justify inferring the consequent of the conditional from the antecedent, which is taken as an assumption. But when the contrary-to-fact assumption is made in the antecedent of the conditional, this assumption, *together with back-ground knowledge that has not been held in abeyance*, entails *the non-applicability* to the case of the very generality that one began by supposing *could* justify inferring the consequent from the antecedent.[4]

Consider another example of a supposed accidental generality, this one offered by Kneale, and slightly more plausible than the dodo example quoted above.[5] Suppose that we know, as a matter of fact, that there never has been, is not now, and never will be, a raven that is not black. We are then justified in asserting as true the generality

(R)     All ravens are black

But, Kneale argues,[6] though this is a true generality, it could not be a law of nature nor sustain the assertion of a counterfactual conditional. For, suppose, first, that, as a matter of fact, no raven has ever, or will ever, live in polar regions; and suppose, second, that we do not know whether dwelling in polar regions affects the colour of ravens, so that so far as we know the descendents of ravens that might migrate to such regions will grow white feathers. Thus, while (R) is true, its truth is a consequence of the "historical accident"[7] that no ravens ever migrate to, and stay to live in, polar regions. Hence, while (R), since true, can be used to predict, it cannot be used to sustain the assertion of the counterfactual conditional

(B)     If something were a raven and an inhabitant of the polar regions, then it would be black.

But since a law does sustain the assertion of such counterfactuals, (R) cannot be a law.

Now, this argument, like the one about the dodos, does not straight off secure Kneale's point. For, if the evidence for (R) is a mere list, as Kneale supposes that it is, then (R) will not, according to the Humean, be worthy of law-assertion. We should therefore not be surprised that it fails to suppose the counterfactual (B).

Kneale supposes us to have full evidence that (R) is true. That precludes the issue of using (R) in predictions. But suppose that we did not have all the evidence that (R) is true; suppose that it was not exhaustively enumerated. In that case (R) would be like

(S)     All swans are white

before 1600. Now, precisely because we know

(K$_1$)     The colour of plummage is not invariant for every species of
            bird

we should hesitate to place full confidence in either (S) or (R). As (S)
proved false, so, in the light of (K$_1$), might (R). Thus, the background
knowledge (K$_1$) casts doubt on whether (R) is true. For this reason we
would not be prepared to use (R) to predict or explain, or, at least, not
with any degree of confidence. In respect of usefulness for predictions,
then, (R) turns out to be much like (I).

It is also similar to the case of (I) when it comes to the issue of the
support of counterfactuals. Since Darwin[8] we know, that is, can law-
assert such things as the following about organisms, including ravens:

(K$_2$)     Members of a species can inhabit a region (= survive and
            reproduce in it) only if their characteristics are fit-making

and

(K$_3$)     The colour of an organism is a characteristic which is fit-
            making only if it does not contrast with the colour of the
            environment.

We also know that

(K$_4$)     In the polar regions, the environment is predominantly white
            for much of the year

and, of course, that

(K$_5$)     The colour black contrasts with the colour white.

From these we can deduce, and law-assert

(L)        For any species, its members inhabit polar regions only if
            they are not black.

Under the assumption

(K$_6$)     Ravens constitute a species of organism

this law sustains the assertion of the counterfactual conditional

(B*)       If something were a raven and an inhabitant of the polar
            regions, then it would not be black.

The assertion of (B*) on the basis of (L) requires us to deny (B). For, the assertion of (B*) requires us to suppose that

> something is a raven
> this thing inhabits the polar regions

These, together with (L) and ($K_6$) entail not only the consequent of (B*):

> this thing is not white

but also the *negation of the rule* (R). Thus, the background knowledge ($K_2$)—($K_6$) that justifies (L) and its application, require one, under the assumption of the antecedent common to (B) and (B*), to reject as inapplicable the generalization used to sustain the assertion of the counterfactual (B). In short, in the case of the accidental generality (R), as in the case of the accidental generality (I), when the contrary-to-fact assumption is made in the antecedent of the counterfactual conditional the assertion of which it is supposed to sustain, *this assumption, together with the background knowledge that has not been held in abeyance*, entails *the non-applicability* to the case of the very generality that one began by supposing *could* justify inferring the consequent from the antecedent.

This is not to say that (R) could not, *with other assumptions*, sustain the assertion of counterfactuals. For example, it could sustain the assertion of the counterfactual

(B**)    If something were a raven and an inhabitant of a polar region of windswept black rock, then it would be black.

But here the antecedent which forms the assumption of the collapsed argument requires, when it is assumed, one to suspend the piece of background knowledge ($K_4$), and without ($K_4$) one cannot derive (L) nor, therefore, assert (B*) which negates both (B) and (B**).

As a final example, consider this ubiquitous one:

(c)    All the coins now in my pocket are copper.

This is an accidental generality because, unlike a law, it cannot sustain the counterfactual conditional

(d)    If this dime were now in my pocket, it would be copper.

It does not sustain this, because the background knowledge

(e$_1$)     For any coin, if it has a certain metallic composition at some time $t$, then it has the same metallic composition at all times

(e$_2$)     This dime was silver at $t_0$

(e$_3$)     Silver and copper are species of metallic composition

(e$_4$)     Whatever is silver is not copper.

These entail, and permit us to law-assert

($\lambda$)     For all times, this dime is not copper.

The law ($\lambda$) sustains the assertion of the counterfactual

(d*)     If this coin were now in my pocket, it would not be copper

which requires us to deny (d), since the assumptions of the antecedent common to (d) and (d*) together with ($\lambda$) requires one to deny (c) which is to provide the connection between the antecedent and consequent of (d). Thus, in the case of this accidental generality too, it turns out that when the contrary-to-fact assumption is made in the antecedent of the counterfactual conditional the assertion of which it is supposed to sustain, this assumption, together with the background knowledge that has not been held in abeyance, entails the non-applicability to the case of the very generality that one began by supposing could justify inferring the consequent from the antecedent.

What these examples argue — and they could be multipled — is that what is characteristic of accidental generalities is that they are at best conditioned generalities and are recognized to be such through our background knowledge of the relevant conditioning factors. If one assumes this background knowledge as well as the generality that is to support the counterfactual conditional, one cannot take it for granted that the contrary-to-fact assumption of the antecedent of the conditional will not conflict with the background knowledge. Indeed, this, I am suggesting, is characteristic of those generalities we call accidental. In effect, what happens simply enough is that one gets oneself into a contradiction. Which one must then resolve. One can continue to make the contrary-to-fact assumption of the antecedent and to use the accidental generality to infer the consequent only if one *also* withdraws some piece of the background knowledge. So far as the counterfactual assertion is concerned, it does not matter which piece of background knowledge is withdrawn so long as the contradiction is resolved. It follows, therefore, that accidental generalities *can* support the assertion of counterfactual conditionals, *provided that* appropriate adjustments are made in the background knowledge that surrounds them and, indeed, is part of what makes them characteristically accidental generalities.

However, such adjustments in our belief-structure are a necessary part of *all* counterfactual reasoning. Consider the conditional

> If this match [which is wet] had been dry, then it would have lit when it was struck.

Here the relevant law would no doubt be

> All dry matches located in an oxygen-containing medium light when struck.

For this to apply we would also have to assume

> This match was located in an oxygen-containing medium.

In addition we assume

> This match was struck.

With these and the contrary-to-fact assumption which is the antecedent of the conditional:

> This match was dry

we infer the consequent of the conditional

> This match has lit

The counterfactual conditional is a compressed version of this inference. Now, we also have as part of our background knowledge (let us suppose) the information that

> This match was lying in the rain.

With this as an antecedent condition and the general law

> Whatever lies in the rain gets wet.

we must conclude

> This match was wet.

The contrary-to-fact assumption is the negation of this. Hence, if we make that contrary-to-fact assumption, we arrive at a contradiction which can be resolved only if some part of our background knowledge is held in abeyance. In general, then, there seems to be no difference between accidental generalities and laws, in that both can be used to support the assertion of counterfactual conditionals, provided certain parts of our background knowledge are also withdrawn.

Thus, in the case of both laws and accidental generalities, if they are used to sustain the assertion of counterfactual conditionals, then, in general, in order to prevent a contradiction from arising, one must suspend, hold in abeyance, some piece of background knowledge over and above the suspension of disbelief with respect to the antecedent of the conditional. The difference is that in the case of laws it is easier, more "natural", to suspend the relevant background knowledge, whereas in the case of accidental generalities it is less easy to make the required additional suspensions. It is easier to suppose that the world is so changed as to permit a generality that is a law to sustain the assertion of a counterfactual conditional than it is to suppose that the world is so changed as to permit a generality that is accidental to sustain the assertion of a counterfactual conditional. This seems to be a matter of two characteristics of accidental generalities: (i) accidental generalities are conditioned generalities, but background assumptions often lawful, and it is easier to suppose that the world is so changed that a conditioned generality is false than it is to suppose that the world is so changed that an unconditional generality is false; and (ii) accidental generalities are often evidentially ill-supported, and it is easier to suppose that the world is so changed that an ill-supported generality is false than it is to suppose that the world is so changed that a scientifically well-supported generality, i.e., a law, is false.

What remains to be done is to articulate in detail these criteria, and provide a *rationale* for them, that is, answer the question, what argues that they are *appropriate* criteria? We shall develop our answer to this question in Chapter III, Section VIII, subsection (iv), below, in the context of a response to those who propose to analyze counterfactual conditionals in terms of "possible worlds".

If this later defence is successful, then it will indeed turn out, as the Humean says, that the criteria for distinguishing accidental and lawful generalities are *contextual*, what we are prepared to do with them, that is, prepared to do with them inferentially in a context of background knowledge and evidential support. In particular, there will be *objectively* no difference in the capacity of a true generality, whether lawful or accidental, to predict, explain, or support the assertion of counterfactual conditionals. What accidental generalities lack that precludes their being used to sustain the assertion of counterfactual conditionals is a certain *context* which laws, in contrast, have. To this extent, Kneale is quite correct: the Humean collapses the distinction between laws and accidental generalities; if true, then objectively what the one can do the other can do. Nonetheless, the point remains, which Kneale overlooks,[9] that for the Humean there is still a distinction to be drawn; only it is contextual rather than objective.

The task remains, of course, as we have said, to detail the principles for the making of the belief-contravening assumptions that are implicitly part of the assertion of any counterfactual conditional. We shall discuss such principles in detail in Chapter III, below.

In the meantime, however, we have, I think, sufficient for our present purposes. For, it would seem that we may reasonably conclude that while we must distinguish between accidental generalities and laws there is no difference between them that prevents the former from yielding scientific explanations when used as premises in arguments of sort (A), provided that the objective conditions for such an argument being an explanation are fulfilled. And this is really all we need for our defence of the empiricist's "deductive-nomological" model of explanation and for our defence of the Humean account of laws and of counterfactual conditionals.

### III. LAWS AS RELATIONS AMONG UNIVERSALS: A RECENT DEFENCE OF NECESSARY CONNECTIONS

For the Humean, what distinguishes a law-like from a non-law-like generality is the preparedness of one who asserts it to use it to predict and explain and to justify asserting subjunctive conditionals. And then, one is *justified* in so law-asserting a generality just in case that the evidence for its truth satisfies certain conditions, roughly that the generality has passed the tests of experimental science, that the evidence has been gathered according to the norms of scientific method. This position has recently been challenged by Tooley,[1] whose case has been endorsed by Armstrong.[2] For Tooley and Armstrong, a generality[3]

(1)        $(x)(Fx \supset Gx)$

is a law if, and only if, there is an unanalyzable atomic relational fact about the properties $F$ and $G$, say

(2)        $\mathcal{N}(F, G)$

such that this fact (2) entails the generality (1).

Now, this alternative to the Humean account is of some interest, both in its own right, and even more so, in its claim to solve problems that the Humean cannot solve.

The aim of this alternative is to account in *objective* terms for the *nomological necessity* of lawful generalities. Where the Humean offers a subjectivist, or, if you wish, emotivist, account of this necessity, Tooley and Armstrong ground it objectively in the fact (2). There is an important sense in which one must ascribe a sort of necessity to any fact on which two (or more) properties are related by a primitive unanalyzable relation. For, if one assumes what seems reasonable, that it is events which are in time, that temporal relations hold among events and not among properties, then a fact like (2) will be *atemporal*. In that sense, such a fact may be said to hold *for all times*. But what holds for all times is, in one sense of 'necessary', *necessary*. Hence, a fact like (2) has to be considered a necessary fact. Tooley and Armstrong thus attempt to ground the nomological necessity of a generality (1) in the objective necessity that attaches to facts like (2).[3a]

However, there is a problem which would seem to be insuperable for the Tooley-Armstrong account of nomological necessity. Clearly, that account can succeed only if (2) *entails* (1). For, only in that case could the necessity of (2) be transferred to (1): a conclusion is necessary only if *both* premises are necessary; if one premiss is contingent, then so is the conclusion. The difficulty is that the premiss

(3)        $\mathcal{N}(F, G) \supset (x)(Fx \supset Gx)$

is not a tautology. At least, it is not a tautology upon any explication of the idea of 'necessity truth' that is based on the customary idea of logical form that appears in functional logic of the sort found in *Principia Mathematica*. Since (3) is not a tautology, one cannot say (2) entails (1). But in that case, even if (2) obtains, *that* fact *in no way guarantees* the truth of (1). Thus, Tooley and Armstrong fail to provide an objective ground for the necessity of (1), if is the case that (1) can be used to state the law.[4]

This may be called the *ontological difficulty* for the Tooley-Armstrong position. There is also an *epistemological problem*. Clearly, if we are to know that (1) is a law, i.e., that it can be asserted to hold with objective nomological necessity, then we must *know* that the ground of this necessity, to wit, the fact (2), does actually obtain. Now, in general, our grounds for asserting an atomic fact is, other things being equal, that it is *presented to one*. One need not deny that we are presented with certain facts in which properties are related by higher order relations. The difficulty is that what science seeks as it attempts to justify its law assertions is *experimental confirmation* of generalities (1); that is, it seeks to obtain *instances* of the generality, and if these instances satisfy certain conditions then that is taken to provide grounds for taking the generality to be a law. It would seem that, if the Tooley-Armstrong position were correct, we should not do this, but rather something else, or, if we did do it, then the aim would be not simply that of collecting instances of the law so as to satisfy certain conditions on what it is for an instance to be confirming.[5] Rather, what we should aim to do is be presented with facts like (2). Once such a fact was observed, we would have the grounds required (or so Tooley and Armstrong claim) for asserting (1) to be a law, nomologically necessary. But, of course, the latter is *not* what scientists do. Rather, they *do* search for confirming instances. So the Tooley-Armstrong position seems, in this epistemological problem, to run into conflict with actual practice in science.

Now, Tooley and Armstrong nowhere face up to the ontological difficulty.[5a] That alone condemns their position, I believe. Of course, to say that is not by itself to defend the Humean position. To do the latter one must examine the problem they believe that their account of laws can handle and the Humean's cannot, in order to see whether there really is a difficulty for the Humean. As we shall argue, however, the problem they raise is in fact solvable by the Humean. So there is, after all, no need here that forces one to resort to their account with its insuperable ontological difficulty. However, Tooley takes up the epistemological difficulty and turns it back on the Humean, producing an argument that if one adopts his view then certain features of scientific research practice can be accounted for, whereas upon the

Humean position they cannot. The argument is ingenious but, as we shall see, also unsound. As it turns out, the Humean can account in a somewhat similar way as does Tooley for the research practices to which he directs our attention. In fact, we shall argue, Tooley fails to see this because he fails to see how the Humean can deal with the problem taken by Tooley and Armstrong to be insolvable for the Humean. So it is to this supposed problem that we must first turn.[6] But we shall also argue that Tooley's approach to the ·epistemological problem cannot be adequate until the ontological difficulty is solved. Thus, however ingenious is his treatment of the epistemological problem, his alternative to the Humean account of laws is a failure.

We are asked to consider a world containing ten different types of fundamental particle.[7] We further suppose that the behaviour of the particles in interactions is a function of the types of the interacting particles. Consider only interactions involving two particles. There are 55 possibilities with respect to the types of the two particles. We now suppose that 54 of these possible interactions have been observed and studied carefully. The upshot is, suppose further, 54 different laws, not interrelated in any way that permits the deduction of some from the others. Finally, we suppose that, given the way particles of types $X$ and $Y$ are currently distributed, it is lawfully impossible for them ever to interact at any time, past, present or future. "In such a situation it would seem very reasonable to believe that there is some *underived* law dealing with the interaction of particles of type $X$ and $Y$."[8] This is indeed reasonable.

Can the Humean account for this reasonable judgment? Tooley claims not. For, he suggests, the only way in which the Humean can deal in a satisfactory manner with generalizations that are vacuously true (i.e., without "positive" confirming instances), is to hold they are laws only if they are derived from generalizations that are not vacuously true.[9] But if this is so, then in the complex situation we above supposed, the judgment that there is an underived law for the $X-Y$ interaction must be unreasonable. The question is, then, whether Tooley is correct in his claim that the Humean has no other way to account for the reasonability of the judgment that *there is* an underived law for the always un-instanced $X-Y$ type of interaction.

Let us see.

There are exactly 10 different sorts of fundamental particle in the universe. Call them $S_1$, $S_2$, ... , $S_{10}$. We can classify them as "basic particles species" by means of the following generic predicate:

$$\mathscr{S}f =_{\text{df}} f = S_1 \vee f = S_2 \vee \ldots \vee f = S_{10}$$

The classes $S_i$ are mutually exclusive:

$$(f)\,(g)\,[\mathscr{S}f \,\&\, \mathscr{S}g \,\&\, f \neq g: \supset (x)\,(fx \supset \sim gx)]$$

and jointly exhaustive of all basic particles:

$$(x)\,(\exists f)\,[\mathscr{S}f \,\&\, fx]$$

In order to make the example somewhat more concrete, let us suppose that the laws concerning interacting pairs are laws concerning the *accelerations* of the interacting particles. Let "$a(x, t)$" assert that particle $x$ has *specific* acceleration $a$ at $t$, and let "$\mathscr{A}\,a$" be the generic statement that asserts that the specific acceleration $a$ is an acceleration. Since the $\mathscr{A}$'s are accelerations, we must assume that the domain is quantifiable in the relevant way.[10] And we must assume, as part of this, that every particle has, at any moment, a unique acceleration. We must further suppose that two particles $x$ and $y$ interact at $t$ when a certain relation $I(x, y, t)$ obtains. What we then have are laws like this describing such interactions:

(i)     $(a)\,(b)\,(x)\,(y)\,(t)\,[S_1x \,\&\, S_2y \,\&\, \mathscr{A}\,a \,\&\, \mathscr{A}\,b \,\&\, a(x, t)$
        $\&\, b(y, t) \,\&\, I(x, y, t): \supset a = R_2^{\$}b]$

(In laws describing interactions among particles of the *same* class we must add a clause to the antecedent that $x \neq y$.) In (i), $R_2^{\$}$ is a *specific* mathematical function. Let us assume that it satisfies certain generic conditions $\mathscr{R}$ (e.g., those similar to those imposed by classical mechanics on force fucntions). From (i) and this assumption that $R_2$ is of the genus $\mathscr{R}$, we immediately *deduce* that

(ii)    $(\exists! r)\,(a)\,(b)\,(x)\,(y)\,(t)\,[\mathscr{R}\,r \,\&\, S_1x \,\&\, S_2y \,\&\, \mathscr{A}\,a \,\&\, \mathscr{A}\,b$
        $\&\, a(x, t) \,\&\, b(y, t) \,\&\, I(x, y, t): \supset a = r^{\$}b]$

Let us now say that $\mathscr{L}\,(f, g)$ obtains just in case a law of the form (ii) holds for $f$ and $g$ in the way that (ii) holds for $S_1$ and $S_2$. In particular, this abbreviates (ii) as

$$\mathscr{L}\,(S_1, S_2)$$

What our physicists discover, as they explore fundamental particle interactions, is that the following laws obtain:

$$\mathscr{L}\,(S_1, S_3)$$
$$\mathscr{L}\,(S_1, S_4)$$
etc.

Discovering these laws then permits him to make the following *induction about laws* concerning fundamental particle interactions.

(iii)      $(f)\,(g)\,[\mathscr{S} f\ \&\ \mathscr{S}\,g\colon \supset \mathscr{L}\,(f,\,g)]$

The scientists then go on to confirm this law *predictively*. They first use this law to make inferences about unexamined interactions. For example, $S_9$–$S_{10}$ interactions have not yet been examined. From (iii) one deduces that the law

(iv)      $\mathscr{L}\,(S_9, S_{10})$

holds. This is to predict that a law like (ii) holds for $S_9$ and $S_{10}$ just as (ii) holds for $S_1$ and $S_2$. This inference guides their research, describing the *form* of the law that obtains for $S_9$–$S_{10}$ interactions. Research proceeds and the scientists confirm the law that $R_{54}^{\bullet}$ gives the functional relation between $S_9$'s and $S_{10}$'s, just as $R_2^{\bullet}$ does for $S_1$'s and $S_2$'s. That is, they discover a *specific* law for the $S_9$–$S_{10}$ interaction parallel to law (i) for $S_1$–$S_2$ interaction. From this *specific* law and the fact that $R_{54}$ is $\mathscr{R}$, we deduce that (iv) obtains. Thus, in confirming the specific law for the $S_9$–$S_{10}$ interaction, one has provided grounds for asserting the law (iv). But (iv) was predicted by (iii). Thus, in confirming (iv), one has confirmed the prediction of (iii). In other words, to use Lakatos' terms, when the law (iii) concerning laws, that is, the *theory* (iii), was originally law-asserted, it was *theoretically progressive*, and subsequent research confirmed laws of the sort it predicted to hold, thereby showing that the theory is also *empirically progressive*.

Now, the generic theory (iii) also predicts that

(v)      $\mathscr{L}\,(S_{10}, S_{10})$

i.e., that *there is* a law of a certain generic form governing the $S_{10}$–$S_{10}$ interaction. But, as it turns out, for some contingent fact, for all $x$, $y$, $t$ such that $x \neq y$, $S_{10}x$ and $S_{10}y$,

$$\sim I(x, y, t)$$

the condition of interaction does not obtain. So there are no positive instances of a specific law for $S_{10}$—$S_{10}$ interactions. Moreover, there are therefore no instances of a specific law that can provide grounds for asserting (iv). Nonetheless, we have, in the empirically progressive theory (iii), good inductive grounds for asserting (v), that is, we have grounds that make it "reasonable to believe that *there is* some *underived* law dealing with the interaction of particles of types"[11] $S_{10}$ and $S_{10}$, even though that specific law cannot be derived from laws that are not vacuously true. The point is, there is nothing in this account of the reasonability of this belief that is incompatible with a Humean account of laws. The crucial move is that which introduces the induction (iii) about laws; that is, what is essential is a generic theory that has shown itself to be empirically progressive. And there is no reason why the Humean should not analyze this law as he analyzes all other laws. In particular, there is nothing in this account that forces one to reject the Humean position and adopt Tooley's.

Or at least, we can see that the Humean can accept one of Tooley's core ideas. This is that, in order to know the uninstanced specific law obtains, one must know *a relation among universals.* This, after all, is what (v) states to obtain. The point is that Tooley construes this as a *primitive* relation. So construed, it confronts the ontological difficulty that appears to be an insuperable objection to the view. But we have shown that one can construe the relation among universals to be a non-primitive relation analyzable in terms of a generic theory, a law about laws. Thus, Tooley is not wrong in looking for a relation among universals, but he is wrong in construing it as primitive. Once it is recognized that the relation can be analyzed in terms of a unifying generic theory, then this idea of Tooley's can easily be accepted by the Humean.

Before turning to the epistemological issue, it is perhaps worth noting that there are real examples from science that satisfy all the general suppositions of Tooley's contrived example. The van der Waals' law for gases is a generic law like (iii). It has a specific instantiations laws describing the behaviour of different real gases. These specific instantiations correspond to the laws like (i) in the Tooley example. In addition, the van der Waals' formula comprehends as another instantiation the ideal gas law, which describes the behaviour of a type of gas that does not — and, indeed, given other laws, cannot — exist. This corresponds to the specific law for the non-interacting types in the Tooley example. As for why the ideal gas law is law-assertible, the answer we propose[12] is just that we have seen the Humean can give in the case of Tooley's example: even in the absence of positive confirming instances, grounds for law-asserting a generalization can exist, deriving from an empirically progressive generic theory, a confirmed law about laws.

As for the epistemological problem, in attempting to come to grips with it, Tooley organizes his remarks around a probabilistic construal of confirmation. In particular, he relies upon Bayes Theorem which states that, if $Pr(H/E)$ is the probability of H given E, then

$$Pr(H_1/E) = \frac{Pr(H_1) \times Pr(E/H_1)}{Pr(E)}$$

$$= \frac{Pr(H_1) \times Pr(E/H_1)}{\sum\limits_{i=1}^{n} [Pr(H_i) \times Pr(E/H_i)]}$$

where the $H_i$ are mutually exclusive and jointly exhaust the set of possible alternatives. For example, we might let $H_1$ be the hypothesis that $(x)(Fx \supset Gx)$. Let $H_2$ be the hypothesis $\sim H_1$. Then $n = 2$: the two hypotheses are jointly exhaustive and mutually exclusive, exactly one of them being true. We suppose that we have observed that $a$ is $F$, but have not yet determined whether $a$ is $G$, that is, whether $a$ supports H or supports $\sim$ H. Let E be the total evidence available, including the fact $Fa$. The problem is to determine how much the fact $Ga$, *if observed*, would contribute to the probability of H. If we let E* be: E & $Ga$, then this problem is simply that of determining

$$Pr(H_1/E^*)$$

or since E* entails E, what is identical,

$$Pr(H_1/E^* \& E)$$

which in turn is also identical to

$$Pr(H_1 \& E/E^*)$$

Thus, by Bayes Theorem we obtain that

$$Pr(H_1/E^* \& E) = \frac{Pr(E^*/H_1 \& E) \times Pr(H_1 \& E)}{[Pr(E^*/H_1 \& E) \times Pr(H_1 \& E)]}$$
$$+ [Pr(E/\sim H_1 \& E) \times Pr(\sim H_1 \& E)]$$

However, E and $H_1$, that is, $Fa$ and $(x)(Fx \supset Gx)$, entail $Ga$, that is, $E^*$. Hence

$$Pr(E^*/H_1 \& E) = 1$$

from which it follows that

$$Pr(H_1/E^* \& E) = \frac{Pr(H_1 \& E)}{Pr(H_1 \& E) + [Pr(E^*/\sim H_1 \& E) \times Pr(\sim H_1 \& E)]}$$

Now, E is given, so we may set

$$Pr(E) = 1$$

Let

$$Pr(H_1 \& E) = m$$

Then

$$Pr(H_1 \& E) = Pr(E) \times Pr(H_1/E) = Pr(H_1/E) = m$$

and

$$Pr(\sim H_1 \& E) = Pr(\sim H_1/E) = 1 - Pr(H_1/E) = 1 - m$$

Finally, let

$$Pr(E^*/\sim H_1 \& E) = k$$

$m$ measures the probability to be assigned to $H_1$ given the evidence E available prior to the completion of the test or experiment on the individual $a$. $k$ measures the likelihood that the crucial fact $Ga$ will obtain even if the hypothesis $H_1$ is false.

Finally, let

$$Pr(H_1/E^* \& E) = p$$

It follows immediately that

$$Pr(H_1/E^* \& E) = p = \frac{m}{m + k(1 - m)}$$

It is easily seen that

$$p_a > p_b \text{ if and only if } k_a < k_b$$

Thus, $p$ increases as $k$ decreases. What this shows is that if $E^*$ is unlikely if $H_1$ is false, i.e., if $k$ is small, then the greater will be $p$, i.e., the greater will be the amount it contributes to the probability of the hypothesis: *evidence that is unlikely if the hypothesis is false more strongly supports that hypothesis*; evidence that would surprise us if the hypothesis is false is stronger than evidence that would not surprise us.

Tooley works with a slightly more complicated example.[13] He considers the two hypotheses

$$S: (x)\,(Fx \supset Gx)$$
$$T: (x)\,(Hx \supset \sim Gx)$$

He then supposes that we have observed $b$ to be both $F$ and $H$ so that

$$E = \text{total evidence including the facts } Fb \text{ and } Hb$$

Individual $b$ will thus provide a test or experiment deciding between S and T when we have observed whether it is $G$ or not. Then

$$E^* = E \,\&\, Gb$$

Finally, to provide himself with a set of mutually exclusive and jointly exhaustive hypotheses (as S and T are not), Tooley considers the set

$$H_1 = S \,\&\, \sim T$$
$$H_2 = \sim S \,\&\, T$$
$$H_3 = S \,\&\, T$$
$$H_4 = \sim S \,\&\, \sim T$$

Of these four hypotheses $H_i$, one and only one must be true.

Now, again, $E^*$ entails E, so that

$$Pr(H_1/E^* \,\&\, E) = Pr(H_1 \,\&\, E/E^*)$$

and hence, by Bayes Theorem,

$$(*) \qquad Pr(H_1/E^* \,\&\, E) = \frac{Pr(E^*/H_1 \,\&\, E) \times Pr(H_1 \,\&\, E)}{\sum_{j=1}^{4} [Pr(E^*/H_j \,\&\, E) \times Pr(H_j \,\&\, E)]}$$

Furthermore, since E is given, we may set $Pr(E) = 1$, in which case

$$Pr(H_i \,\&\, E) = Pr(E) \times Pr(H_i/E) = Pr(H_i/E)$$

We must now compute $Pr(H_i/E)$, for each $i = 1, \ldots, 4$. Begin with $H_3$:

$$Pr(H_3 \,\&\, E) = Pr(H_3/E) = Pr(S \,\&\, T/E) = 0$$

since E entails that $\sim S \vee \sim T$. Now consider $H_1$:

$$\begin{aligned}
\Pr(H_1 \ \& \ E) = \Pr(H_1/E) &= \Pr(S \ \& \ \sim T/E) \\
&= \Pr(S/E) - \Pr(S \ \& \ T/E) \\
&= \Pr(S/E) - 0 \\
&= \Pr(S/E)
\end{aligned}$$

If we let

$$\Pr(S/E) = m$$

then

$$\Pr(H_1 \ \& \ E) = m$$

Similarly, if we let

$$\Pr(T/E) = n$$

then

$$\Pr(H_2/E) = n$$

Finally, for $H_4$,

$$\begin{aligned}
\Pr(H_4 \ \& \ E) = \Pr(H_4/E) \\
&= 1 - [\Pr(H_1/E) + \Pr(H_2/E) + \Pr(H_3/E)] \\
&= 1 - (m + n)
\end{aligned}$$

For the denominator of (*), we also need the quantities $\Pr(E^*/H_i \ \& \ E)$. But these are easily obtained. Since E and S jointly entail $E^*$, we have

$$\Pr(E^*/H_1 \ \& \ E) = 1$$

And since E and T entail $\sim E^*$, we have

$$\Pr(E^*/H_2 \ \& \ E) = 0$$

Since $H_2 = S \ \& \ T$, and $\Pr(E^*/T \ \& \ E) = 0$, it follows that

$$\Pr(E^*/H_3 \ \& \ E) = 0$$

Finally,

$$\begin{aligned}
\Pr(E^*/H_4 \ \& \ E) = \Pr(E^*/\sim S \ \& \ \sim T \ \& \ E) \\
&= k
\end{aligned}$$

where $k$ measures that the likelihood that the crucial fact $Ga$ will obtain even if both hypotheses S and T are false. Substituting in (*), we obtain

$$Pr(H_1/E^* \& E) = \frac{1 \times m}{(1 \times m) + (0 \times n) + (0 \times 0) + [k \times (1 - (m+n))]}$$

$$= \frac{m}{m - k(1 - m - n)}$$

However,

$$\begin{aligned} Pr(H_1/E^* \& E) &= Pr(S \& \sim T/E^* \& E) \\ &= Pr(S/E^* \& E) \times Pr(\sim T/E^* \& E) \\ &= Pr(S/E^* \& E) \times 1 \end{aligned}$$

since $E^*$ entails $\sim T$. It follows that

$$Pr(S/E^*) = p = \frac{m}{m + k(1 - m - n)}$$

It is easily seen that, as before,

$$p_a > p_b \text{ if and only if } k_a < k_b$$

Hence, $p$ increases as $k$ decreases, so that *in a situation of competing, or contrary hypotheses, evidence that supports one hypothesis and eliminates its contrary, supports that hypothesis more strongly the more unlikely it is when both hypotheses are false.* In particular, the increase in $p$ will be smallest when $k$ is greatest. In the latter case, i.e., when $k = 1$, then $p$ is least, and specifically,

$$p = \frac{m}{1 - n}$$

And even here, $p > m$ provided only that neither $m = 0$ nor $n = 0$.[14] Thus, *even evidence that is antecedently most probable anyway, and therefore of the weakest sort, will nonetheless succeed in raising the probability of an hypothesis in a conflict situation provided only that each of the hypotheses has a prior probability $> 0$.*

It is also easily seen that

$$p_a > p_b \text{ if and only if } n_a > n_b$$

Hence, *in a situation of competing hypotheses, evidence that supports one hypothesis and eliminates its contrary, supports that hypothesis more strongly the greater the antecedent likelihood of its competitor.*

Finally, Tooley notes how evidence provided by survival in conflict situations can raise the probability of hypotheses to quite high values.[15] If $k = \frac{1}{2}$ and $m = n$, then $p = 2m$. Thus, a few observations can raise the probability of an hypotheses, provided that those observations are well chosen. In the worst possible case, where $k = 1$ and $n = 0$, then we have $p = m$ so that in such a case a confirming instance will not raise the probability of an hypothesis. If $k = 1$, then $p > m$, i.e., the probability of the hypothesis is increased provided that both $m > 0$ and $n > 0$. If $n = 0$, then $p > m$ provided that $k < 1$ *and $m > 0$*. In any case, however, *for an hypothesis to have its likelihood increased through confirmation in a test situation, then that hypothesis must have a non-zero probability, and, other things being equal, its likelihood will be greater if its competitor also has a non-zero probability.*

What does this amount to? It is clear that Tooley takes his example of conflicting hypotheses S and T to be a *typical example that may be generalized as a description of experimental research.* Thus, what holds for this example, specific details aside, must hold for all experimental set-ups. Now, we just saw in the example that if the posterior probability of an hypothesis is to increase, then that hypothesis at least must have a non-zero prior probability and that it is better that its competitor(s) also have a non-zero prior probability. Thus, on the assumption that Tooley's example is typical, we may conclude that *experimental research aimed at increasing the probability of laws makes sense only if one can always assign a non-zero prior probability to the hypotheses being tested.*

However, as Carnap has made clear,[16] if one is dealing with an infinite universe, then it is hard to see how one can be justified in assigning a non-zero probability to a generalization, given evidence concerning only a finite number of instances. In order to be able to assign a non-zero prior probability to generalizations about particulars, one must be able to assume that *it is at least probable that there is some such generalization that is true* — this has been called the Principle of Determinism[17] — and furthermore, if experiments, which eliminate some hypotheses are to lead the probabilities to converge on the true generalizations, then one must also assume that *it is at least probable that the true generalizations are among the members of a given limited set of possibilities* — since a set of possible hypotheses can be specified by listing properties that are then taken as possibly conditioning each other, this has been called the Principle of Limited Variety.[18] Carnap assumed that a limited set of properties is given,[19] and thus assumed the Principle of Limited Variety — though he never attempted to justify this assumption. But he never assumed a Principle of Determinism, and was thus forced to conclude that all generalizations have a zero prior probability in any circumstance and therefore that their posterior probability is always zero also. In spite of Carnap's ingenious reaction to this,[20] this has always been taken as a damning criticism of Carnap's system of inductive logic.[21]

Tooley aims to respond to this situation, to as it were save an inductive logic from the problem that Carnap uncovered. He argues that, *upon his view of laws, non-zero prior probabilities can be assigned to generalizations.* He reasons as follows. In a system like Carnap's,[22] all atomic facts are assigned a non-zero prior probability. But two universals exemplifying the unanalyzable relation of causal necessitation, as in (2)

$$\mathcal{N}(F, G)$$

is an atomic fact. Hence this fact will have a non-zero prior probability. But this fact entails the generalization (1)

$$(x)(Fx \supset Gx)$$

Now, if $f$ entails $g$, then the probability of $g$ is greater than or equal to the probability of $f$. So, since (2) has a non-zero prior probability, (1) will also have a non-zero prior probability. Upon Tooley's view of laws, generalizations thus receive a non-zero prior probability, and Carnap's problem is solved.

This account of research does *not* presuppose that we know which facts like (2) obtain. All that it requires is that each such possible state of affairs receive a non-zero prior probability. And this, Tooley suggests — he doesn't really argue the point — can be established by the same *a priori* considerations that lead one to assign a non-zero probability to any atomic state of affairs about individuals.

This is ingenious. It avoids the epistemological difficulty we raised above, that research aimed at providing evidence justifying the assertion as laws of generalizations about particulars, does not proceed by looking for facts like (2) about universals but by experimentation among particulars. Not only that, but it further argues that such experimentation makes no sense if facts like (2) are excluded, while it does make sense if facts like (2) are admitted. And this, Tooley further suggests, provides grounds for adopting his view of laws.

The case, however, is not sound.

In the first place, one may ask just what this relation $\mathcal{N}$ is. Although we are certainly acquainted with relations among universals — e.g., "higher than by a third" among sounds, and "darker than" among colours — it would seem that we are *not* acquainted with this special causal relation $\mathcal{N}$. On this point we must, I think, agree with Hume, who examined in detail in the *Treatise*[23] the claim that we are acquainted with such an objective causal tie, and concluded that such claims are mistaken. Tooley's proposal thus violates the general principle of empiricism, the Principle of Acquaintance that nothing is to be admitted into one's ontology unless one is acquainted with it or it is of a kind or type with which one is acquainted.[24] The latter clause is, of course, important, for it allows one to include in one's ontology such things as sounds inaudible to the human ear: one is not acquainted with those sounds but they are of a kind, namely, sounds, with which one *is* acquainted, and one can therefore admit them into one's ontology.

It would seem, indeed, that one *could not be acquainted* with such facts as (2). And if we assume, as seem reasonable, the anti-Platonist Principle of Exemplification,[25] that properties and relations do not exist apart from entities that have those properties or stand in those relations, then it follows that one could not be acquainted with the relation $\mathcal{N}$. If Tooley is correct, then $\mathcal{N}$ *cannot* satisfy the empiricist's Principle of Acquaintance. Now recall Tooley's evidence E, which consists of the data about particulars *Fb* and *Hb*. It seems that acquaintance with such data, while not necessarily incorrigible, can provide us with grounds for taking such data to be certain, i.e., for assigning Pr(E) = 1. The considerations that permit Tooley to assign Pr(E) = 1 would seem to be equally sound with respect to other facts known by acquaintance. Acquaintance should therefore be able to provide us with grounds for assigning a probability of 1 to facts like (2). And since facts like (2) are supposed to entail generalizations like (1), it would follow that acquaintance with relational facts among universals can provide us with grounds for assigning probabilities of 1 to generalizations about particulars. But if that is so, then there is no need for experimentation in such cases to try to find data about particulars that will raise the probability of the generalizations. However, scientific practice indicates otherwise: no generalization about particulars is law-asserted unless it has been subjected to experimental and observational tests. If we take, as Tooley insists we should, such practice seriously, then acquaintance can never provide us with grounds for assigning a probability of 1 to facts like (1). But for the empiricist, there can be no difference in kind between acquaintance with universals and acquaintance with particulars.[26] So, if acquaintance can provide grounds for taking facts about particulars to be certain, it should also be able to provide grounds for taking facts about universals to be certain; and since the consequences of taking such facts as (1) to be certain are at variance with scientific practice, it follows that we cannot be acquainted with such facts. It would therefore seem that Tooley's own view requires him to assert that one is never acquainted with facts like (2), nor, therefore, with the primitive nomological relation $\mathcal{N}$. Thus, Tooley's own view requires him to hold that the primitive nomological relation is inconsistent with empiricist methodology.

In fact, however, Tooley suggests that it is not necessary to be acquainted with $\mathcal{N}$ in order to admit it into one's ontology. Rather, one can refer to it by means that one refers to, e.g., the sound that the dog heard that was inaudible to me, to wit, by means of *definite descriptions*. He speaks of nomological relations among universals as being *theoretical relations*, where a theoretical relation is referred to, not by a name, but by means of the apparatus of variables and quantifiers, i.e., in effect by definite descriptions.[27]

The idea is one with which we are by now familiar. Let $\mathcal{G}$ represent the genus of being a species of *germ*. Then, $\mathcal{G} G_1$, represents that the species $G_1$ of microscopic organism is a species falling within the genus $\mathcal{G}$, i.e., that it is a species of germs. Let $D$ be a sort of disease. Then

(a)        $(\exists f)\,[(x)\,(fx \supset Dx)]$

represents that there is a property $f$ the presence of which in $x$ is sufficient for the presence of $D$. But (a) is trivially true: $D$ itself is such an $f$.[27a] In contrast, to say that there is a species of germ that causes $D$ is non-trivial. The latter would be represented by

(b)        $(\exists f)\,[\mathcal{G} f \,\&\, (x)\,(fx \supset Dx)]$

If we have confirmed laws like this:

(c)        $(x)\,(G'x \supset D'x)$
           $(x)\,(G''x \supset D''x)$

where

$$\mathscr{G} G'; \mathscr{G} G''$$

and where $\mathscr{D}$ is a genus of diseases such that

$$\mathscr{D} D'; \mathscr{D} D''$$

then we might well generalize to the generic law about laws, i.e., theory

(d)        $(g) [\mathscr{D} g \supset (\exists f) [\mathscr{G} f \& (x) (fx \supset gx)]]$

Assuming that

$$\mathscr{D} D$$

we can deduce from (d) that (b) obtains. If a theory like (d) is *empirically progressive* then that can provide grounds for asserting (b) even though the entities (b) asserts to exist *have never been observed.*[28] The point is that (b) asserts that these entities are of a kind, $\mathscr{G}$, that has been observed. Thus, though we are not acquainted with what (b) asserts to exist, we can still admit such entities into our ontology because they are of a kind with which we are acquainted.

Tooley proposes to apply this idea to his nomological relation $\mathscr{N}$. No doubt he would like to use a generality like

(e)        $(\exists r) [r(F, G) \rightarrow (x) (Fx \supset Gx)]$

where the symbol '$\rightarrow$' indicates *entailment*. But this, like (a), is trivial. For, we can always define

$$R^*(f, g) =_{df} (x) (fx \supset gx)$$

and this automatically satisfies (e). We therefore need to place restrictions on $r$ in (e) in order to avoid trivialization. Let us call them $\rho$. Thus, we need, not (e), but a generalization like

(f)        $(\exists r) [\rho r \& [r(F, G) \rightarrow (x) (Fx \supset Gx)]]$

Just as, on the basis of (b), one can speak of *the* germs that cause disease $D$, so, on the basis of (f), if it were available, one could speak of *the* nomological relation that grounds the truth of the generalization that $F$'s are $G$'s.

However, it does not seem that any generalization like (f) is available. In the *first* place, one asserts (b) only because one has available a theory like (d) that justifies its assertion. Tooley provides us with no such theory, nor can I imagine what a plausible version of such a theory would be like. But in any case, in the *second* place, such a theory could be counted as antecedently confirmed, or as empirically progressive, only if one had antecedently confirmed *other instances* of the form

$$\rho r \ \& \ [r(F, G) \rightarrow (x)(Fx \supset Gx)]$$

just as we accept (b) because we have confirmed other instances (c) of that form. Now, the only constraints $\rho$ that Tooley tells us about are that $r$ be primitive and that its obtaining between two universals entails the corresponding generality among particulars.[29] The problem is that entailment relation: *we are acquainted with no unanalyzable relation among universals that entails a corresponding generalization among particulars.* The properties and relations with which we are acquainted are all *self-contained* in just the sense that no state of affairs of the form

$$\mathscr{R}(F, G)$$

with which we are acquainted somehow so reaches beyond itself to all particulars that it guarantees the truth of

$$(x)(Fx \supset Gx)$$

To this extent, logical atomism follows once we accept the empiricist's Principle of Acquaintance.[30] This is, of course, but another way of posing what we earlier called the ontological difficulty with Tooley's position. What we now see is that this difficulty cuts deep enough to prevent Tooley from avoiding the problem of acquaintance with the nomological relation by referring to the latter by means of a definite description use of which is justified by prior acceptance of some scientific theory.

Moreover, unless the ontological difficulty is solved, Tooley's appeal to a nomological relation to solve Carnap's problem simply won't work. Tooley proposes that (1)

$$(x)(Fx \supset Gx)$$

has a non-zero prior probability because, one, (2)

$$\mathscr{N}(F, G)$$

entails (1), and, two, (2) has a non-zero probability. In particular, since the prior probability of (1) is zero unless (2) obtains, it follows that no experiment can ever raise the probability of (1) above zero unless there is some fact *not* about particulars that can probabilify (1) to a degree greater than zero. The relevant fact, since it can't be one about particulars, must be one about universals, as Tooley says; and it is hard to see how a contingent relation between (1) and (2) can permit the latter to probabilify the former, so that the relation must, as Tooley also says, be that of entailment. So, if Tooley's solution to the epistemological problem is to work, he must hold that (2) entails (1). That being so, unless Tooley can solve the ontological difficulty, his proposed solution to Carnap's problem remains unsuccessful. On the other hand, if we are right, then, on the basis of the empiricist's Principle of Acquaintance, we can conclude that there is no reason to suppose that Tooley can solve the ontological difficulty. And since he recommended his account of laws as superior to the Humean's precisely because it could account for the research process as he described it, we may further conclude that he has given us no reasons for preferring his view to that of the Humean.

But *is* it impossible for the Humean to account for research practice as described by Tooley? This practice consists of treating hypotheses as being made more probable as a result of experiments. This practice makes sense only if non-zero prior probabilities can be assigned to the hypotheses being put to the experimental test. Can the Humean assign non-zero prior probabilities to hypotheses in a test situation? It would seem that he can provided that he has available something analogous to the Principle of Determinism and of Limited Variety which would enable him to determine prior to the experiment a set of hypotheses among which the true can be found.

Return to our germ example. The law (b) states that [1] *there is* a cause for disease $D$ and that [2] this cause is to be found *among the factors* $\mathcal{G}$. It is up to research to identify which of these factors [2] is the cause that [1] asserts to exist. Experiment must eliminate the false hypotheses. As experiment proceeds, the true hypothesis will be that one that survives the experiments. Clearly, [1] and [2] function in just the way that the more abstract Principles of Determinism and of Limited Variety function. Hence, if the Humean can give grounds for accepting a law like (b), then he will be able to assign non-zero probabilities to the hypotheses tested in an experimental set-up. And that in turn will enable the Humean to account for scientific practice Tooley asserts he cannot.

The question is, therefore, whether the Humean can ever have reasons for accepting a generic law like (b). However, the answer to this would appear to be affirmative. For, as we saw, if we have a theory like (d), then that can provide grounds for accepting a law like (b) as a guide to research. Nor is our disease example particularly artificial. As we saw in Chapter I, Section II, the axioms of classical mechanics function just like (d) in justifying the assertion of laws like (b) to guide research. We saw there, as an example, how that the theory told Coulomb that *there is* a force function for electrical phenomena, and it was then his research task to find this force that his theory asserted to be there. Again, the van der Waals' law asserts that *there are* coefficients describing the behaviour of real gases; it is then the research task to find what these are specifically for given real gases. Or, to use Tooley's own example about interacting fundamental particles, we might have a generic theory like (iii) that enables us to deduce the form the specific laws for unexamined interactions must have; it is then the research task to discover specifically what law of this form it is that governs the about-to-be-explored type of interaction. Thus, provided we have a generic theory or law about laws, the Humean can account for the research practices that Tooley describes. And it would seem that *such theories are acceptable provided that they are empirically progressive.*[31] Of course, problems remain. As we saw in our discussion above, in Chapter I, Sec. II, the acceptability of a generic theory depends upon evidential support being transmitted from observational data via the consequence and converse consequence relations, and this latter raises real problems.[32] These, though, are merely technical problems. For, the discussion of Newton's theory in Chapter I, Sec. II shows clearly enough that, *however this evidential support is transmitted, empirically progressive generic theories do guide empirical research.*[33] *This is all that the Humean needs in order to make his case.* It follows that the Humean can, after all, do what Tooley asserts he can't do, and what, ironically enough, Tooley himself cannot do.

It is, of course, true that the *prior* probabilities the Humean must assign are assigned on *a posteriori* grounds. In *this* sense, the response we have suggested that the Humean can make to the epistemological difficulty raised by Tooley is *not* a reply to Carnap's problem, of how to assign *a priori* non-zero prior probabilities to generalizations of fact. And if Tooley demands a Carnap-type response on the part of the Humean to the epistemological difficulty, then we have failed to satisfy him. But there are really two issues: (1) *Can* Carnap be satisfied? that is, can we provide non-zero *a priori* prior probabilities for statement of law? And (2) *Must* Carnap be satisfied? As for (1): If Hume is correct, then we *cannot* give *a priori* probabilities at all, and in particular we cannot give them *a priori* for laws. For, there is always a logical gap between sample and population that precludes any such assignment of *a priori* probabilities. Hence, the issue of trying to find non-zero *a priori* probabilities for laws does not arise.[34]

For the Humean, the *prior* probabilities can never be *a priori* probabilities. However, it is unfair to demand the latter of the Humean — if that is what Tooley is doing — for the reason that the Humean has already argued that *such a demand is impossible to fulfill.* As John Stuart Mill insisted in his discussion of the Principle of the Uniformity of Nature, only an *a posteriori* justification can be given for the maximal principles that guide research.[35] But *that* is no reason why such principles cannot guide research and be used as a basis for assigning prior probabilities to hypotheses being put to a test in an experimental situation, that is, assigning *a posteriori* prior probabilities. And these *can* do what the Humean asks of them, namely, provide a solution to the epistemological difficulty Tooley raises, since the answer to the second of the two questions cited above is negative: we need not satisfy Carnap. Of course, if Hume is correct, then we *cannot* satisfy Carnap, and therefore need not do so.[36] But there is a less problematic point to be made here also. Tooley issues a challenge to the Humean: account for the use, which he (Tooley) describes, of the probability calculus in research practice. This challenge was what we called the epistemological difficulty. Tooley correctly describes research practice — at least, we were prepared to grant him that. But what he did not anywhere defend was his assumption that that practice, when it employed prior probabilities, was employing *a priori* prior probabilities. In that sense, then, no case was made that Carnap must be satisfied. We may take it, then, that all that the Humean must do in order to meet the epistemological difficulty, is to show how the required prior probabilities can be introduced, rather than show, as Tooley assumes, how *a priori* prior probabilities can be introduced. And *this* we have shown that the Humean *can* do, once he admits that there are generic theories that can permit the assignment of *a posteriori* prior probabilities to untested laws at the specific level. One may speculate that Tooley fails to see that the Humean has an answer to the epistemological difficulty because he fails to note the possibility — quite compatible with the Humean account of laws — of generic theories, laws about laws. In overlooking the role such laws play is research, however, Tooley is not alone. In fact, not only do many opponents of empiricism overlook such laws, so do many of its defenders.[37] But if we are right, both in our discussion in Chapter I, Section II, and in our discussion of Tooley, then it is essential to empiricism to recognize both their existence and the role they play in guiding research.

## CHAPTER THREE

# POSSIBLE WORLDS: A DEFENCE OF HUME

> All possible worlds lie within the actual one.
> Nelson Goodman
> *Fact, Fiction and Forecast*

## I. INTRODUCTION

Upon the thesis we are defending, explanation is by means of deduction from laws. In particular, since statements like

(1)     *c* causes *e*

are explanatory, we are defending the thesis that these statements involve a deduction from laws.[1] One particular attempt to so construe (1) that it does *not* involve the assertion of a law is of particular interest in the present context. This is the position of David Lewis.[2] This position not only separates causal statements like (1) from laws, but also denies the Humean account of laws, construing laws as *mere* matter-of-fact regularities, as not involving what the Humean insists must be involved in any analysis of laws, to wit, the *idea of necessity*.

Lewis, in contrast to Hume, separates counterfactual or subjunctive conditionals from laws. This means, of course, that Lewis must provide an alternative account, one not based on laws, of the assertibility of subjunctive conditionals. This alternative account is in terms of the idea of *possible worlds*.[3] Possible worlds are related to each other by the relation of *similarity*. The counterfactual conditional

(2)     If A were then B would be

is said to be true in the actual world provided that, in those possible worlds that are most similar to ours and in which A holds, B also holds. Then, using this notion of counterfactual implication, Lewis analyzes the notion of causal dependence.[4] The notion of being a law is analyzed independently of the idea of counterfactual conditionals. All laws are regularities, and specifically are those regularities that can be fit into the neatest or simplist axiomatic development.[5] Using this notion of law, Lewis goes on to develop the idea of one event being nomically dependent upon another with such nomic dependence carefully distinguished from causal dependence. Of course, one of the things that can make two worlds similar is that the same laws hold in each, so that there is a close connection between laws and counterfactuals and between nomic and causal dependence. Nonetheless, according to Lewis, two worlds differing in laws might still be more similar than two worlds sharing laws, so that a counterfactual conditional might be true by virtue of holding in the world most similar to ours, even though the regularities that would, in our world, permit the inference of consequent from antecedent, do not hold in that possible world, and in particular do not lawfully connect antecedent and consequent. The contrast to the Humean, then, is nearly complete: For the latter, a law is not a mere regularity; that concept also involves the idea of necessity; and it is laws in this sense that the Humean then uses to account for causal language and counterfactual conditionals. Clearly, if one is to defend the deductive-nomological model as basic to explanation, then we have to say something about Lewis' position.

## II. THE POSSIBLE WORLDS ACCOUNT OF COUNTERFACTUALS

Lewis states the essentials of his position thusly:

"*If kangaroos had no tails, they would topple over*" seems to me to mean something like this: in any possible state of affairs in which kangaroos have no tails, and which resembles our actual state of affairs as much as kangaroos having no tails permits it to, the kangaroos topple over.[1]

'Possible state of affairs' is fleshed out to become 'possible world.' If we suppose that we have a complete description of the actual world, then we obtain a description of a non-actual but possible world by altering that description. If the alteration is slight, the possible world is similar to ours, i.e., to the actual world; it is more similar to the actual world than the possible world described when the alteration is more than slight. A non-modal sentence is true in a possible world just in case that it is entailed by the description of that possible world, or, in the case of a sentence containing quantifiers, just in case that its quantifier-free development is entailed by the description of the possible world.[2] If we represent the counterfactual (2) by

(3)      $A \to B$

then this is said to be true, i.e., true in the actual world, just in case that

$A \supset B$

is true in every member of a class of possible worlds. This class of possible worlds is picked out by the relation of *similarity*.

But more fundamental than the relation of similarity is the more generic notion of *accessibility*.

Lewis has us conceive a set of possible worlds as constituting a kind of filled sphere.[3] The world $i$ is at the centre. Those possible worlds that are *accessible* from the world $i$ constitute the filled sphere, which is called the "sphere of accessibility" of $i$. Once the notion of a sphere of accessibility is available, Lewis gives truth conditions for modal sentences.[4] If Ø is a non-modal sentence, and N is the modal operator *necessary*, then

$N\,Ø$

is true in world $i$ if and only if Ø is true in every possible world in the sphere of accessibility $S_i$ assigned to $i$. Defining the strict implication

$A \to B$

to mean, as usual,

$N(A \supset B)$

it follows that the latter is true at $i$ if and only if

$A \supset B$

is true in every possible world in $S_i$. If we define a "P-world" to mean "possible world in which P is true", then we can equivalently say that $A \to B$ is true in $i$ if and only if B is true in every A-world in $S_i$, i.e., in every A-world accessible from $i$. Note that if no A-world is accessible then $A \to B$ is *vacuously true*. We obtain *different kinds of necessity* according to the accessibility relation we choose. Corresponding to *logical necessity*, we assign to each world $i$ as its sphere of accessibility $S_i$ the set of *all* possible worlds. Thus, the logical strict implication $N(A \supset B)$ is true at $i$ if and only if B is true at all A-worlds whatever.

In this case, there are no inaccessible A-worlds to be left out of consideration. Corresponding to *physical necessity*, we assign to each world $i$ as its sphere of accessibility $S_i$ the set of all possible worlds in which the laws of nature prevailing at $i$ hold. Thus, the physical strict implication $N(A \supset B)$ is true at $i$ if and only if B is true at all those A-worlds in which hold all the laws that hold at $i$. Possible worlds in which there hold laws contrary to those in $i$ would be inaccessible from $i$; they would be outside the sphere of accessibility of $i$.

Of course, neither of these notions is unproblematic until the notion of "possible world" is further elucidated. Presumably that would suffice for the case of logical necessity. And in the case of "physicl necessity" one needs in addition a clarification of the concept of "natural law."

Further notions of necessity can also be defined, by specifying different accessibility relations to define different spheres of accessibility. The crucial one is, of course, that which defines counterfactual implication. In this case, the accessibility assignment is determined by similarity of worlds: the sphere of accessibility of $S_i$ for a world $i$ is the set of all worlds that are "similar to at least a certain fixed degree"[5] to the world $i$ — where the relevant similarity is "overall similarity, with respects of difference balanced off somehow against respects of similarity."[6] Then, if (3)

$$A \rightarrow B$$

or $N(A \supset B)$ is taken to be a sort of strict conditional, then it is true at $i$ if and only if

$$A \supset B$$

is true throughout $S_i$; i.e., if and only if B holds in all A-worlds similar to at least that degree to $i$.[7] Again, if no A-world is accessible, then $A \rightarrow B$ will be vacuously true.

But in fact, this won't do for counterfactual implication.[8] Consider this pair of counterfactual conditionals

> If one of us walked on the lawn, no damage would be done; but if everyone walked on the lawn it would be ruined.

Here we have the pair

$$A \rightarrow B$$
$$(A \, \& \, A') \rightarrow \sim B$$

both of which are true. Now, if we accept the above account of the truth conditions for counterfactual implication, then the first of these is true if and only if B is true in every accessible A-world. And the second is true if and only if $\sim$B is true in every accessible (A & A')-world. It follows that, since both are true, no (A & A')-world can be accessible; i.e., if the first is true, then the second can only be vacuously true. But if a strict conditional is vacuously true, then so is any other with the same antecedent. Hence, from the truth of the above two counterfactuals, one can infer the truth of

> If everyone walked on the lawn then the cow would have jumped over the moon.

This, however, is *not* true; or, at least, while the first pair may both be justifiably asserted, that fact does not justify asserting the third counterfactual. And as a consequence the construal of counterfactual implication as a strict conditional must be rejected as mistaken.

Lewis solves this difficulty by noting that similarity — unlike the accessibility relations defining logical and physical necessity — admits of degrees. Hence, where accessibility is defined by the similarity relation, some possible worlds are more accessible from a centre $i$ than others. Similarity thus defines not just *a* sphere of accessibility for $i$, but a whole set of *nested spheres*, one for each degree of similarity.[9] Actually, 'degree' is misleading. One does not need a metrical notion, only a comparative notion of similarity: if $x$ is more similar to the world $i$ than is $y$, then it is closer to the centre. Possible worlds that are equally similar form a subsphere about the centre $i$, and the whole set of possible worlds that are in any way similar to $i$ are structured by the similarity relation into a sphere made up of a nested sequence of subspheres, in the sense that, for any two subspheres, one wholly contains the other. Thus, if a possible world $x$ belongs to a subsphere $S$, then it belongs to every subsphere in the sequence which is larger than $S$. Hence, if $x$ is nearer the centre than is $y$, then at least one subsphere contains $x$ but not $y$, and none contains $y$ but not $x$. Note also that there may be possible worlds that do not occur in the sphere of accessibility assigned to the world $i$.

Lewis uses this idea to propose that the *counterfactual* conditional (3)

$$A \rightarrow B$$

not have truth conditions defined, as in the case of a strict conditional, relative to a single sphere of accessibility, but relative rather to the sequence of such spheres defined by the similarity relation. Specifically, he proposes that (3) be considered as true if and only if
> either (i) no A-world belongs to any subsphere in the set about $i$
> or (ii) some subsphere does contain at least one A-world, and

> A $\supset$ B is true in every world in that sphere.

Lewis calls such a conditional a "variably strict conditional." In case (i), there are no A-worlds in the sphere about $i$; in this case, A $\rightarrow$ B is vacuously true at $i$. In case (ii), A $\rightarrow$ B is true at $i$ if there is a subsphere throughout which A $\supset$ B is non-vacuously true. If a sub-sphere is said to be "P-permitting" just in case that it contains at least one P-world, then we may express case (ii) as asserting that A $\rightarrow$ B is true at $i$ if there is an A-permitting subsphere in which every A-world is a B-world.

On this account, A $\rightarrow$ B can be true at $i$ in spite of there being any number of (A & ~B)-worlds, just so long as *each* of the latter is further from $i$, that is, less like $i$, than is some (A & B)-world.

Moreover, for a given A there may be no *smallest* A-permitting subsphere.[10] For, it could be the case that, given any A-world $j$ other than the world $i$, there is some other A-world which is closer to the centre $i$ than is $j$. If there always were a smallest subsphere, then, as Lewis says,[11] he could "make the truth conditions for counterfactuals simpler" in the non-vacuous case. For, he could say that A $\rightarrow$ B is true if A $\supset$ B is true in *the smallest* A-permitting subsphere.

In this scheme, we can construct spheres around *any* world $i$. For the simplist cases, $i$ is the actual world; that is, we are interested in conditionals that are relative to *our* possible world, which is to say, relative to the actual world. But often it is useful to consider spheres about some nonactual world, e.g., for purposes of analyzing complex counterfactuals like "If Grant had been drunk at Appomattox, then it would have been the case that if the Federal cavalry had been kept in reserve Lee would have won the battle." For our purposes, however, we may ignore these cases, and restrict ourselves to the sequence of spheres that centres on the actual world.

In we now consider the pair

> A $\rightarrow$ B
> (A & A') $\rightarrow$ ~B

we can see how the problem that confronted construing counterfactual implication as a strict conditional has been avoided.[12] Taking it to be a strict conditional, we had to choose one of the nested spheres about $i$, $S_i^1$, $S_i^2$, $S_i^3$, ..., to be *the* sphere of accessibility about $i$ and no choice was right. $S_i^1$ was right for the first, but not the second. $S_i^2$ was right for the second, but not for the first, nor for a third

$$(A \& A' \& A'') \rightarrow B$$

and so on. Now, however, construing counterfactual implication as a variably strict conditional, it is not necessary to choose: the several spheres are all there as subspheres in the nested set; $S_i^1$ is there to make the first of the counterfactuals true, $S_i^2$ to make the second true, $S_i^3$ to make the third true, and so on.

Such is Lewis' account of counterfactuals.[13] As we said above, however, the whole scheme, ingenious as it may be, remains problematic until the crucial notions of *possible world* and *natural law* are further elucidated — and to this list we must now add the notion of *similarity* among possible worlds. We shall look at these in turn.[14]

### III. HOW POSSIBLE ARE POSSIBLE WORLDS?

Lewis is among those who now seek to specify *truth-conditions* for counterfactual conditionals.[1] This is in contrast to the Humean position that we have been defending.[2]

For the Humean, the *assertion* of the counterfactual conditional

(4)        If $a$ were $F$ then it would be $G$

is to be construed as a *complex assertion* involving the argument

(5)        $(x)(Fx \supset Gx)$
           $Fa$
           _____
           $\therefore Ga$

where the first premiss is *law-asserted*, the second premiss is merely *supposed* rather than asserted, and the conclusion is *asserted to follow* from those premisses. The counterfactual statement does not *itself* literally have truth-conditions. Rather, *it is indicative sentences alone that have truth-conditions*, and so, in the case of the counterfactual assertion in the subjunctive, the bearers of truth-values are the indicative sentences that are implicitly part of the complex assertion into which the assertion of the counterfactual to analyzed. If an indicative sentence is true, then its having the truth-value justifies its assertion. What justifies the assertion of a subjunctive conditional is a more complex matter, according to the Humean. For Lewis, in contrast, there is no essential difference between subjunctives, at least so far as concerns counterfactuals, on the one hand, and indicatives, on the other: both are bearers of truth-values, and in each case it is truth that justifies assertion.

The Humean can, naturally, claim a certain plausibility for his account. It is the Humean's claim that a generality is implicitly law-asserted that is, perhaps, the hardest claim to defend. At any rate, that is where Lewis disagrees with the Humean. Nonetheless, the Humean's claim is not implausible. For example, to take a similar case, not in the subjunctive, the assertion

> Since he arrived before her, and she arrived before Roland, he must have arrived before Roland.

is agreed by most as most reasonably construed not as a single sentence with truth-conditions of its own, but as an *enthymeme*, that is, as a *series* of sentences, one asserted only implicitly (namely, the assertion of the transitivity of "before"), two asserted as premises, and the third asserted to follow from the two explicit and the one implicit premiss. The question is, does the plausibility that attaches to this example carry over, as the Humean asserts it does and as Lewis asserts it does not, to the case of subjunctive or counterfactual conditionals? The question can be answered only by considering the arguments for and against the two alternatives. Moreover, the argument, in the end, must be *normative*: what *ought* people to mean when they assert counterfactuals? Thus, the mere fact that in normal verbal behaviour no law-assertion is explicit when subjunctive conditionals are asserted might very well turn out to be irrelevant to the Humean's case. For, if he can successfully argue that such a law-assertion *ought* to be there, then any assertion of a subjunctive conditional, *if it is justified, must* contain, implicitly at least, a law-asserted generality, and if it doesn't, then it is unjustified. In fact, it can be argued that misguided attention to the niceties of describing overt verbal behaviour while ignoring the normative issue can lead to unwarranted criticisms of the deductive-nomological model of explanation.[3]

But the difficulty with Lewis' position does not lie here, but rather in its conflict with the basic empiricist framework of those who defend the Humean and deductivist positions. A counterfactual like (4) *is a statement, not about actualities, but about possibilities. It is not about how things are but how they might be.* For the empiricist, there is *this* world, the actual one, and no other: besides what *is* the case, there are not along side them a whole host of further facts, *possibilities* along side the actualities, *might be's that are.* This is why subjunctive conditionals cannot have truth-conditions in the way indicative sentences do: what would make counterfactuals true could only be possibilities, and they simply are not real in the way actualities are. Hence, for the Humean, subjunctive conditionals have assertion conditions but not truth-conditions. However, when Lewis asserts that subjunctive conditionals have truth-conditions, he intends it quite literally: *its truth-conditions are possibilities and such possibilities exist in just the way actualities do.* When he asserts that A → B is true just in case B is true in the possible A-worlds most similar to ours, he means to say that just as it is a *fact* in our world that makes A true in our world (if it is) so it is a *fact* in a possible world that makes it true in that world (if it is). What makes A true in our world is a fact in our world; what makes A true in some other possible world is a fact in that world. Both facts exist in exactly the same sense. Or rather, both possible worlds — this one, which is our world (i.e., the one actual to us), and the other possible world (which is actual to those in it) — exist in exactly the same sense. For the empiricist, this sort of discourse, taken uncritically, is essentially meaningless.

The essential point has been made by Quine:

Wyman's overpopulated universe is in many ways unlovely. It offends the aesthetic sense of us who have a taste for desert landscapes, but this is not the worst of it. Wyman's slum of possibles is a breeding ground for disorderly elements. Take, for instance, the possible fat man in that doorway. Are they the same possible man, or two possible men? How do we decide? How many possible men are there in that doorway? Are there more possible thin ones than fat ones? How many of them are alike? Or would their being alike make them one? Are no *two* possible things alike? Is this the same as saying that it is impossible for two things to be alike? Or, finally, is the concept of identity simply inapplicable to unactualized possibles? But what sense can be found in talking of entities which cannot meaningfully be said to be identical with themselves and distinct from one another?[4]

Quine's point is, of course, not that there are no answers to such questions, but that the concepts used to pose them are so radically unclear, puzzling and problematic that it is just not clear how one is to go about answering such questions. What is required is that the crucial concepts be *explicated*, rendered unproblematic. Philosophers have said many things about the possible, the necessary, and related concepts. For example, many have held that necessary truth is what is true in all possible worlds. But what is a possible world? The skilful dialectician can, like Quine, raise problems that are apparently unsolvable. In this way, most of the things philosophers have said about possibility and necessity remain puzzling and problematic. The task that confronts the empiricist is that of making sense of such discourse. *To do that is to so explicate the crucial concepts that they are no longer problematic and so that we can see a clear sense in the problematically stated claims of traditional philosopher.*[5]

Lewis, in contrast, claims such notions need no explication. Specifically, we can take the notion of a *possible world* as a *primitive concept*[6] — it is sufficiently unproblematic as to require no further analysis[7] — and then use it to define other modal notions like "necessary", etc.

In fact, he suggests, there is, Quine notwithstanding, no distinction between the actual and the possible: all possible worlds exist, and the actual world is simply that possible world that is *actual to us*, that is, it is the possible world that we are in. According to Lewis, we ought to construe 'actual' as a token-reflexive term like 'now': just as "now" means "the moment that is simultaneous with the utterance of 'now'" so "actual" means "the possible world in which the utterance of 'actual' occurs". "Actuality" is always relative to a speaker. This world is actual to me; another possible world is actual to Hamlet. Actuality is not a property that distinguishes one possible world from all the others but is a relation between an utterance and the possible world in which the utterance is made.[8] Where Quine and the empiricist will somehow distinguish between the actual, which exists, and the merely possible, which does not exist,[9] Lewis insists that there is no such distinction, that all possible worlds exist in exactly the same way, and that all possible worlds are equally actual, that is, actual in the same relative sort of way.

This position leaves the crucial notion of "possible world" undefined and unexplicated. The claim is that explication is not[2] required because the notion is already sufficiently clear in its philosophical use that one can recognize that it is implicitly defined by certain axioms,[10] and that once these are articulated it becomes evident that the problems Quine poses do not in fact arise.[11] Among the axioms that Lewis proposes as implicitly defining the crucial notion are

$$(P_1): (x)(y)(Ixy \supset Wy)$$
$$(P_2): (x)(y)(z)(Ixy \ \& \ Ixz: \supset y=z)$$

where

$Wx = x$ is a possible world
$Ixy = x$ is in $y$

so that these axioms assert that

$(P_1)$: Nothing is in anything except a possible world
$(P_2)$: Nothing is in two possible worlds

If these axioms are accepted, then it is indeed clear that at least some of the difficulties that Quine raises are avoided. Quine asks how many possible men are in the doorway at which he is looking. If Lewis' axioms are accepted, then this question cannot be asked. For, the doorway is actual — it is in *this* world — while the possible men are in *other* worlds. However, the actual door has *counterparts* in other possible worlds — doors that are similar, but not identical to it. In other possible worlds, the counterparts of this door will have men in them, and these can be counted. And of course, how many are in the counterpart will vary from possible world to possible world. Quine can raise the problems he does because he confuses possible worlds with each other and does not specify exactly which one it is that he is asking his question about.[12]

This, naturally, requires *another primitive notion* "$Cxy$" = "$x$ is a counterpart of $y$",[13] which is implicitly defined by such axioms as:

$(x)(y)[Cxy \supset (\exists z)(Ixz)]$
   = whatever is a counterpart is in a world
$(x)(y)[Cxy \supset (\exists z)(Iyz)]$
   = whatever has a counterpart is in a world
$(x)(y)(z)[Ixy \ \& \ Izy \ \& \ Cxz: \supset z=x]$
   = Nothing is a counterpart of anything else in its world
$(x)(y)[Ixy \supset Cxx]$
   = Anything in a world is a counterpart of itself.

This relation is not transitive, nor symmetric; something in one world may have more than one counterpart in another; two things in one world may have a common counterpart in another; there are worlds in which there is something in the one that is the counterpart of nothing in the other; there are worlds in which there is something in the one such that nothing is the counterpart of it in other world.[14] All this is so because "counterpart" is a *similarity* relation, and therefore does not have a neat set of structural properties.[14a]

Your counterparts resemble you closely in content and context in important respects. They resemble you more closely than do the other things in their worlds. But they are not really you. For each of them is in his own world, and only you are here in the actual world. Indeed, we might say, speaking casually, that your counterparts are you in other worlds, that they and you are the same; but this sameness is no more a literal identity than the sameness between you today and you tomorrow. It would be better to say that your counterparts are men you *would have been*, had the world been otherwise.

The counterpart relation is a relation of similarity. So it is problematic in the way all relations of similarity are: it is the resultant of similarities and dissimilarities in a multitude of respects, weighted by the importances of various respects and by the degrees of the similarities.[15]

One may well wonder, however, whether the concept of "possible world" is sufficiently clear that it is self-evident that its use conforms to the axioms Lewis lays down.

In the first place, it is not at all clear whether Quine's problem has been avoided. He raised problems concerning the possible state of affairs consisting of some possible but non-actual man being in relation to *this actual* doorway. Lewis lays down axioms that preclude this way of speaking. Nonetheless, merely saying so does not establish the illegitimacy of Quine's way of speaking. Indeed, one wants to insist that there are many possibilities that obtain with respect to *this* doorway.[16] The philosophical task, surely, is to clarify the problematic notion, so explicate it that we can see that a clear sense *does* attach to these claims about possibilities with respect to this actual doorway. Again, consider a person confronting a choice situation. There are two possibilities before him. If Lewis is right, one individual has the one possibility before him, while it is another individual that has the other possibility before him. To be sure, the two individuals are counterparts, but since contrary facts hold of them, they exist in different possible worlds and are, therefore, *literally different*. But this is, surely, not how we ordinarily conceive of choice situations:[17] we think of individuals as having alternative futures which are possible for them as the *very same* individuals.[18] This discourse may, indeed, as Quine insists, be problematic, but it is problematic in that peculiar philosophical way that demands clarification rather than dismissal. Lewis, alas! simply abandons the philosophical task, and lays down axioms that dismiss rather than clarify the problematic modes of discourse.

Another aspect of Lewis' axiomatization of "possible world" that remains problematic is his dismissal of the idea that there is a *real difference* between actuality and possibility. Consider a case of perceptual error: a piece of white paper is under a red light, and I perceive that the paper is red. This is a *mis*-perceiving, but nonetheless the state of affairs that appears to me is that of the paper being red. This state of affairs is a *possible* state of affairs. On Lewis' account this possible state of affairs must be construed as an actual state of affairs in another possible world.[19] However, one wants to insist, surely, that what I perceive is *not* an *actual fact in another world* but rather *a possible fact in this world*. Moreover, what I perceive is a possibility about *this* piece of paper which is actual rather than a fact about its counterpart in another world.

One could get around this by making "actual" into yet another primitive concept. Lewis' axioms for his indexical 'actual' could still be used:

$$(\exists x)[Wx \ \& \ (y)(Iyx \equiv Ay)]$$
= Some world contains all and only actual things
$$(\exists x)(Ax)$$
= Something is actual.

This raises many problems of its own, however. Though the axioms just laid down suggest otherwise, "actuality" cannot be treated as a property among properties. If it is a fact that this is red, then that state of affairs has the property of actuality. But now consider the state of affairs of this being red being actual. Does this also have the property of actuality? Clearly, a vicious regress looms.[20] As Bergmann has pointed out, this regress can be stopped only by granting the simple "property" of actuality an extraordinary status.[21] That is, rather special and delicate axioms must be laid down for its use. Bergmann, alone among the recent friends of possibilities, has attempted to explore the special status that must be granted to "actuality" if the latter is to be treated as a simple unanalyzable property, or, rather, "property".

Bergmann also faces up to another problem that Lewis completely fails to deal with. Frege made the point long ago that "... one can never expect basic propositions and theorems to determine the reference of a word or symbol."[22] If we say

$$(x)(Fx \supset Gx)$$

is (part of) the implicit definition of '$F$' and '$G$', that fact in no way determines what it is that those predicates refer to. The most that is achieved is that, since we wish the implicit definition to be true when we assign designata to '$F$' and '$G$', we have placed non-syntactical restrictions on the interpretations we can assign to those predicates.[23] Nonetheless, the formula remains *uninterpreted* until designation rules are assigned to the predicates, coordinating these linguistic items to *perceivable properties* of things.[24] Only after such designation rules are assigned does the formula become a sentence *about the observable world*. The same must be said for such terms as 'possible world' or 'actuality.' Lewis takes these as primitive, and lays down axioms for them. But he does not interpret these concepts. We must say, therefore, that until he provides us with an interpretation of these terms, the formulae containing them must be reckoned as uninterpreted and meaningless, at least so far as the empiricist is concerned. It is Bergmann's great virtue to argue that the "modes" of "actuality" and "potentiality" (i.e., "possibility") are in fact, on some occasions, presented to one.[25] (Bergmann calls these "properties" "modes" to mark the special status they must be granted.)[26] In this respect one must judge Bergmann's ontology of the possible to be incredibly more articulate than that of Lewis, who is content to present us with mere axiomatics. Nonetheless, even at its most articulate, as in the case of Bergmann, an ontology which includes possibles along with actuals strikes the empiricist as odd.

For our purposes, however, there is one feature which is most damning to any attempt to explain counterfactuals in terms of truth-conditions that obtain in real possible but non-actual worlds.

According to Lewis, the counterfactual (3)

$$A \rightarrow B$$

is true just in case B is true in the nearest A-world.[26a] Hence, to *discover* whether the counterfactual is true, we must first *locate* the nearest A-world, and then *discover* whether B is true in it or not. For indicative sentences, one specifies truth-conditions, and then finds out whether they are worthy of assertion by discovering whether those truth-conditions are actual or not. Lewis holds that counterfactuals have truth-conditions in exactly the same way. So, presumably, exactly as with indicative sentences, one must *discover* whether a counterfactual is true by *discovering* whether its truth-conditions obtain. But just how does one *locate* an A-world, that is, a possible world other than ours in which A is true (as it is not in ours)? Clearly, it is not like locating my lost watch. For, my lost watch is in the actual world. And if Lewis is correct, possible worlds are wholly separate from each other: nothing in one is in any other. In fact, how can I ever get out of the actual world so as to locate some other possible world? And if I do locate an A-world how do I go about exploring it in order to find out whether B obtains in it? I do have rules for getting about in the actual world, but, to be frank, I have no idea how to go about exploring a non-actual world. Indeed, it would seem that, if Lewis is right, then if I do get to the non-actual world so that I can explore it, then I am no longer me, but rather my counterpart. So how do *I* — or *we* — or even Lewis — find out these things? To put it in a sentence: if counterfactual conditionals are literally about non-actual possible worlds, then there is no way in which the truth of such statements can be verified.[27] (Bergmann avoids this difficulty by making potentialities and actualities all part of the one world that exists.)[28]

There is little, then, that can recommend Lewis' odd and rather inarticulate ontology of possible worlds as a foundation for any account of counterfactual conditionals. It does not follow, however, that one must reject what he says about such conditionals. Rather, one must do what he eschews, to wit, explicate the basic idea of 'possible' or 'possible world'. Once this notion has been explicated it might well turn out that there is much to be said for Lewis' analysis of counterfactuals.

Now, Humeans and other defenders of the deductive-nomological model have in fact defended a fairly standard explication of the idea of 'possibility.' Briefly put, a state of affairs $S$ is possible just in the sentence '$S$' is a non-contradictory well-formed sentence of the empiricist's language. The latter sort of language consists of a logical framework — essentially that of ordinary formal logic (propositional logic, lower functional calculus and often — though it is not relevant for our purposes — higher functional calculi capable of expressing arithmetic) — to which are added non-logical constants (individual constants and predicate constants, both monadic and relational) which are interpreted into observable things, properties and relations.[29] 'Necessary,' 'possible', etc., receive *syntactical* explications. These can be connected to the idea of "possible worlds" as follows. (The device is Carnap's, deriving from Wittgenstein's *Tractatus*.)[30] Let $D$ be a conjunction of sentences each of which is either atomic or the negation of an atomic sentence, and let every atomic sentence occur once, either negated or unnegated, in $D$. Then $D$ is called a *state description. Each state description is a description of a possible world.* More precisely, the notion of state description *explicates* the notion of a description of a possible world.[31] And the set of all possible worlds is given by the set of all possible state descriptions, where the notion of "all possible" is the *completely non-problematic* notion of *combinatorial possibility*:[32] one considers all possible combinations of terms that are consistent with the formation rules of the language.[33] Once the idea of "possible world" has been explicated in terms of this non-problematic notion, the rest of the picture falls quickly into place. For a state of affairs $S$ to be true in or to hold in the possible world described by $D$ is for '$S$' to be entailed by $D$, where *entailment* receives the usual syntactical explication. For a state of affairs $S$ to be necessary, true in all possible worlds, is for '$S$' to be entailed by all state descriptions. For $S$ to be impossible is for '$S$' to be a well-formed sentence entailed by no state description. And so on. Among the necessities is that at least one and at most one state description is true. In all this there is no question of the literal reality of possible worlds. There is only one world, namely, the actual world, that is, the world described by *the* true state description. The other possible worlds are as it were but syntactical variations on this actual world.[33a]

Lewis himself has argued that this explication will not do. In the first place, he asserts, we *do* quantify over possible worlds in ordinary language: *there are* many ways that things could have been besides the ways they actually are.[34] This is certainly true, but as it stands it is irrelevant to the issue. What is important is not *how* we speak but *what philosophic sense* we are to make of that way of speaking. No one proposes to challenge Lewis' naive belief that there are possibilities.[35] That is not at issue. Rather, what one insists upon is that this belief requires *analysis*, that is, *explication*, and to explicate is to clarify it, not challenge it. Lewis has every right to his naive belief that there are possibilities — it is, after all, true! — but that is not the same as his far from naive realistic analysis of these possibilities. What the empiricist, with his syntactical explication of "possibility", is challenging is not the former but the latter.[36]

Lewis has a second argument against this explication of "possible worlds."[37] It is, he claims, either circular or false. If we explicate "possibly $S$" as "'$S$' is a consistent sentence" and then explain 'consistent sentence' as one that *could be true* or one that is *not necessarily false*, then the theory is circular, presupposing the very ideas it proposes to explain. On the other hand, if we explain 'consistent sentence' as one that comes out true under some interpretation of the non-logical vocabulary, then the account is false, since it construes as consistent, or as possible, does not exclude as impossible, such impossible states of affairs as

($\alpha$)      This pig is a sheep

or, what is the same,

($\beta$)      All pigs are sheep

As for the first charge, that of circularity, then this is not correct. It is true that "possibility" is explicated in terms of *combinatorial possibility*. But there is a difference: the former is philosophically problematic, the latter is not. Hence, to use the latter to explicate the former is *not* circular. Once the shift is made to combinatorial possibility, then, in effect, one has shifted to explicating "possible" as "true under some interpretation of the non-logical constants." The question is, are ($\alpha$) and ($\beta$) to be construed as "impossible." If so, the notion remains unexplicated. The explication is being challenged as not capturing all of the traditional notion that it ought to capture. Such judgments are hard indeed to adjudicate. In this case, however, there are relevant considerations that cannot be omitted. If we explicate "possible" as "true under some possible interpretation" then "necessarily true" turns out to mean "true under all possible interpretations of the non-logical constants." That means that in "necessary truths" the non-logical words occur, as Quine put it, at most vacuously. The logical words alone are essential to the truth of necessary truths. Similarly, "necessary falsehoods" are false by virtue of their logical form, that is, the order and arrangement of the logical words, alone. This explicates one of the traditional criteria for "necessary truths (falsehoods)", namely, they are statements the truth (falsify) of which is a matter of their form rather than content. Even in its unexplicated form, this criterion picks out such statements as

Either $a$ is $F$ or $a$ is not $F$

as necessary truths. Statements like this are obviously different in kind from ($\alpha$) and ($\beta$), the falsify of which is *not* determined by their logical form *alone*: ($\alpha$) and ($\beta$) are indeed false, but the non-logical words, the descriptive predicates, clearly do not occur vacuously.[38] One may therefore reasonably conclude that if ($\alpha$) and ($\beta$) are impossible, as Lewis holds, then they are so in a sense different from that which the empiricist aims to elucidate through his syntactical explication. What Lewis is proposing is a univocal account of "possible" and of "necessary", and there is no ground whatsoever to think this condition on an explication is reasonable.[39] In fact, Lewis himself rejects it when he himself distinguishes logical from physical necessity. And indeed, one wonders why Lewis does not, what seems reasonable, construe

No pigs are sheep

as a law of nature, and therefore ($\alpha$) and ($\beta$) as physically impossible.

I think we may conclude that Lewis' objections to the empiricists' syntactical explication of the notion of "possible worlds" are unsuccessful. We therefore have a concept of "possible world" that is acceptable to the empiricist. Once this is in hand, there is no reason why one could not mount a defence of Lewis' account of counterfactuals that is fully compatible with the empiricist strictures of the Humean and the defenders of the deductive-nomological model. What is crucial to Lewis' account of counterfactuals is the idea of being "true in a possible world". This is no longer understood realistically, as it is by Lewis. But that way of taking it is philosophically problematic. At least, so the empiricist argues. However, the empiricist can also explicate the idea of being "true in a possible world" to his own satisfaction, and once he has done that the whole of Lewis' apparatus becomes available.

On the other hand, even if the empiricist can make sense of Lewis' account of counterfactuals, it does not follow that this explicated theory is true. It is the latter, of course, that the Humean disputes. But before turning to this — which is the real issue that concerns us — we must look at two more of Lewis' problematic notions, those of "law of nature" and of "similarity" among "possible worlds."

## IV. NOMIC AND CAUSAL DEPENDENCE

Lewis adopts a view of laws in which the latter are *mere* regularities. There is no element of necessity in laws, as there is for Hume and the Humean. Nonetheless, Lewis, as he must, wishes to distinguish regularities which are laws from those which are not. The view he adopts is what we reckon those regularities to be laws that we can organize deductively into a theory. In his terms,

> . . . a contingent generalization is a *law of nature* if and only if it appears as a theorem (or axiom) in each of the true deductive systems that achieves a best combination of simplicity and strength. A generalization is a law at world *i*, likewise, if and only if it appears so a theorem in each of the best deductive systems true at *i*.[1]

Several virtues are proposed for this definition.[2] (1) It explains why the property of lawhood is not just a matter of the generality of the statement. That is, it doess serve to distinguish generalities which are laws (*propter hoc*) from those that are accidental (*post hoc*). (2) It explains why being a law is a contingent property. (3) It explains how we can know a generalization true by exhausting its instances — e.g., Bode's "Law" — while we do not know whether it is a law. (4) It explains why we have reason to take the theorems of well-supported theories provisionally as laws. (5) It explains why lawhood has seemed a vague and difficult concept: our standards of simplicity and strength, and the proper balance between them, are only roughly fixed. *All these virtues are shared by Lewis' view of laws and the Humean's.* Lewis adds another which, if sound, distinguishes his view as justified and the Humean's as not. This is the point that his account (6) "explains why *being* a law is not the same as being regarded as a law — being projected, and so forth — and not the same as being regarded as a law and also being true. It allows there to be laws of which we have no inkling."[3]

We shall evaluate Lewis' account of laws, and his criticism of the Humean below. Right now, however, we are simply trying to clarify his views to see if they fit within the empiricist framework of the Humean and the defender of the deductive model of explanation. There is no question that this criterion of lawhood is one that is compatible with empiricism. That being so, the disagreement here is not between the empiricist and the anti-empiricist (as was the disagreement about the reality of possible worlds), but rather one within the empiricist ring.

With the notion of *law* at hand, Lewis has succeeded, in a way that can be accepted by the empiricist, in elucidating the concept of *physical necessity* earlier defined thusly: the physical strict implication $N (A \supset B)$ is true if and only if B is true at all those A-worlds in which hold all the laws that hold in the actual world.

Physical strict implication is defined in terms of *laws*. Counterfactual implication is defined in terms of *similarity* among possible worlds. *Since these two criteria need not coincide, lawful necessity and counterfactual dependence do not always coincide.* Here, of course, we have located the crucial point of disagreement with the Humean. Before enlarging on this point, however, we had better introduce some more of Lewis' definitions, specifically those relating to causation.

For two propositions, A and B, if we assume the occurrence of A is possible, then the counterfactual conditional

(*)         if A were to obtain then B would obtain

is true (at the actual world), according to Lewis, if there is a possible world in which

($W_1$)     A obtains
            B obtains

both hold that is more similar to the (actual) world than any possible world in which

($W_2$)     A obtains
            B does not obtain

hold. If (*) is true then B is *counterfactually dependent* upon A.[4]

Counterfactual dependence is necessary but not sufficient for causal dependence. If causal dependence is to obtain, events must be located in *families*. A family of non-compossible events $\{e_i\}$ is *causally dependent*[5] upon a second family of non-compossible events $\{c_i\}$ if and only if, for each *i*, $e_i$ depends counterfactually on the corresponding $c_i$, i.e., if and only if, for each *i*,

            if $c_i$ were to occur then $e_i$ would occur

is true. As for single events, *e* depends causally on *c* just in case both

            if *c* were to occur then *e* would occur
            if *c* were not to occur then *e* would not occur

are true. If neither *c* nor *e* is actual the second is automatically true, so causal dependence will depend on the truth of the first; if *c* and *e* are actual the first is automatically true, and causal dependence will depend on the truth of the second.

Since Lewis characterizes a law at world *i* as a generalization at world $i$ that appears as a theorem in each of the best deductive systems true at $i$, it follows that L is a law in the actual world only if L is a true generalization.

A proposition B is *counterfactually independent*[6] of the family $\{A_i\}$ of non-compossible propositions if and only if B would hold no matter which of the A's were true, i.e., if and only if, for all $i$,

if $A_i$ were to obtain then B would obtain

is true. Such a counterfactual conditional will be true just in case there is a possible world for which

(W$'_1$)     $A_i$ obtains & B obtains

that is closer to the actual world than any possible world in which

(W$'_2$)     $A_i$ obtains & B does not obtain

Finally, the family $\{C_i\}$ of non-compossible propositions *depends nomically*[7] on the family $\{A_i\}$ if and only if there is a non-empty set L of true law-propositions and a set F of true propositions of particular fact such that L and F jointly (but F alone does not) entail all the material conditionals

$A_i \supset C_i$

Suppose the C's depend nomically on the A's in virtue of L and F. Suppose, further, that all members of L and F are counterfactually independent of the A's. From the latter, we know that, for each $i$,

($\alpha$)      if $A_i$ were to obtain then L and F would obtain

is true. So there is a world (W$'_1$) closer to the actual world than any world (W$'_2$). In a (W$'_1$)-world L and F obtain. Since C's are nomically dependent on A's, L and F entail '$A_i \supset C_i$'. So this material conditional obtains in (W$'_1$). Since $A_i$ also obtains in (W$'_1$), so does $C_i$. Hence, so long as the counterfactual ($\alpha$) is true, any world in which

$A_i$ obtains & $C_i$ obtains

is closer to the actual world than one in which

$A_i$ obtains & $C_i$ does not obtain

That is, so long as ($\alpha$) are true, the counterfactuals

if $A_i$ were to obtain then $C_i$ would obtain

are all true. Thus, if the C's depend nomically on the A's in virtue of L and F then the C's depend counterfactually on the A's. We may regard nomic dependence as accounting for counterfactual dependence.

Nonetheless, there may well be counterfactuals that are true though there is no law L that, together with an F, connects the antecedent to the consequent.[8]

Similarity and difference of worlds with respect to their laws is a major factor in estimating the similarity and difference of worlds. We may expect among the worlds very similar to ours there to be worlds which share our laws rather than worlds in which our laws hold only as accidental generalities, or are false, or are replaced by contrary laws. Thus, the laws of our world will tend to hold in the subspheres that, on Lewis' scheme, as it were surround our world. And so, since A → B is true in our world just in case that B is true in the nearest A-world, it is more than likely that the laws of our world will hold in that nearest A-world, and that these laws entail A ⊃ B, i.e., that B is true in that A-world. But we could still have worlds which are almost like ours save that a *small* violation of one of the laws occurs, an event that is accounted for by no law in that world, a little miracle. In such a case, the violated law is not replaced by a contrary — which would be a much more radical change in laws. Indeed, it may be that the violated law can be treated as imperfect, holding only under a specific condition that permits the one exception. On the other hand, worlds with the same laws as ours, and where the same sort of event as the miracle (i.e., its counterpart) occurred, would have to differ from ours in initial conditions and in the whole history from those initial conditions to the event in question. It may very well be that the world in which the small miracle occurred was more similar, overall, to ours than the nearest world in which the laws of our world hold. "Laws are very important, but great masses of particular fact count for something too; a localized violation is not the most serious sort of difference of law."[9] If the laws in our world entail A ⊃ ~B, B may occur as a small miracle in an A-world, and that A-world be *overall* the A-world most similar to ours. In that case the counterfactual implication A → B would, according to Lewis, be true in our world, even though the laws of our world establish that there can be no nomic connection between the two events. If A and B are events that are members of families of non-compossible events, and if A → B is true, then B is causally dependent on A. So, it is possible to explain an event causally where it is not possible to explain it nomically, by subsumption under laws.

Given that the notion of "possible worlds" can be explicated in a way satisfactory to the empiricist, this separation of causal and nomic explanation is Lewis' crucial disagreement with the Humean and the defender of the deductive-nomological model of explanation. Before we can examine this separation critically, however, it is necessary to explore in more detail the notion of similarity among possible worlds upon which so much of Lewis' account depends.

## V. SIMILARITY AMONG POSSIBLE WORLDS

We can most easily bring out some relevant unclarities in Lewis' notion of inter-world similarity by examining a recent discussion of Lewis' view by A. Rosenberg.[1]

Rosenberg makes three main points about David Lewis' theory of causation. The first concerns the notion of similarity among possible worlds. The second concerns whether Lewis can defend the interrelated notions that he develops, of counterfactual independence and of nomic dependence. The third concerns whether the advantage Lewis claims for his theory over what he (Lewis) calls a "regularity theory" of causation, viz., that his solves what is called "the problem of effects", really does exist.

I shall begin with a few preliminary remarks, then comment on each of Rosenberg's points, but in reverse order.

### 1

Rosenberg's second and third points are successful only if certain assumptions are made about the nature of laws. These assumptions conflict, I believe, with the actual logical structure of most scientific laws. The assumptions impose a simplistic model of the logical form of laws, where, as I see it, in order to handle the complexities involved in counterfactual inference it is precisely the complexities of logical form to which one must attend. The assumptions derive, I suspect, from that philosophical methodology employed by Hempel[2] and Carnap[3], the methodology that, leaving aside complications concerning modalities, requires all statements to be expressable within the straight-jacket of the formalism of the lower functional calculus. However appropriate is the use of this formalism in an initial gambit — and I do not deny that it can be — we must not let it control our vision of science and of scientific laws. Carnap and Hempel have provided us with a very bad example, I am afraid.

The particular logical feature of certain laws I wish to cite is what we have earlier called their *generic* nature.[4] Thus, the Law of Inertia asserts that

For any *kind* of mechanical system, *there is* a force function that relates the accelerations of objects in that system to the circumstances in which the objects are located.[5]

For that kind of mechanical system we call planetary, the force function is that of the Law of Gravity; for that kind we call spring balances, the force function is that of Hooke's Law; and so on. The Law of Inertia makes a generic assertion about all specific kinds falling within the genus of mechanical system; and it says about each kind that *there is* a force function that accounts for motions within that kind of system; but it does not assert *specifically* what force function it is that holds in each kind. The Law of Inertia quantifies over kinds of system and over force functions; it cannot be construed as quantify·ᵢg over only individuals, i.e., it cannot be expressed if one insists the language of science is in its essence the LFC. I shall argue that Rosenberg's second point in particular is successful, only if he ignores that there are generic laws.[6]

2

One of the virtues Lewis claims for his account of causation is that it can solve the "problem of effects." We will return to this point again later. Right now we will develop it far enough to evaluate Rosenberg's remarks on the topic.

Suppose that two events $c$ and $e$ occur, that $c$ causes $e$, and that $e$ does not also cause $c$. We are further to suppose that, given the laws and some of the actual circumstances, $c$ could not have failed to cause $e$. It would seem to follow that if the effect $e$ had not occurred, then its cause $c$ would not have occurred. This is a counterfactual conditional of the sort (*)

If A were to obtain, then B would obtain

and it would therefore follow that $e$ does, after all, cause $c$, contrary to the supposition that this reverse causation does not hold. This is what Lewis calls "the problem of effects". According to Lewis, the solution of this problem is

... flatly to deny the counterfactual that causes the trouble. If e had been absent, it is not that c would have been absent. . . . Rather, c would have occurred just as it did but would have failed to cause e. It is less a departure from actuality to get rid of e by holding c fixed and giving up some or other of the laws and circumstances in virtue of which c could not have failed to cause e, rather than to hold those laws and circumstances fixed and get rid of e by going back and abolishing its cause c. . . . To get rid of an actual event e with the least overall departure from actuality, it will normally be best not to diverge at all from the actual course of events until just before the time of e.[7]

Rosenberg criticizes this solution to the problem of effects by means of a counter-example.[8] This counter-example consists of three possible worlds, a, b and c. Each world consists of one object o, and two simple properties P and Q. An event occurs when o has a property at a time. The temporal chains of events on a, b and c are summarized in the following table.

|   | $t_g$ | $t_h$ | $t_i$ | $t_j$ | $t_k$ | $t_l$ |
|---|---|---|---|---|---|---|
| a | P | Q | P | Q | P | Q |
| b | P | Q | P | — | P | Q |
| c | P | Q | — | — | P | Q |

Rosenberg holds that the only laws that hold in world a are of the form

(L)  $(x)(j)(i)[j = S^\bullet i \rightarrow [P(x, t_i) \rightarrow Q(x, t_j)]]$
$(x)(j)(i)[j = S^\bullet i \rightarrow [Q(x, t_i) \rightarrow P(x, t_j)]]$

Here, "$S^\bullet i$" represents "the immediate successor of $i$". Rosenberg assumes for the sake of the argument that the laws (L) are counterfactually independent of the events in world a. It follows that o being Q at $t_j$ in a is counterfactually dependent upon, and therefore nomically dependent upon o being P at $t_i$ in a. Does a being Q at $t_j$ cause o to be P at $t_i$? To give a negative answer, Lewis denies the counterfactual conditional

(**)  if o being Q at $t_j$ were not to occur then o being P at $t_i$ would not occur

(**) is true if and only if some possible world in which neither

o is Q at $t_j$
o is P at $t_i$

hold is more similar to the actual world than any possible world in which

> $o$ is Q at $t_j$

does not hold while

> $o$ is P at $t_i$

In Rosenberg's world $c$ the former hold, in world $b$ the latter. Lewis denies (**) by holding worlds like world $b$ are more similar to the actual world $a$ than worlds like world $c$. Rosenberg argues that this solution to the problem of effects is unsuccessful because world $c$ is in fact more similar to world $a$ than is world $b$. His reason is this: laws (L) hold in worlds $a$ and $c$, but not in $b$; hence world $b$ differs from world $a$ in both history and laws, while world $c$ differs from would $a$ only in history and not in laws.[9]

Rosenberg's counter-example will not work, for two reasons.

In the *first* place, he has not described *all* the laws that hold in world $a$. It is clear we must add at least

$$(L') \quad (x)(j)(i)[j = S^\bullet i \rightarrow [Q(x, t_j) \rightarrow P(x, t_i)]]$$
$$(x)(j)(i)[j = S^\bullet i \rightarrow [P(x, t_j) \rightarrow Q(x, t_j)]]$$

If we again assume these are counterfactually independent of the events in $a$, then it follows from these laws that $o$ being P at $t_i$ in $a$ is nomically dependent upon $o$ being Q at $t_j$ in $a$. Thus, the supposition Lewis requires to get his problem of effects to be a problem, namely, that $c$ causes $e$ but not conversely, is not fulfilled in Rosenberg's world $a$.

In the *second* place, Rosenberg's worlds $a$ and $c$ do not share the same laws. In addition to the rules just mentioned, the law

$$(L'') \quad (x)(i)[P(x, t_i) \oslash Q(x, t_i)]$$

where '$\oslash$' is the exclusive "or", holds in world $a$. Clearly, it does not hold in world $c$ — nor in world $b$. So Rosenberg's claim that $c$ is more similar to $a$ than is $c$ because the same laws hold in both is not justified. Moreover, the laws (L) do *not* hold in $c$; the second is falsified by the sequence from $t_h$ to $t_i$. Similarly, the sequence from $t_j$ to $t_k$ falsifies the second of the laws (L').

Three other comments are relevant. *One.* The object $o$ in worlds $b$ and $c$ is very odd indeed. In world $b$, no basic properties are present in $o$ at $t_j$. On these occasions, $o$ is, if not bare, then at least nude. For myself, I am strongly inclined to accept the metaphysical principle, sometimes called the Principle of Exemplification, that there are no nude particulars.[10] This fact, this deep metaphysical fact, places limits upon what is lawfully possible. It would seem to exclude Rosenberg's worlds *b* and *c* as real possibilities. Does he really wish to hold that it is possible that at certain times for some objects only negative predications will be true of them?

*Two.* In any case, negative events, events in which an object does *not* have a basic property at a time, do require explanation.[11] To cite the old example, it is not unreasonable to seek an explanation why the gardner did *not* water the lawn. When Rosenberg asserts laws (L) hold in world *c* he is clearly ignoring the two wholly negative events occurring at $t_i$ and $t_j$. Note that in world *a*, once (L″) is recognized to hold in it, the negative events all receive an explanation.

*Three.* I would argue that, in a *deterministic* world *all* events, possible as well as negative, require an explanation. (It is not necessary to add clauses about conjunctive, disjunctive, etc., events. If for each basic property, we know whether it holds or does not hold of an object at a time then we can deduce which other truth-functionally complex predications are true of the object at that time — cf. what Carnap called Q-predicates.[12])

Finally, if Rosenberg really believes that the laws (L) adequately describe what happens in world *c*, then he has a view of what counts as an adequate lawful description that is, I think, incompatible with what science actually aims to discover. Science aims to discover laws that give deterministic explanations of individual processes. I think this is so, but even if it is not, we may assume, with Lewis, for the sake of the argument, a deterministic world. In explaining a process one should have a law that enables one, given a set of initial conditions, to deduce the state of the system at any future or past time. Compare, here, Newton's law explaining the process the solar system undergoes. I refer, of course, to the ideal of process knowledge. The laws (L) do not do this for the process object *o* undergoes in world *c*: they do not enable us to deduce the state of the system at times $t_i$ and $t_j$. Rosenberg thus overlooks a crucial feature that laws would exemplify in a deterministic world.

3

In the above discussion, we assumed, with Rosenberg, that the laws (L) were counterfactually independent of the events in world *a*. But

Rosenberg does, of course, challenge this assumption: this is the second of his three main points.

Lewis requires laws to be generalizations true of a world. Rosenberg argues that

> Since laws restrict actualities at a world, but not also possibilities, if two deterministic worlds are identical in histories — in the chains of events that occur "in" there — then they must be identical in their laws.[13]

and that

> ... since on Lewis' view laws describe only actualities at a given world, if the laws of two worlds differ then their histories must also differ.[14]

From these considerations he deduces the two principles [15]

(A)     For any two possible worlds, if they share the same history then they share the same laws.

(B)     For any two possible worlds identical in initial conditions, if they do not share the same history, then they do not share the same laws.

(B) has the immediate consequence

(B*)    For any two possible worlds, if they do not share the same history but do share the same laws then they differ in initial conditions.

Rosenberg argues that if we have an L and family of propositions $\{A_i\}$ describing the occurrence of particular events, then "no set of laws like L can be shown to be counterfactually independent of a family of propositions like the A's." [16]

If L is counterfactually independent of the A's, then each of

($\beta$)      if $A_i$ were to obtain, then L would obtain

is true. Rosenberg denies at least one ($\beta$). He takes $A_1$ and supposes $A_1$ is false. Rosenberg then denies

($\beta'$)      if $A_1$ were to obtain, then L would obtain.

To do this, he argues there is a possible world

($W_2''$)    $A_1$ obtains & L does not obtain

that is closer to the actual world than is the possible world

(W$_1''$)    A$_1$ obtains & L obtains.

He compares the three worlds as follows.

| | Crucial event | Laws | Initial conditions | History prior to crucial event | History subsequent to crucial event |
|---|---|---|---|---|---|
| actual | ~A$_1$ | L | C | Hp | Hs |
| W$_2''$ | A$_1$ | other than actual | C | Hp | Hs |
| W$_1''$ | A$_1$ | L | other than actual | — | — |

The first row and the first two columns are assumptions. The entries in the last three columns of the second row are constructions designed to give a world (W$_2''$) as close to the actual world as possible: Rosenberg asserts that "Lewis would choose a world with an *almost* perfect match in history as the closest."[17] The entry in the third column of the third row is by deduction from (B*). The history of (W$_1''$) diverges in an least two places from the actual, that of (W$_2''$) at only one. Hence the possible world (W$_2''$) is closer to the actual world than the world (W$_1''$). ($\beta'$) is therefore false, and L cannot be counterfactually independent of the A's.

Clearly, this argument is successful only if Rosenberg's assertion is acceptable, that "Lewis would choose a world with an *almost* perfect match in history as the closest" — that is, would so choose no matter which particular L and A's we were considering. For, if there were exceptions — if there was just one case of L and A's in which, for all A$_i$ that are not actual, (W$_1''$) was closer than (W$_2''$) to the actual world — then Rosenberg would not have established laws like L could not be counterfactually independent of the family of A's. Are there any such exceptions? Lewis clearly thinks so:

. . . similarities in matters of particular fact trade off against similarities of law.[18]

Why does Rosenberg think not?

The answer lies in a principle that Rosenberg believes Lewis is committed to. This is the principle

(R)      Worlds that differ in law must differ in history and this difference in history exhausts their nomological differences.[19]

If differences in history *exhaust* nomological differences, then $(W_2'')$ will always be closer than $(W_1'')$ to the actual world. Historical similarity will exhaust nomological similarity, and the closer a world is in its history to the actual world will *ipso facto* also be a world nomologically more similar, more similar in its laws, to the actual world. No similarity in law between $(W_1'')$ and the actual world could move the former closer to the actual world than any world like $(W_2'')$ that was more similar in its history to the actual.

Now, if we take this argument of Rosenberg seriously then much more follows than he concludes: one can infer not just that no law in counterfactually independent of the relevant A's, but that *no* C is counterfactually dependent upon its A. C is counterfactually dependent on A just in case the counterfactual conditional

(*)      If A were to obtain than C would obtain

is true. And this is true just in case there is a possible world in which

$(W_1)$      A obtains & B obtains

hold is closer to the actual world than any possible world in which

$(W_2)$      A obtains & B does not obtain

hold. Every $(W_1)$ world contains two deviations from the actual, and corresponding to each such world there is a $(W_2)$ world containing only one deviation from the actual. If histories alone were relevant, as (R) requires, then one could *never* find a $(W_1)$ world closer to the actual than all $(W_2)$ worlds. In that case, (*) is always false, and no C is ever counterfactually dependent on an A. Histories alone cannot judge of overall similarity among possible worlds.

Could considerations of law ever lead one to judge, as in Rosenberg's example, that a $(W_1)$ world is closer to the actual than a $(W_2)$ world? I think so. Consider the set of laws

(G)      $y = f(x)$
$\phantom{(G)\quad}z = g(y)$
$\phantom{(G)\quad}x = h(z)$

and the following histories:

| | | | | | |
|---|---|---|---|---|---|
| actual | $x_1$ | $y_1$ and so ~ $y_2$ | $z_1$ | $x_1$ | $y_1$ |
| $W_2'''$ | $x_1$ | $y_2$ | $z_1$ | $x_1$ | $y_1$ |
| $W_1'''$ | $x_2$ | $y_2$ | $z_2$ | $x_2$ | $y_2$ |

Column 1 is the column for initial conditions and also the history prior to the crucial event, since the second column is that of the crucial event. Columns 3, 4 and 5 are the histories subsequent to the crucial event. Rosenberg, as we saw, relying upon (R), would have Lewis judge that ($W_2'''$) is closer to the actual world than is ($W_1'''$). Yet the actual world and ($W_1'''$) have the functional laws (G) in common, while whatever laws hold in ($W_2'''$), since $y$ takes on two values for a single value of $x$, these laws cannot include a functional law relating $x$ and $y$. Such a significant difference in laws may well lead one to judge ($W_1'''$) is more similar than ($W_2'''$) to the actual world, even though the actual world and ($W_1'''$) have quite different histories.

This sort of consideration in judging similarity of possible worlds seems pretty reasonable. On the other hand, if (R) is acceptable then this is ruled out, and all counterfactual and so all causal dependence disappears. But *is* (R) acceptable? If taken in isolation, there is nothing to recommend (R). Rosenberg himself deduces it from other premises, specifically from (A) and (B). The first conjunct of (R) is simply the transposition of (A), so it is acceptable. In any case, it is not this conjunct that prevents ($W_1''$) from ever being closer than ($W_2''$) to the actual world. It is the second conjunct of (R) that causes the trouble. Now, we can assert that differences in history exhaust nomological differences only if, in general, a difference in history implies a difference in laws. Rosenberg, it seems, believes this follows from (B). But it does not. (B) is more restrictive. (B) asserts that *if* two worlds have the same initial conditions *then* a difference in history implies a difference in laws. The crucial second conjunct of (R) can be inferred only if we also accept

(B†)     if two worlds have different initial conditions, then a difference in history implies a difference in laws.

However, (B†) is in fact false. For one can find possible worlds that differ both in initial conditions and in history and yet have the same laws. And once we allow this, differences in history do not exhaust differences in law, and we allow that, in some cases at least, $(W_1'')$ may be closer than $(W_2'')$ to the actual world, and therefore also allow that, for some L and A's, L can be counterfactually independent of the A's.

Consider the law

(F)        $y = f(x)$

In World 1, $x$ takes on the value $x_1$; this is the initial condition of World 1. In this world, $y$ takes on the value $y_1 = f(x_1)$. The history of World 1 consists of the values $x_1$ and $y_1$ being exemplified in it. In World 2, the initial condition is $x_2$ and the history consists of $x_2$ and $y_2 = f(x_2)$ being exemplified in it. (F) is, of course, very simple. We could complicate it so as to cover the process feature of development over time by means of the more complicated example

$$x(t') = f_1[x(t), y(t)]$$
$$y(t') = f_2[x(t), y(t)]$$

This would enable us to construct complete histories of worlds over time. But with the simpler (F) and Worlds 1 and 2 the conclusion to be drawn is clear enough: worlds can differ in both initial conditions and in history and yet have the same laws explaining what occurs in them.

(F) is, of course, a whole body of laws. $x$ and $y$ are both generic concepts. The species in $x$, $x_1$, $x_2$, etc., cannot be co-exemplified; that is, the event of $x_i$ being exemplified in an object at $t$ is non-compossible with the event of $x_j$ being exemplified in that object at $t$, for all $i \neq j$. The same holds for the species of $y$. The species in both genera satisfy the condition of forming a measurable dimension. (F) quantifies over species. It says something about *each species* within the genus $x$ and $y$ species related by the function $f$. (F) states in brief form the body of specific laws

($F_1$)     For any object, if it is $x_1$ then it is also $y_1 = f(x_1)$
($F_2$)     For any object, if it is $x_2$ then it is also $y_2 = f(x_2)$
. . . . .   . . . . . . . . . . . . .

In World 1, it is the specific law ($F_1$) that holds; in World 2, it is the specific law ($F_2$). But in both, the generic law (F) holds. Worlds that differ in initial conditions and in history differ in specific laws, and if this was all there was to it then ($B^\dagger$) and therefore (R) would be acceptable. But worlds that differ in specific laws may have generic laws in common, and it is this possibility that falsifies ($B^\dagger$). Rosenberg wrongly accepts (B$^\dagger$), I suspect, because he attends only to the specific laws, ignores the generic. Like Carnap and Hempel, he tends to ignore those aspects of science that can't be fit into the LFC, those aspects that go beyond the specific to the more abstract, generic levels. It is perhaps this generic feature of laws like (F) that Lewis is hinting at when he offers his definition of causal dependency in terms of families of non-compossible events.

4

The term 'short' is not vague; it is relative. To say John is short is not to say something vague about John; it is to say something relative. To say John is short is to say he is shorter than, say, Peter, or less in height than some other standard. Except, of course, when we say John is short we don't explicitly mention the standard; that standard is left implicit in the context. In most quotidien contexts, the implicit standard is such that John is short if he is 4' 11 3/4" or less. In other contexts, however, the implicit standard might be other than this. If John is 6' tall and one says of him that he is short, one has not necessarily said something false. This would be so, for example, if he was a basketball player and one was using the average height of basketball players as the standard. Nor need the shift to a non-normal standard be made explicit; the context might very well make it clear that this is the standard one is using. What one is saying when one says John is short is unclear only when the context does not make clear the standard being used. Such unclarity is to be rectified by asking the speaker to make explicit what standard he is using.

"$x$ is far from $y$" is like "$x$ is short": it is relative in the same way. When one says $x$ is far from $y$ implicitly one is saying $x$ is farther from $y$ than $w$ is from $z$. The latter, of course, presupposes that "farther from" admits of degrees: to say $x$ is farther from $y$ than $w$ is from $z$ is to say that the distance from $x$ to $y$ is greater than the distance from $w$ to $z$. Here, one is asserting a relation, "greater than", of two other relations, viz., the distance from $x$ to $y$ and the distance from $w$ to $z$. "$x$ is far from $y$" is relative because the standard "the distance from $w$ to $z$" may vary from context to context.

Rosenberg compares "$x$ resembles $y$" to "$x$ is short".[20] But it is closer to "$x$ is far from $y$". Rosenberg suggests "$x$ resembles $y$" has the same sort of implicit structure as "$x$ is far from $y$", and suggests the former amounts to "$x$ resembles $y$ more than $z$". The latter involves a comparison of two relations: "the resemblance of $x$ to $y$ is greater than the resemblance of $x$ to $z$." So we are back to "the resemblance of $x$ to $y$". We have got to where we started: with the need to define "$x$ resembles $y$" in a way that admits of degrees.[20a]

This locates the real problem, however. It is not, as in the case of "$x$ is short" or "$x$ is far from $y$", that there is an implicit standard to be made explicit, but rather that resemblance is always in respect of something: $x$ never *just* resembles $y$ but only in respect of something — in respect of colour, or of shape, or, in the case of possible worlds, in respect of (at least) either laws or histories. "$x$ resembles $y$" is like "$x$ is worth doing" rather than "$x$ is short". $x$ may be worth doing, i.e., more worth doing than other things, on the scale of inclinations, on the scale of prudence, on the scale of duty, or, worth doing, as one says, "everything considered," that is, worth doing on some overall scale which results from somehow combining these other scales. "$x$ is worth doing" is in this way ambiguous, and it is left to the context to disambiguate it. Of course, if the context doesn't do that job, then exactly what it is that is being said is left unclear, an unclarity to be rectified by asking the speaker just which scale of values it is he is using.

Lewis wants "$x$ resembles $y$" as applied to possible worlds to be like "everything considered, $x$ is worth doing". It seems to me to be more like "$x$ is worth doing": sometimes one scale is meant and sometimes another, and (perhaps) sometimes an overall scale.

In particular, one world may be closer to another along a scale that reckons similarity in terms of similarity of more generic laws. Consider the van der Waals' law, a generic law that (let us assume) describes the behaviour of all actual gases. This law states that

> For each (kind of) gas there are unique constants $a$, $b$ such that for each sample $x$ of that (kind of) gas

$$\left( p(x) + \frac{a}{(v(x))^2} \right) (v(x) - b) = R T(x)$$

There is a specific law for each actual gas $G_i$, which states that

> For each sample $x$ of $G_i$,

$$\left( p(x) + \frac{a_i}{(v(x))^2} \right) (v(x) - b_i) = R T(x)$$

In the actual world there are many possible values of $a$ and $b$ in the van der Waals' formula that correspond to no actual gas. Suppose the actual world contains gases $G_1, \ldots, G_n$. Now consider world A: In this world there are all the gases $G_1, \ldots, G_n$, and the specific laws for these are as in the actual world. But in addition, world A contains a gas $G_{n+1}$ the specific laws of which violates the van der Waals' formula. The specific laws that obtain in the actual world all obtain in world A, but world A and the actual world differ at the level of generic laws. As for world B, it contains gases $G_1, \ldots, G_{n-1}$, the specific laws of which are as in the actual world, and it contains gas $G_n$ but for $G_n$ the specific law, while conforming to the van der Waals formula, has constants slightly different from those it has in the actual world, say $a_n$ and $b_n + \delta$. Thus, world B differs from the actual world in its specific but not in its generic laws. On a scale that judged similarity in terms of specific laws, world A would likely be the closer to the actual. On a scale that judged similarity in terms of generic laws it would be world B that would likely be the closer to the actual. If we now consider the counterfactual conditional

> If this sample were to have pressure $p_1$ and volume $v_1$ then it would have temperature $T_1$

this might turn out to be true if we used the scale of similarity based on similarity of specific laws and false if we used the other scale. Novels, I think, are, as Alice Kaminsky has argued,[21] best construed as bodies of what are, in effect, counterfactual conditionals. Science fiction novels are often judged veracious on what seems to be a scale based on the similarity of specific but not generic laws: these novels often deal with worlds in which the specific patterns are all the same save for certain extreme cases (e.g., faster-than-light travel), which extreme cases, however, serve to establish a radical dissimilarity from the actual world at the generic level. In satire, on the other hand, all sorts of specific level changes are made but there must remain similarity at a more generic level. The criteria of success are based on (among other things) two different sets of criteria of judging inter-world similarity.

We have two scales, then. Sometimes we seem to use one, sometimes the other. Perhaps we sometimes use a weighted combination of both. Certainly, I see no reason why we shouldn't use such scales as we please, so long as our meaning is clear. There is no need, it seems to me, to agree with Lewis that there is and ought to be one scale of inter-world similarity.

In fact, there are others who adopt just a position as this while working within what is essentially a Lewisian framework. For example, Stalknaker[22] has defended a possible-world semantics of counterfactuals that does not differ in any essentials from Lewis.[23] In Stalknaker's account,[24] in order to specify the truth-conditions for A → B one must pick out by means of a "selection function" $f$ from among all possible A-worlds that A-world that is the closest to the actual world.[25] The conditions that Stalknaker imposes on any selection function ensure that it will be a similarity relation.[26] But, without expanding on the point, he allows there to be several selection functions, that is, several similarity relations each of which might pick out a different A-world as most similar to ours.[27] The cost is a "pragmatic ambiguity" in counter-factuals that can be eliminated only by making explicit the selection function that remains only implicitly specified in most contexts.[28] Lewis' counterfactuals are not in this way ambiguous, but only at the cost of an implausible account of our judgments of interworld similarity.

## VI.  LEWIS' REGULARITY VIEW OF LAWS

Lewis offers us a criterion to distinguish generalizations that are laws from those that are not: a generalization is a law just in case that "it appears as a theorem (or axiom) in each of the true deductive systems that achieves a best combination of simplicity and strength."[1] Does this in fact achieve what it sets out to do?

Some philosophers hold views that separate explanation and prediction. Others adopt views that hold predictive capacity is not a sufficient condition for lawfulness.[2] All hold, however, that predictive capacity is a necessary condition for lawfulness, and further that laws (though perhaps not correlations) are sufficient for explanation, even if they are not necessary. Nor is it unreasonable that the property of being usable to predict is deemed important. After all, if only from the view point of our pragmatic interest, this feature can render laws as things worth knowing. But be that rationale as it may,[3] this capacity to be used successfully in predictions is generally taken to be *a* defining condition for laws, and for present purposes that suffices.

This permits us to make our question about Lewis' criterion of lawfulness more precise: does appearing in an axiom system confer upon a generalization the capacity to be used successfully in predictions? does our belief that a generalization can be fit into some sort of axiom system render it reasonable to take the generalization to be one usable in predictions? does it make it reasonable to law-assert the generalization?

It would seem, however, that Lewis' condition is neither necessary nor sufficient for predictive capacity nor, therefore, for lawfulness.

In the first place, it is not *sufficient*. For, we can easily construct axiomatic systems that encompass non-laws. For example, consider the non-law

> Every Canadian father of quintuplets in the 20th century is a Canadian of French ancestry.[4]

This is a non-law since it is clear that it be unreasonable to use this to justify betting about the next set of quintuplets to be born in Canada in this century, which is to say that it would unreasonable to treat this generalization as having any predictive capacity. Nonetheless, this non-law can be deduced from the following two axioms:

> Every Canadian father of quintuplets in the 20th century is named 'Dionne'.

> Every Canadian male named 'Dionne' is of French ancestry.

Or take another example, Goodman's notorious

> All emeralds are grue

where $x$ is grue just in case that either $x$ is green and examined before $t$ or $x$ is blue and examined after $t$ (where $t$ is later than now). This is hardly to be reckoned by lawlike, yet it can be deduced from the axioms

> All emeralds are beryls containing chromic oxide.
> All beryls containing chromic oxide are grue.

The point is, of course, that axioms confer predictive capacity on a generalization only if they themselves have that capacity. That is, axioms can transform a generalization into a law only if they themselves are laws. And, as our examples show, appearing as an axiom, that is, *as a premiss in a deduction*, does not confer the status of a law on a generalization.

Two points might be made in response. It might be suggested that not any axiom system will do; it must be one that is *simple* and has *strength*, i.e., scope. However, simplicity *by itself* cannot even confer truth, let alone predictive capacity. As for scope, it is always fairly easy to find a theory or axiom system that encompasses what we *take* to be laws. We can always obtain a theory with sufficient scope to encompass what we *think we know*. N. R. Campbell made this point some time ago.[5] *The question is whether such a theory has sufficient scope to include laws that we do not know.* That is, the theory must be the theory that is *objectively* the best, encompassing as many of the generalizations that, objectively, obtain, and not just those that, subjectively, we believe we have that we have evidence for. So the question becomes that of estimating whether a given axiomatic organization of laws is, or is a good approximation to, the theory that, objectively, is of maximal scope. Capacity to organize known laws is no criterion, since it is easy to do that, as Campbell says, yet hard indeed to find a good theory.[6] Nor can the criterion be the objective capacity to organize laws, since that is just what is at issue. Rather, the criterion is, again as Campbell says, the capacity to lead us to the discovery of new, i.e., hitherto unknown, laws.[7] *If a theory is successful at prediction, or, in Lakatos' terms, is empirically progressive, then that constitutes grounds for holding that the theory is of sufficient scope to include not only known but also unknown laws.*[8] The axioms of a theory are supposed to confer the status of lawhood on the generalizations they encompass in their deductive net; but only some theories do this, to wit, those with scope; moreover, the criterion for having sufficient scope to confer the status of lawhood is that the axioms themselves have predictive capacity; hence, axiomatization alone does not confer predictive capacity; to the contrary, predictive capacity determines which axiomatizations are to be reckoned good theories. *Lewis' criterion of strength or scope thus presupposes rather than accounts for the capacity for successful predictions that is the mark of lawhood.*

A second response to the counterexamples, given above, to the idea that axiomatization will confer the status of lawlikeness, could be that the examples may be rejected because we *do not believe* that *these mini-theories will fit into any deeper or more comprehensive deductively organized structure of generalization.* Now, we certainly do not believe that, but the question is, why? What grounds do we have? Surely the answer is that we do not believe these generalizations will fit into a system of laws precisely because they are *not laws*. These axiomatic systems do not confer law status; they do not do so because we believe that they do not fit into any deeper theory; and we do not believe that they fit into any deeper theory because they are not laws. Again, the appeal is circular.

It seems, therefore, that Lewis' criterion for being lawlike is not sufficient. Neither is it *necessary*. For people had many law beliefs, many of them quite reasonable — e.g., fire burns, food nourishes, etc.[9] — long before there was any idea of what an axiomatic system looks like, let alone any idea of how such laws might be fit into one. Thus, the question of whether a law belief is reasonable can be raised without raising the question whether the generalization can be fit into some deductive structure of laws. Or, in other words, the latter is not necessary to the former.

We may conclude, therefore, that Lewis fails to provide a criterion that distinguishes lawful generalizations from non-laws.

What of the supposed superiority of Lewis' criterion to the Humean's? We noted above that Lewis[10] claimed several virtues for his account, and all but one were shared with the Humean's account.[11] What of the remaining one? The idea was that Lewis can and the Humean cannot explain why *being* a law is neither the same as *being regarded* as a law, nor the same as being regarded as a law and true. Presumably, the Humean cannot make this distinction. Now, one must distinguish between a generalization being *lawlike* and a generalization being a *law*: a law is a lawlike generalization that one is *justified* in treating as lawlike. For the Humean, a generalization is lawlike just in case it is regarded as a law, that is, as Lewis says,[12] just in case it is actually "projected", used in making predictions. One will be *objectively justified* in treating the generalization as lawlike just in case that the generalization is true, for that is a sufficient condition for its being used to successfully predict. But, since we can know only a sample and not the population, it is *never possible* to know if the assertion of the generalization is objectively justified.[13] Hence, objective truth alone cannot serve as the criterion by which *we*, who are not omniscient, distinguish laws from generalizations which are *merely* lawlike. The criterion to which the Humean appeals to distinguish lawlike generalizations which are laws from those which are, e.g., mere prejudices, is that *a generalization is to be reckoned* — so far as we can tell — *as a law just in case that it is lawlike and the (fallible) evidence supporting its assertion has been gathered in accordance with the rules of the scientific method.*[14] Thus, for the Humean there *is*, as there is for Lewis, a difference between being a law and being regarded as a law, and between being a law and being regarded as a law and also true. The alleged superiority of Lewis' position over the Humean's does not exist.

Lewis glosses his claim with the remark that his position, in contrast to the Humean's, "allows there to be laws of which we have no inkling."[15] But the Humean can say as much. For him, to say that there are laws of which we have no inkling is to say that: there are objectively true generalizations, of which we have no inkling, and objectively true observable facts, of which we have no inkling, but such that if one were to observe the facts in conformity to the rules of the scientific method then they would justify one's law-asserting the generalization. These are existence claims — about generalizations and observable facts — and a counterfactual claim about them. The history of the growth of scientific knowledge, i.e., the fact that scientific research has *regularly* pushed back the frontiers of human knowledge, together with the evident gaps in our knowledge, and the evident fallibility of our judgments, suffice to justify the existence claims and the counterfactual about them. Thus, the Humean, as much as Lewis, can claim, what seems reasonable, that there are "laws of which we have no inkling."[16]

## VII. ARE THERE COUNTERFACTUAL CONDITIONAL THAT INVOLVE NO LAWS?

Once Lewis' theory is purged of its real possibles, and rendered, through explication of the modal concepts, compatible with the empiricism of the Humean, the crucial distinction between his view and that of the Humean lies in the fact that Lewis' account allows, where the Humean's does not, the assertion of counterfactuals where no law is available to connect antecedent to consequent. Of course, Lewis allows, as we have seen, that nomic dependence (necessity due to laws of nature) *often* explains causal dependence and the truth of counterfactual conditionals. But *not always* — which is where the difference lies. The problem is that the difference is sufficiently small to make adjudication difficult.

If the Humean is correct, then to assert a counterfactual conditional is to assert some appropriate lawful generalization which, however, is not made wholly explicit in the context. If Lewis is correct then to assert a counterfactual conditional is to assert something concerning inter-world similarity — on (we have now argued) some appropriate scale of similarity which, however, is not made wholly explicit in the context. It is difficult to decide between such claims. Perhaps Lewis does mean what he says he means! And if so, then he can and may continue to do so! Lewis has a perfect right to let language mean what he wishes it to mean. Similarly, of course, the Humean can continue to mean what he means. And in any case, most often the two will coincide, as Lewis makes clear, so that *usually* there will be no disagreement when a counterfactual is asserted. As for most people other than the self-conscious Lewisian or the self-conscious Humean, they probably haven't thought about the matter, and so have no clear idea of *exactly* what they do mean.

However, there is an indirect route by which we can perhaps get at what people *ought to mean* when they assert a counterfactual conditional. This is by considering their *purposes*. It may be that by considering purposes, or the point of using such language, we can discover that they ought to mean what the Humean says they (for the most part, anyway — with the possible exception of David Lewis) *do* mean; or, alternatively, that they ought to mean what Lewis says they do (for the most part) mean. That is, by answering the normative question, and assuming that people normally do as they ought (perhaps, indeed, unreflectingly and from learned habit alone), then we can answer the descriptive question of who — Lewis or the Humean — provides the more adequate account of normal discourse.

Moreover, there is one point of disagreement between Lewis and Humean. Though small, it can, perhaps, be made into a test case. Nonetheless, even if this test case shows that Lewis is wrong, and even if the Humean can in fact handle it, it does not follow that the Humean is correct. For this latter, one needs the argument from norms. Thus, in the end the Humean must base his case not so much on actual usage but on considerations of how language *ought* to be used, that is, ought to be used *relative to the cognitive interests that men have*. This dictates our strategy. We shall first look at the area of disagreement. The upshot will be that Lewis' account is mistaken. We shall also argue that the Humean is able to account for the facts of usage that refute Lewis' position. We shall then argue that, given the usual purposes of counterfactual discourse, the Humean's is the position that ought to be adopted.

The problematic area is that of counterfactual conditionals that turn out to have *true antecedents*.[1] For Lewis, the following two inference patterns are valid, given the truth-conditions that he proposes for counterfactual conditionals:

(v1)       $\dfrac{A\ \&\ \sim B}{\therefore \sim(A \rightarrow B)}$

(v2)       $\dfrac{A\ \&\ B}{\therefore A \rightarrow B}$

As for (v1), if the antecedent of $A \rightarrow B$ is true and its consequent is false, then there is no A-permitting sphere around our world throughout which $A \supset B$ holds, since it fails at the actual world which belongs to every A-permitting sphere; hence $A \rightarrow B$ false in this case. As for (v2), if the antecedent of $A \rightarrow B$ is true, and the consequent also, then there is an A-permitting sphere around the actual world, namely, the one that has this world as its only member; so $A \rightarrow B$ is true when both A and B are true.

We will consider (v1) first.

Now, $A\ \&\ \sim B$ is logically equivalent to $\sim(A \supset B)$. Hence, (v1) is logically equivalent to

$$\dfrac{\sim(A \supset B)}{\therefore \sim(A \rightarrow B)}$$

And since $\sim q \supset \sim p$ is logically equivalent to $p \supset q$, this last is logically equivalent to

(v3)       $\dfrac{A \rightarrow B}{A \supset B}$

Thus, if (v1) is valid, then so is (v3). And this in turn guarantees that if (v1) is valid, then so is *modus ponens*

$$\dfrac{A \rightarrow B}{\qquad A \qquad}$$
$$\therefore B$$

and *modus tollens*

$$A \rightarrow B$$
$$\underline{\sim B}$$
$$\therefore \sim A$$

which is precisely what one expects of a *conditional*.[2]

If, however, that is how, almost by definition, a *conditional* behaves, then the Humean had best be in agreement with Lewis on the validity of (v1). Not surprisingly, of course, he is. On the other hand, and equally without surprise, however, the Humean disagrees on how to interpret (v1). Since the Humean does not grant that counterfactuals literally have truth-conditions, he cannot quite so freely move from falsity to negation, as does Lewis: there is no literal falsity to move from! For the Humean, to say that A → B is false is to say that one is not justified in asserting it. Hence, for the Humean, what (v1) means is that if

$$A \& \sim B$$

obtains then that establishes that one *cannot* be justified in asserting

$$A \rightarrow B$$

But this is clearly so for the Humean. For, he takes it that a necessary condition for asserting A → B is that there be a true statement of law L such that

$$L$$
$$\underline{A}$$
$$\therefore B$$

is valid. But a valid argument is such that it *cannot* be that the premises are all true and the conclusion false. Hence, if this argument is valid, and A is true and B false then the "law" L, *whatever it may be, cannot* be true.[3] Thus, if A & ~ B holds, then it is *impossible* for one to be justified in asserting A → B upon the Humean account. So the Humean, as well as Lewis, is prepared to accept (v1); and its immediate consequences that *modus ponens* and *modus tollens* inferences are valid with counterfactual conditionals. The Humean, like Lewis, insists that counterfactual conditionals are *conditionals*.

The disagreement is with respect to (v2), to which we now turn.

For the Humean, to say that (v1) is *valid* is to say that A & ~ B is *logically sufficient* to render the assertion of A → B unjustified. Analoguously, for the Humean, to say that (v2) is valid is to say that A & B is *logically sufficient* to justify the assertion of the counterfactual conditional A → B. But this is not so for Humean: two statements of individual fact can never justify the assertion of the law that is also part, implicitly, of what is asserted when the counterfactual is asserted.

The Humean and Lewis thus disagree on the acceptability of inferences in accordance with the pattern (v2). It is this disagreement that we will try to use to decide between them, or, to be more explicit, try to show that Lewis is mistaken.

If the Humean is correct, then we cannot expect that when speakers are prepared to assert.

A & B

they are also prepared to assert the counterfactual conditional

A → B

Lewis is also prepared to agree that, normally, when persons are prepared to assert the former they are not also prepared to assert the latter. That is, Lewis and the Humean are agreed that this is the pattern found in our *normal linguistic habits*. But they disagree on the reason why this pattern exists. For the Humean, it is because the inference is invalid. For Lewis, it is for other reasons: the inference is indeed valid but for these other reasons it would be "odd" to assert A → B if one is prepared to assert A & B. Lewis thus defends the validity of (v2) in the face of *apparently* contrary linguistic habits, and he does this by invoking an explanation, other than that based on invalidity, for why the pattern is "odd", contrary to normal linguistic habits.[4] Now, *in principle*, this sort of move is perfectly legitimate. It does not follow, however, that the *specific* case that Lewis makes is sound. Indeed, we shall argue that it is not. But in order to make this point it will be useful to look at a related case where this sort of move *is* legitimate.

Recall our discussion of Chapter II, Section I, above. Consider *modus ponens*

(a1)      If A then B
          A
          ——————
          ∴ B

and let us ask to what use it may be put. One use is that of convincing someone else to the truth of B. In order to do that the other person must be in the epistemic condition of *not* knowing that B is true. If, however, he is in the further epistemic condition of knowing both that A and that if A then B, then it is possible to use the argument of *modus ponens* to prove to him that B must be true. Another use of *modus ponens* is in *reasoned* predictions. If we are in the epistemic position of not having observed B, but have observed A and know that if A then B, then we can predict that B. That is, prior to the occurrence of B we have reasoned grounds for expecting its occurrence. There are other cases, but we need not enumerate them: the basic pattern is clear. *Modus ponens* is, because valid, often useful in providing grounds for believing that its conclusion is true. It can do this only if its premisses are true. For, if they are not, the argument is unsound and incapable of proving its conclusion true. Two other subjective conditions must obtain concerning the epistemic state of the person for whom the argument is to give grounds for believing the conclusion. The first of these is that the person must know (have reasons for believing) that the premisses are true. Otherwise, no matter how good the argument was objectively in establishing the truth of its conclusion, the person would have no grounds for taking it to be one that did so. The second epistemic condition is that the person *not* know that the conclusion is true. For, if the person *already knows* that the conclusion is true, then he has no need of the *modus ponens* argument: the argument will be useless to him, at least for the purpose of bringing it about that he knows the conclusion to be true.

Suppose we can construe the 'if-then' of the *modus ponens* (a1) as the horseshoe of material implication. (a1) is therefore

(a2)     $A \supset B$
            $A$
            _____
            $\therefore B$

Now, $A \supset B$ is true if either A is false or B is true or both A and B are true. This has important consequences for the utility of (a2) in debates and predictions. For, if (a2) is to be useful in for such purposes, then we must have grounds for believing that the major premiss is true; and if those grounds are a knowledge that one of the possible truth-conditions obtains, then (a2) is useless in debate and prediction: the three mentioned conditions cannot simultaneously be fulfilled. If we affirm the major premiss because we know the antecedent to be false,

then the condition that the argument have true premisses is not fulfilled.
If we affirm the major premiss because we know the conclusion to be
true, then the second epistemic condition is not fulfilled. The same
holds for the final case where we affirm the major premiss because we
know that both antecedent and consequent are true.

Similar considerations will show that if we construe *modus tollens*

> If A then B
> ~ B
> ――――――
> ∴ ~A

as having a material conditional for its major premiss:

> A ⊃ B
> ~ B
> ――――――
> ∴ ~A

then this argument form, too, cannot be used in debate and prediction
to provide grounds for believing its conclusion to be true.

This is not to say that there might not be other uses to which such
arguments might be put. For example, we might very well affirm

> Hitler was a good man ⊃ I'm a monkey's uncle

on the basis of knowing that both the antecedent and the consequent
are false. If such a conditional were affirmed then it would be the
expectation of its user that the hearer would use the evident falsity of
the consequent and *modus tollens* to infer the falsity of the antecedent.
But it would not really be a question of *proving* that the antecedent is
false: if there was any real question about the matter, then the falsity of
the consequent would place the truth-value of the conditional itself in
question, and it could not, after all, be used to *prove*, by *modus tollens*,
the falsity of its antecedent. Rather, asserting such a conditional is a
special rhetorical device whereby one can *emphatically* deny the
antecedent. The purpose of such a conditional is not the epistemic one
of proof and evidence but the rhetorical one of emphasis.[5]

We may conclude, then, that in contexts of debate or prediction — in
general, in contexts of giving reasons for conclusions for which we
otherwise lack evidence, the assertion of a conditional is not merely the
assertion of a material implication but contains a *moment of necessity*
that marks the conditional as one that is *stronger than* a material

implication and is asserted on the basis of evidence other than that of a simple knowledge of the truth-value of its antecedent and consequent.[6]

It follows that in contexts of debate and prediction, the assertion of a material conditional would always be "odd". This oddity would not be due to the falsity of the conditional. Nor would it be due to the fact, claimed by some,[7] that the material conditional is not *really* a sense of 'if-then'. After all, since it satisfies both *modus ponens* and *modus tollens*, it has as much right to be a sense of 'if-then' as any other concept that satisfies those inference patterns. But be that as it may, the point is that the material conditional *could be* asserted, so far as its logic is concerned, but *would not be* asserted in such contexts precisely because it could not do what one would want a conditional to do in contexts of debate and prediction, namely, prove its conclusion true. This is why its use would violate normal linguistic habits. So we have accounted for the "oddity" of the material conditional in much of normal discourse without concluding that there is something wrong with it from the point of view of logic.

In the context of debate, the moment of necessity — the "must" — that marks the normal conditional off from the ("odd") material conditional may be one of several varieties. It might be logical:

If $p$ and if $p$ then $q$, then $q$ *must* be true

or it might be definitional:

If he's a bachelor then he *must* be unmarried.

In the context of (reasoned) prediction, however, the *must* can be provided by only one thing: a law-asserted generalization. (That may provide the relevant connection in the context of debate also.) In the context of a reasoned prediction, then, the assertion of a normal conditional involves the law-assertion of a generality. For, if it did not, the conditional could not be used to yield, via *modus ponens* and *modus tollens* arguments, predictions about otherwise unknown events.

The preceding remarks can hardly be said to be the end of the matter, of course.[8] But for our present purposes they suffice, however. The point is that the Humean wishes to extend this account of predictively useful conditionals to the case of counterfactual conditionals: the counterfactual conditional is stronger than the material conditional because it implicitly involves the law-assertion of a generality. Lewis, in contrast, while maintaining the greater strength of the counterfactual conditional denies that the assertion of one involves the assertion of a law.

Moreover, Lewis wants to argue that the assertion of a counter-factual conditional in a context in which both its antecedent and consequent are true is "odd" in much the same way in which the assertion of a material conditional in such a context would be "odd". A counterfactual conditional in such a context *could be* asserted — after all, (v2) is valid — but *would not be* asserted because it "could serve no likely conversational purpose that would not be better served by separate assertions of [antecedent and consequent]."[9]

But oddity is not falsity; not everything true is a good thing to say. In fact, the oddity dazzles us. It blinds us to the truth value of the sentences, and we can make no confident judgements one way or the other. We ordinarily take no interest in the truth value of extreme oddities, so we cannot be expected to be good at judging them. They prove nothing at all about truth conditions.[10]

These remarks are just. Unfortunately, however, Lewis does not tell us what these conversational purposes are that could be served as well by asserting A & B as by asserting A → B. Nonetheless, two very likely possibilities come to mind, and in both cases Lewis' point is sound.

One purpose is that of *conveying information* about particular facts and events. Now, if P entails Q but not conversely, then there is a good sense in which P is *informationally stronger* than Q. In this sense, A is informationally stronger than A ∨ B. And in the same sense A & B is stronger than A ⊃ B. Now, according to Lewis, A → B says something about the A-worlds that are most similar to ours, viz., that is them A ⊃ B is true. But when A & B holds then the A-world most similar to ours, i.e., the actual world, is the actual world itself. Thus, when A & B holds in our world, A → B makes an assertion about this world, and all that it asserts is that A ⊃ B is true; or, in other worlds, when A & B holds, it is informationally stronger than A → B. Thus, the purpose of conveying information would be better served by asserting A & B than by asserting A → B. Relative to this purpose, then, it would be "odd" to assert A → B when one can assert A & B.

A second purpose is the one we have already noted, that of provid-ing grounds for asserting the consequent. But if we know that A & B, we already know that B, and so there would be no point in asserting A → B. Relative to the purposes of prediction or debate, then, it would be "odd" to assert A → B when one can assert A & B.

Lewis provides a second reason why it would be "odd" to assert A → B on the basis of A & B. It would be odd for one to assert, say, that if the sky were blue then the grass would be green, "because he is using the counterfactual construction with an antecedent he takes to be true, though this construction is customarily reserved for antecedent taken to be false."[11] This is certainly true. Nonetheless, Lewis offers no reason why the subjunctive construction *is* reserved for antecedents that are not known to be true. The Humean, of course, has an answer to this, in his complex analysis of the assertion of subjunctive conditionals. But he has less need of the point than Lewis, since, unlike the latter, the Humean does not have to appeal to such facts in order to account for our not asserting A → B when we assert A & B. In any case, however, because Lewis does not explain why the subjunctive construction is customarily used as it is, the point of the defence of his position is weaker than he seems to think.

What these points establish about the oddity of asserting A → B when we assert A & B is that *if* Lewis' account of counterfactuals is adequate, *then* he can account for the fact that it is contrary to normal linguistic practice to assert a counterfactual when its antecedent and consequent are both true. But it provides no direct support for Lewis' account as opposed to that of the Humean. For, as we saw, the Humean also can account for the relevant linguistic practices. What Lewis would like to have, of course, is some positive evidence from linguistic practice that his view is true. And this he attempts to provide. The context is, of course, not a straight-forward one of giving information or of making reasoned predictions, but one that is somewhat more complex, and to that extent non-normal — as, naturally, it would have to be, given Lewis' arguments about the oddity of asserting A → B on the basis of A & B.

The context that Lewis presents is that where one person is *correcting* another. Here the assumption is that one person has the purpose of conveying information and that the other person is asserting that this purpose is not being fulfilled, that *mis*-information is being conveyed. This sort of case is thus parasitic upon the normal use of language to convey information. That is what makes it out of the ordinary, not normal. But it is also a matter of the second person raising the issue of whether the counterfactual asserted by the first is true or false — or, if the Humean is correct, whether the assertion of counterfactual is justified or unjustified.

Lewis considers this case:

> You say: 'If Caspar had come, it would have been good party.' I reply: 'That's false; for he *did* come, yet it was a rotten party.' . . . [The] reply seems perfectly cogent. In [it], I correct your false belief that Caspar was absent, manifest in your use of the counterfactual form; but I do this while expressing my overall disagreement . . . with your conditional assertion. Moreover, I justify my disagreement . . . by giving an argument. The argument is abbreviated but its presence is signalled by the word 'for' in my reply.[12]

The argument to which he alludes is, of course, (v1):

$$\frac{A \ \& \ \sim B}{\therefore \ \sim(A \to B)}$$

As it stands, Lewis' sample of discourse is plausible enough. But, since both Lewis and the Humean accept the validity of (v1) (though interpreting the argument somewhat differently), it is neutral between the two: if it supports the one that it equally supports the other.

The second case is this:

> You say: 'If Caspar had come, it would have been a good party.' . . . I reply: 'That's true; for he *did* [come], and it *was* a good party. Your didn't see him because you spent the whole time in the kitchen missing all the fun.' [This] reply [too] seems perfectly cogent.[13]

In this case the false belief that Caspar was absent is again corrected, while overall agreement with the counterfactual is expressed. Moreover, a reason is given for agreeing with the counterfactual — a reason marked by the occurrence of 'for'. This reason consists in asserting both the antecedent and the consequent of the counterfactual. That is, it is of the form (v2)

$$\frac{A \ \& \ B}{A \to B}$$

The reason seems cogent, and if (v2) were valid then that would account for its cogency. This makes it plausible to suppose that (v2) really *is* valid.

It is now the turn of the Humean to do some explaining. He must explain why, in the conversational context that Lewis presents, the response that is made is a *natural* thing to say — why it is *not* an "odd" thing to say — in spite of the fact that (v2) is (if the Humean is correct) invalid, that is, in spite of the fact that A & B alone is not sufficient to justify the assertion of A → B.

It would seem, however, that the Humean can do this fairly easily if he interprets the response along the following lines: You have asserted that A → B; the data A & B support what you say, not directly and demonstratively by entailing it (as Lewis maintains), but by supporting, as confirming evidence, the law that is part of the assertion of the counterfactual. That the response so interpreted would be a natural one is due to the fact that the situation being talked about (i.e., in Lewis' example, the party) provides a *test situation* for the law that is involved in the assertion of the counterfactual. The antecedent A obtains (Caspar comes to the party), and we know that it obtains prior to our knowing whether the consequent B (the party was good). We know that A but do not know that B. But using the relevant law, we infer that B will obtain since A does. If it turns out that B does come about, the law will have been (further) confirmed; and if B does not come about, then the generalization that led us to expect it will have been falsified. If B does come about then we shall have become that much more justified in law-asserting the generalization that is part of the counterfactual, and therefore that much more justified in asserting the counterfactual. And if B does not come about and the generalization is falsified, then the assertion of the counterfactual will have been shown to be unjustified, or, what is the same, it will have been shown that one ought to deny rather than affirm that counterfactual. In asserting your counterfactual you knew nothing about this, since you did not know that Caspar was at the party, nor that there was great fun at it too. But I do. And in correcting you on these facts it is equally *natural* for me to also point out that these facts not only don't dispute your claim to have been justified, in the epistemic situation you were in before I corrected you, asserting the counterfactual, but in fact strengthen your claim to have been justified in that assertion. What I point out is that, even though you were in error on one point, that error not only does not invalidate everything you claimed but in fact adds further support to that remaining portion of it. It is, surely, *natural* to make both points so as not only to correct but to pinpoint the correction. After all, is not the end of rectifying error better achieved in that way?

I conclude that the Humean can account for what was said in the fragment of discourse that Lewis provided. It follows that Lewis cannot claim that conversation as support of his position against that of the Humean.

Lewis makes a stronger claim that the Humean, however. Lewis is committed to the position that the inference from A & B to A → B is *always* leading from truth to truth, even in those cases where it would be odd to make such an inference. In contrast, the Humean is not so committed. For the latter, whether A & B supports the assertion of the counterfactual A → B will depend upon the law that is (implicitly) involved in the assertion of the latter. Thus, if the Humean is correct then there will likely be *some* cases in which A & B is true while A → B ought to be denied, i.e., where one is not justified in asserting the counterfactual. In contrast, Lewis is committed to the position that there is *no* case in which A & B is true and A → B ought to be denied. If, therefore, we can find cases where A & B are true yet where A → B ought to be denied, then we shall have shown Lewis' position to be mistaken, and shall have provided a little more support for the Humean.

Consider the following case.[14] You say: "If Caspar had come to the party, it would have been a good one." I hear you say this, and know that Caspar did come to the party. So I can correct you on that point. I also know that it was a good party. So, if Lewis is correct, then I ought not to dispute the counterfactual you asserted. However, suppose that I also know Caspar sufficiently well to know that he drinks too much, that he tends to get nasty when he is drunk, and that he in fact ruins most parties that he attends; suppose I also know that he almost ruined this one, and that he would have had it not been for the extraordinary tact and tolerance of the hostess. That is, what I know is that it was a good party *despite* the fact that Caspar came. And *that* knowledge, *surely*, justifies my *disagreeing* with the counterfactual you asserted. That is, the knowledge that I have justifies me in denying A → B even though I also know that A & B. This being so, Lewis' claim that (v2) is valid must be wrong.[15]

On the other hand, the Humean has no difficulty in analyzing the situation; for him, everything falls right into place.

For the Humean, A → B is a sentence that may be used to make different assertions in different contexts.[16] This is because the law that is implicitly asserted when A → B is asserted may vary from context to context. In one context the assertion of A → B may have to be analyzed so as to involve the argument (f₁)    Whenever an A then a B

      A
_____

    ∴ B

where the first premiss is implicit rather than explicit and is law-asserted, and the second premiss and the conclusion are either both denied or neither asserted nor denied. But in another context the assertion of A → B may have to be analyzed so as to involve a more complicated argument, perhaps somewhat like the following:

(f₂)    Whenever a ~C then, if an A then a B

    ~ C

     A
_____

    ∴ B

where the first two premisses are implicit rather than explicit, the first being law-asserted and the second asserted, and the third premiss and the conclusion either both denied or neither asserted nor denied. Indeed, this second sort of case is no doubt the more common, as Lewis himself points out when defining nomic dependence of A on B as occurring in case "there are a non-empty set $\mathscr{L}$ of true law-propositions and a set $\mathscr{F}$ of true propositions of particular fact such that $\mathscr{L}$ and $\mathscr{F}$ jointly imply (but $\mathscr{F}$ alone does not imply)"[17] the material conditionals A ⊃ B.

In the situation we are envisaging, where the party was a good one despite Caspar's attendance, you have asserted the counterfactual "If Caspar had come to the party, it would have been a good one." Your assertion will, according to the Humean, be analyzed along the lines of (f₁) or perhaps (f₂). But what I know about Caspar is that the following generalization is true: in the absence of special conditions (C) his attending a party (A) is sufficient for it to be a bad party (~B); or, to put it more briefly,

(g)    Whenever a ~C then, if A then ~B

Furthermore, I also know, again what you don't know, that A & B. These particular facts and (g) yield, via the argument

(f₃)    (g)

    A

    B
_____

    ∴ C

the conclusion the special conditions C are present rather than, as usual, absent. Since the special conditions C are *normally* absent, then (g) *normally* permits one to assert the counterfactual A → ~B, where this assertion is analysed into

$(f_4)$         (g)

~ C

A

—————

∴ ~B

with the first two premisses implicit only, the first law-asserted, the second asserted, and the third premiss and the conclusion both not asserted. But the occurrence of A & B establishes the assertion of the counterfactual A → ~B is not justified. It does so, not by falsifying the law (as in the case where (v1) applies), but by eliminating (one of) the assertions of particular fact that must be part of what is implicitly asserted if the law (g) is to establish a connection between the antecedent and the consequent of the counterfactual. What I am entitled to assert, in the light of my knowledge (g) and knowledge that A & B, is *not* the counterfactual A → ~B. For, the fact that, by $(f_3)$, C obtains, means that the argument $(f_4)$ that I would assert were I to assert the counterfactual would be unsound and my assertion unjustified. Instead, all I can assert is the counterfactual ~C → ~B, that is, "if the party had been normal, then it would have been ruined", where this assertion is analyzed into

$(f_5)$         (g)

A

~C

—————

∴ ~B

with the first two premisses implicit only, the first law-asserted, the second asserted, and the third premiss and the conclusion both supposed And I can add that the special conditions that are normally absent were fortunately present and that fact saved the party. But besides this, my knowledge justifies a certain response on *my* part towards *your* assertion of the counterfactual A → B. For, if we have a law and a normally obtaining sort of particular fact that between them mean that one is normally justified in asserting A → ~B, then we do *not* have a law and set of particular facts that could justify the assertion of A → B; indeed, in such a case we *cannot* have such a law and set of particular facts, since the law would have to be *logically contrary* to the law (g) that we *do* have. But if we cannot have such a law, then we cannot be justified in asserting A → B, that is, we must deny that counterfactual. In short, given what I know, to wit, the lawful generalization (g) and the particular fact A & B, then, if the Humean's account is correct, then one is fully justified in denying the counterfactual that you asserted. The Humean can therefore account without difficulty for the fragment of ordinary discourse that (we have argued) falsifies Lewis' claim that A → B is true (worthy of assertion) *whenever* A & B is true.

Another example might be helpful. Let '*Px*' represent that a person takes a certain poison. Let '*Bx*' represent that he dies. Let '*Ax*' represent that he takes a certain antidote. Assume that in the majority of cases taking this antidote is sufficient to prevent dying. Assume, finally, that we know that *there are* certain conditions the absence of which is both necessary and sufficient to prevent death when the antidote is taken, but that we do not know *specifically* what they are. Let '*C̄x*' describe the presence of those conditions, with the bar indicating that we have only a determinable rather than a determinate or specific description of them. In these assumptions we have let the following generalization be law-assertible:

$$(h) \qquad (x)[Px \supset [Ax \supset (\sim Bx \equiv \sim \bar{C}x)]]$$

Since we do not know *specifically* what the conditions $\bar{C}$ are, it follows that we cannot have direct confirming data for the law; we cannot identify instances of $\bar{C}$, identification of which is a necessary condition for identifying a confirming instance. But a background theory about various kinds of poisons and how they work might very well provide indirect reasons for law-asserting the generalization. Moreover, since we cannot know *antecedently* for whom the antidote will be effective, we cannot use the law to *predict* about a poisoned person whether the antidote will be effective or not. This would seem to make it impossible to use the law to intervene effectively when someone is poisoned. To a point this is so. But only to a point. The law is still useful, for it can be used to *statistically estimate* the frequency of $\bar{C}$ in the population, even when we are unable to determine specifically who has the crucial property. Consider a person *j* such that

$Pj$
$Aj$

Then suppose that $j$, alas, dies:

$$Bj$$

It follows, given the law, that

$$\bar{C}j$$

obtains. We can thus know that $\bar{C}j$ obtains, even when we can't identify such individuals independently of the law. If we consider the group of people who are both $P$ and $A$, then the frequency of those people who are $B$ will also be the frequency of those who are $\bar{C}$. And if the group of people who are both $P$ and $A$ is a representative sample from the population, then the relative frequency with which they are $\bar{C}$ will likely be close to the relative frequency of $\bar{C}$ in the population as a whole.

Now let us suppose that the frequency of $\bar{C}$ is quite low. In that case the *normal* thing to expect when the antidote is administered to a person is that he will recover.[18] If Caspar is a person who has been poisoned, then we can, in this context, reasonably assert the subjunctive conditional:

($c_1$)    If Caspar were given the antidote, then he would recover.

This would be analyzable into the argument

($f_6$)              (h)
                     $Pc$
                  $\sim \bar{C}c$
                     $Ac$
                  _____
                  $\therefore \sim Bc$

where the first three premisses are only implicit in what was said, the first premiss is law-asserted, the second premiss asserted on the basis of, say, direct observation, the third premiss asserted but with a degree of certainty proportionate to the statistical likelihood of its obtaining, and with neither the last premiss nor the conclusion asserted. *Because the evidence is such that we are subjectively justified in asserting the counterfactual we use that as a justification for intervening in the process in order to achieve certain purposes that we have — in this case, that of saving Caspar's life: we administer the antidote. The counterfactual now becomes a reasoned prediction. The law that supports the counterfactual assertion yields a reasoned prediction when we intervene in the process in such a way as to make the antecedent of the conditional true.* Moreover, if Caspar survives the poison, then ($f_6$) constitutes a sound deductive-nomological explanation of why that happened. *The law that supports the assertion of the counterfactual permits us to predict what will happen when we intervene to make the antecedent true and yields an explanation for the success of the intervention.* Provided, of course, that the intervention is successful — that is, in this case, provided that ~ $\bar{C}c$ does obtain objectively, that Caspar is *not* one of the infrequent persons who are $\bar{C}$.

Unfortunately, it turns out that Caspar is an exception: he is one of those persons in which the normally absent conditions are present:

$$\bar{C}c$$

*Subjectively*, given the evidence available, we were justified in asserting the counterfactual ($c_1$). But given facts that we had no means for antecedently determining whether they obtain, it turns out that *objectively* we were not justified in asserting ($c_1$).

Let us suppose, finally, that Caspar actually dies:

$$Bc$$

For Lewis, since we have

$$Ac \,\&\, Bc$$

one could truly assert (though perhaps it would be odd to do so) the counterfactual

($c_2$)      If Caspar were given the antidote, then he would die.

In the situation we are supposing, the response to an assertion of ($c_2$) would be complicated. It would be something like this: "Yes and no. Yes, we *can* say that *now*, that is, now that Caspar's died. But that is only because we know something now that we didn't know before. So also No, because *earlier* we had no reason to think that Caspar was abnormal and that he would die despite being given the antidote." This conversational fragment tells against Lewis, not by yielding a straightforward case where A & B is asserted and A → B denied. Rather, it challenges what is claimed when it is claimed that (v2) is valid, namely, that the assertion of A & B is *logically sufficient* for the assertion of A → B. The response to ($c_2$) shows that one is agreeing with the assertion of A → B not *merely* because A & B obtain, but because *some other information* has become available. This reply is relevant only if A & B is *not* logically sufficient for A → B. And since it *seems* to be a cogent response, it directly calls into question Lewis' analysis. Moreover, if the Humean is correct, then the response *is* cogent.

Information becomes available after Caspar dies that could not be known to us *prior* to the administration of the antidote. This is the information that Caspar was one of the infrequent $\bar{C}$ persons. Once Caspar dies the following inference provides good grounds for asserting that $\bar{C}c$:

$(f_7)$      (h)
         $Pc$
         $Ac$
         $Bc$
         $\overline{\phantom{Bc}}$
         $\therefore \bar{C}c$

Once this information has been acquired, it can be used to provide a *deductive-nomological explanation*[19]

$(f_8)$      (h)
         $Pc$
         $\bar{C}c$
         $Ac$
         $\overline{\phantom{Bc}}$
         $\therefore Bc$

of why Caspar died *despite* being given the antidote. Moreover, this explanation is precisely the argument into which the counterfactual conditional $(c_2)$ should (if the Humean is correct) by analyzed. This explanation $(f_8)$ shows why one can agree with the assertion of $(c_2)$, while the need to rely on $(f_7)$ to discover the truth of a necessary premiss shows why we could not have been justified in asserting $(c_2)$ earlier. And, of course, precisely because $(c_2)$ is to be analyzed into $(f_8)$, the Humean can, as Lewis cannot, account for the cogency of the response to $(c_2)$ that occurred in the conversational fragment given above.

We must be careful in our conclusions. If Lewis wishes to mean what he says he means when he uses a counterfactual conditional, then he may do so. Yet if our discussion is correct, then Lewis' account does not square with an important part of ordinary usage.[20] That usage occurs when counterfactuals are used in a definite context, for a definite purpose. That context is defined by the end of *intelligent interference* in natural or social processes.[21] We confront a situation which is not-A. We want the situation to be B. In this context we reason: if it were A then it would be B. On the basis of this claim, we conclude that we can bring about B by bringing about A. We now bring about A. At this point B is not yet achieved. The counterfactual has ceased to be that and has become a prediction that, since A has occurred so will B. And when B comes about, the prediction becomes the explanation that B occurred because A occurred. The counterfactual becomes a prediction when its antecedent becomes (through our efforts) true, and this becomes an explanation when we observe, as they come to be, the facts that make its consequent true. Now, if this is to be reasonable, then the prediction must be more than guesswork: it must be a *reasoned prediction*; otherwise the interference could hardly be reckoned to be intelligent. The prediction, that the consequent B will come about, will be avoiding any element of guesswork only if it is based on grounds that establish that, since A occurred, B *must* occur. We make a reasoned prediction of B given A only if it is based on *known connection* between A and B, a connection that establishes that if A occurs then B *must* occur. In particular, it must be based on the existence of a connection rather than on an estimation of interworld similarity that might obtain even when no connection exists. Nor can the conjunction A & B alone be logically sufficient for the prediction, because conjunction is not a connection that permits prediction. But how are we to understand this 'must' of reasoned predictions? As we argued in Chapter I, Section I, the only plausible way to understand that 'must' of reasoned predictions is in terms of the prediction being based on a law-asserted generalization.[22] However, in the context of reasoning about the intelligent interference in a process, a counterfactual is transformed into a prediction by bringing about the facts that make its antecedent true. All that is involved is a change from withholding assent to the antecedent to affirming it. That being so, *the law that makes the prediction a reasoned prediction will have also to be present in the assertion of the counterfactual.*[23] Similarly, the prediction can become an explanation when one succeeds in bringing about the facts that make the consequent true only if the deductive-nomological model of explanation is correct.[24] Thus, *the discourse — counterfactual reasonings, predictions, and explanations — that occurs in the context defined by the end of intelligent interference or potential interference can achieve that*

*end only if one accepts the Humean analysis of counterfactual con-
ditionals and the deductive-nomological model of explanation and
prediction.* In short, *in contexts defined by our pragmatic interests in
intelligent interference or potential interference in natural and social
processes, the semantics of counterfactuals that we* OUGHT *to adopt is
that proposed by the Humean.*[25]

This, by the way, accounts for the close connections between counterfactual and causal discourse — a connection acknowledged even by
Lewis, who proposes to explain the latter in terms of the former, as we
saw.[26] One can achieve one's goals only by employing the *means* that
are appropriate to those ends. But the relationship of means to ends is
simply the relation of cause to effect. When counterfactual reasoning
occurs as we deliberate how to achieve our ends, it is not surprising,
then, that it connects up with causal language. What Lewis' account of
counterfactuals fails to do, is explain the connection of both to *reasoned
prediction.* And once the connection is seen, and is seen to be rational
relative to the end of intelligent interference, then we recognize that
the *laws* that make reasoned prediction possible must be part also of
the counterfactual and causal discourse. I.e., once we recognize the
relevance of reasoned prediction, and that laws *ought* to be part of
what is asserted in such contexts, then we recognize that it is the
Humean's analysis of counterfactuals, and not Lewis', that we *ought* to
accept.

Another context in which counterfactuals are used is that defined by
the end of *excusing* people from moral responsibility for certain things.
For example, you might say:

   (e$_1$)     If Caspar had done A, then B would not have happened.

The idea would be to establish Caspar was responsible for B having
occurred. That this idea involves the assertion of a *connection* between
A and B can be brought out best, perhaps, by noting a possible
response:

   (e$_2$)     No: if Caspar had done A, B would still have occurred.

What you assert is

        $A \to B$

and what I assert is

        $\sim A \to B$

Since, clearly, ~ A and B are both true, (e$_2$) will, according to Lewis, be true. And, if his account is correct, then (e$_1$) will be true just in case B holds in the nearest A-world. So, for Lewis, (e$_2$) makes a claim about *this* world, and (e$_1$) makes a claim about *another* possible world. In fact, for Lewis, it is possible for both (e$_1$) and (e$_2$) to be true! But surely this is wrong; surely, as my response indicates, I am in fact *disagreeing* with you.

In fact, there seem to be *three* difficulties that arise on the account Lewis would give of this conversational fragment. In the *first* place, (e$_1$) and (e$_2$) are clearly both making claims about the *same* world, namely, the actual world: you are trying to establish that Caspar is responsible in *this* world, and I am trying to establish that he is not. *Secondly* — and this is much the same point — both of us are talking about Caspar, whereas if we take Lewis' views on possible worlds seriously, then I am talking about Caspar, the actual Caspar, whereas you are not talking about Caspar but about his counterpart in another possible world. As we said above,[27] Lewis' introduction of counterparts seems to conflict with assumptions implicit in discourse in which we ascribe moral responsibility. Moreover, *thirdly* — what is important in the present context — the point I am making when I assert (e$_2$), Lewis notwithstanding, is *not* exhausted by the conjunction ~A & B; rather, I am *also* saying that A is irrelevant to B, that the two events are *not connected*.[28] I'm saying that, no matter whether A occurred or not-A, B would still have occurred. And it is *this* which makes (e$_2$) a denial of (e$_1$). Precisely because there is no connection between antecedent and consequent, you cannot be justified in your assertion of (e$_1$); precisely because there is no connection between A and B, we ought to deny what you claim: A → B. Lewis can perhaps allow that if both (e$_1$) and (e$_2$) are true, then B is *counterfactually independent* of A.[29] But such counterfactual independence occurs *only if* (e$_1$) and (e$_2$) are *both true*. The point is that to reply to an assertion of (e$_1$) by asserting (e$_2$) is *a way of denying* the former counterfactual. (e$_1$) ascribes a responsibility to Caspar, and (e$_2$) excuses him, *denies* that he is responsible. For Lewis, asserting (e$_2$) in the context of an assertion of (e$_1$) establishes causal independence only if (e$_1$) is true, i.e., *ought to be asserted*; whereas from our example it seems clear that asserting (e$_2$) establishes causal independence by denying (e$_1$), i.e., by establishing that (e$_1$) *ought not to be asserted*. This makes no sense on Lewis' account of counterfactual and causal dependence, but of course finds its natural place in the Humean account.

Perhaps, though, in the case of discourse defined by the end of excusing people, the relevant connection is something other than the laws that the Humean insists upon. However, it is clear enough, from the conversational fragment above concerning Caspar, that what is relevant is the issue of what the consequences would be upon *interfering* or *failing to interfere* in certain ways in certain natural or social processes. That is, this context is simply a special case of the one we discussed above, concerning intelligent interference in processes. The considerations advanced above for why the Humean analysis ought to be adopted in contexts defined by the end of intelligent interference apply equally to the context of excuses.

In fact, I believe that the goal of intelligent interference pervades the greatest part, at least, of the discourse in which counterfactuals occur. Counterfactuals and causes go together, as we have, with Lewis, insisted. But knowledge of causes is always knowledge of at least *potential* means. It is knowledge of what we would have to do were we to want to interfere in the process to bring about various ends. Of course, such knowledge can be sought out of idle curiosity as well as out of some definite pragmatic interest in application. Nonetheless, it remains knowledge that tells us how we ought to interfere (or not interfere) *if* we have certain ends. It is in this way that the goal of intelligent interference determines the context in which the greatest part of our causal and counterfactual — and explanatory — discourse occurs. And since, as we have argued, in such contexts counterfactuals *ought* to be used so as to fit the Humean analysis, it follows that there is little room for speaking in the way Lewis recommends. I conclude, therefore, that it is the Humean and not Lewis who gives the correct analysis of counterfactual and subjunctive conditionals. *At least for contexts defined by the goal of intelligent interference, there are, or ought to be, no counterfactuals that involve no laws.*[30]

This context determines other features of the counterfactuals that are appropriate to it. Among these, the one most worth noting — we shall return to it below, in more detail — is the requirement that the *antecedent be consistent.* Thus, we cannot have a counterfactual like

$$(A \ \& \ \sim A) \rightarrow B$$

But neither can we have a counterfactual like

$$(A \ \& \ C) \rightarrow B$$

in a context where one also law-asserts

>Whenever A then ~C

The latter guarantees that A and C are, in Goodman's phrase, not *contenable*,[31] that is, their joint assumption is not consistent with the laws of nature that one asserts in the context. *It is a necessary condition for the assertibility of a counterfactual in a context defined by the goal of intelligent interference that the antecedent of the counterfactual be logically self-consistent and be consistent with (known) laws of nature.*[32] For, if the antecedent were self-inconsistent or self-uncontenable, then it would be impossible that it be brought about; no means could effect it. But interference is possible only on the basis of *realizable* states of affairs; prediction makes no sense where what is offered as the grounds for the prediction cannot be realized; explanation is always in terms of what does happen, and, since what does happen *can* happen, unrealizable states of affairs are irrelevant. Thus, *in contexts defined by the goal of intelligent interference that its antecedent is either logically or lawfully unrealizable is a sufficient condition for denying a counterfactual.*[33]

## VIII. THE CASE FOR LEWIS' ANALYSIS EXAMINED

We have just argued that Lewis' account of counterfactuals is mistaken. It follows that the arguments he offers in support of it must be unsound. This, however, should be shown rather than merely asserted. To do this, we shall proceed in a series of steps. *First*, we shall look at claims that certain features of the logic and semantics of counterfactuals are better accounted for by Lewis' analysis than the Humean's. *Second*, we shall look at claims that certain features of causal discourse are better handled by Lewis' account that the Humean's. *Third*, we shall look at an example of a complex causal process Lewis calls "pre-emption" which he claims his analysis can handle but the Humean's cannot. *Fourth*, we shall discuss the problem of what Goodman has called "cotenability."[1] Finally, *fifth*, we shall examine a particularly intriguing example that Lewis claims refutes the Humean's position.

### (i)  *Semantics of Counterfactuals*

For the Humean, the counterfactual

>$Fa \rightarrow Ga$

is *ambiguous.*[2] Depending on the context, its assertion may be analyzed
be the argument

$$(x)(Fx \supset Gx)$$
$$Fa$$
$$\overline{\qquad\qquad}$$
$$\therefore Ga$$

or by the argument

$$(x)[(Fx \ \& \ Hx) \supset Gx]$$
$$Ha$$
$$Fa$$
$$\overline{\qquad\qquad\qquad}$$
$$\therefore Ga$$

or by any number of other possibilities. If we take *semantics* to be the
determination of meaning in the sense of *truth-conditions*, then this
ambiguity is clearly *semantical.* For, in different uses of the sentence
different generalizations will be law-asserted, and different generalizations do, of course, have different truth-conditions.

Such ambiguity is not uncommon in language. For example,

> *a* is tall

is semantically ambiguous in just this way. For, implicit in the context is
a *standard*; it may be analyzed into

> *a* is tall relative to Wilt Chamberlain

or into

> *a* is tall relative to the average man

or into

> *a* is tall relative to the King's dwarf

or by any number of other possibilities. In different uses of the sentence
'*a* is tall', different standards will be involved, and whether what is
asserted is true or false will depend on the standard used. Since truth-
conditions, i.e., semantics, depend on context, the sentence '*a* is tall' is
semantically ambiguous.

Nonetheless, while such ambiguity is common, it does produce a
messiness, a lack of simplicity, when one attempts to provide an
analysis of the logical structure of what is said: from the viewpoint of

logic, it would indeed be nice if grammatical form and logical form coincide. They, alas, do not, but because their coincidence *is* the ideal[3] it is a virtue of any analysis of a mode of discourse that it minimize this gap. In particular, then, from this point of view an analysis that represents certain sentences as semantically unambiguous is to be preferred to one that represents those sentences as semantically ambiguous.

Just this virtue has been claimed by Stalnaker to render a Lewis-type analysis of counterfactuals superior to that of the Humean: on the latter counterfactuals are semantically ambiguous, on the former they are not.[4] To be sure, Stalnaker admits, there is a pragmatic ambiguity, but that is less of a problem to logical analysis than a semantic ambiguity.

We may see whether this claim is justified by looking at the semantical analysis. Upon the Stalnaker-Lewis semantics for counterfactuals, when one asserts a counterfactual

$$A \rightarrow B$$

one employs a selection function $f$ that asigns to A an A-world $\alpha$ and the counterfactual is true just in case that B is true in $\alpha$.[5] The selection function is *a* similarity relation. But, as we saw above,[6] several similarity relations are possible. Which one has been selected will depend on the context. Counterfactuals are therefore ambiguous, but the ambiguity is said to be pragmatic rather than the more difficult semantic.[7] This claim is justified by an analogy.

The truth conditions for quantified statements vary with a change in the domain of discourse, but there is a single structure to these truth conditions which remains constant for every domain. The semantics for classical predicate logic brings out this common structure by giving the universal quantifier a single meaning and making the domain a parameter of interpretation. In a similar fashion, the semantics for conditional logic brings out the common structure of the truth conditions for conditional statements by giving the connective a single meaning and making the selection function a parameter of the interpretation.[8]

As argument, this is totally unpersuasive.

Consider the case of a universally quantified sentence like

They all bark

This certainly has, in one sense of 'meaning', a meaning;[9] but nonetheless, its meaning in the sense of *truth-conditions* is *not* fully specified. For, whether or not this sentence is true or false will depend upon the universe of discourse that is specified implicitly in the context in which it is used: in the universe of dogs, the sentence is true, but in the universe of mammals it is not. The sentence is therefore semantically ambiguous in just the way that

> $a$ is tall

is semantically ambiguous. And similarly, the sentence

> $(x)(Fx)$

is semantically ambiguous until the universe of discourse is specified. We do of course have the rule of interpretation.

> Sentences of the form '$(x)(Fx)$' are true if and only if every instance of '$Fx$' is true in the universe of discourse

This ensures that there is a "common structure" to the meanings assigned to sentences of the form '$(x)(Fx)$', but any sentence interpreted *solely* by it remains ambiguous in its truth-conditions until the universe of discourse is specified. Thus, it is simply false to say, as Stalnaker does, that a rule like this so assigns a "single meaning" to the universal quantifier that sentences containing it are semantically unambiguous.

In exactly the same way, counterfactuals are, on the Lewis-Stalnaker account, semantically ambiguous. The counterfactual A → B is true just in case that B holds in the A-world chosen by the selection function $f$. This rule does guarantee that there is a "common structure" to the meanings assigned to sentences of the form A → B, but any sentence interpreted by this rule will remain ambiguous in its truth-conditions — semantically ambiguous — until the selection function $f$ is specified.

In fact, exactly the same sort of "common structure" can be displayed by the Humean. He has only to lay down the rule that A → B is true just in case that there is a selection function $f$ that assigns to A a set of laws $\mathscr{L}$ and individual facts $\mathscr{F}$, and that B is entailed by A, $\mathscr{L}$, and $\mathscr{F}$, and not by A, $\mathscr{F}$ or A & $\mathscr{F}$ alone.

Nor should it be suggested that the Lewis-Stalnaker account is a better account of common structure because of a greater simplicity due to there being only a few relevant interworld similarity relations that can yield the selection function they need, whereas there are incredible many such selection functions for the Humean owing to the incredibly many possibilities for selecting relevant $\mathscr{L}$'s and $\mathscr{F}$'s. For, interworld similarity depends upon (among other things) which laws hold. As it is admitted, "It is ... the structure of inductive relations and causal connections which makes counterfactuals ... true or false".[10] Thus, until one has specified what one takes to be laws, one has not picked out a possible A-world as the one in which the truth of B is to ascertained; nor, therefore, has one completely determined the truth-conditions of the sentence one is using until one has chosen an $\mathcal{L}$ — and, for that matter, also an $\mathcal{F}$, since some particular facts will, surely, be more relevant than others in judging of interworld similarity. Thus, the Lewis-Stalnaker account of the semantics of counterfactuals is neither better nor worse than that of the Humean: "simplicity and systematic coherence"[11] in respect of semantics do *not* judge the possible-worlds analysis to be better than that of the Humean.

A second virtue claimed by Stalnaker for the possible-worlds analysis,[12] is that it accounts for the fact that one normally denies

$$A \rightarrow B$$

by asserting

$$A \rightarrow \sim B^{13}$$

Since the former is true just in case that B holds in the nearest A-world, and the latter is true just in case $\sim$B holds in that world, and since $\sim$B holds only if B doesn't, it follows that if A $\rightarrow$ $\sim$B is true then A $\rightarrow$ B is false. This is neat enough, but it counts against the Humean only if the latter lacks this virtue. But it doesn't. For the Humean, to assert A $\rightarrow$ $\sim$B is, among other things, to law-assert a generality that establishes that being $\sim$B is regularly consequent upon being A, at least provided that certain other individual facts $\mathcal{F}$ obtain. But if being $\sim$B is regularly consequent upon being A, then being B is *not* regularly consequent upon being A.[14] However, one can assert A $\rightarrow$ B only if one can law-assert that being B is regularly consequent upon being A. Hence, if one is justified in asserting A $\rightarrow$ $\sim$B then one is not justified in asserting A $\rightarrow$ B, i.e., one is justified in denying the latter. In short, for the Humean also, one can deny A $\rightarrow$ B by asserting A $\rightarrow$ $\sim$B.[15]

Another virtue that has been claimed for the Lewis-Stalnaker analysis of counterfactuals is that it allows what have been called "semifactuals"[16] — sentences of the form

Even if A, (still) B

with a true antecedent and a false consequent, e.g.,

> Even if you were asleep all morning, you would still be tired.

— to be given an analysis precisely parallel to that given to counter-factuals. The counterfactual

> If A were, then B would be

is analyzed as

> A → B

and the semifactual

> Even if A, B

is analyzed as

> A → B

Both are true just in case B holds in the nearest A-world. The fact that the consequent B is false (i.e., false in the actual world) for the counterfactual, but is true in the case of the semifactual is irrelevant to the truth-conditions.[17] In contrast, the Humean distinguishes counter-factuals and semifactuals: the former assert while the latter deny a lawful connection between antecedent and consequent.[18] This separate analysis is, it is said, objectionable.

A separate and non-conditional analysis for semi-factuals is necessary to save the 'connection' theory of counterfactuals in the face of anomalies we have discussed, but it is a baldly *ad hoc* manoeuvre. Any analysis can be saved by paraphrasing the counter-examples. The possible worlds theory . . . avoids this difficulty by denying that the conditional can be said, in general, to assert a connection of any particular kind between antecedent and consequent.[19]

Causal connections *do*, of course, determine truth-values of counter-factuals and semifactuals, even for the possible-world analysis, but for this analysis they do so by determining interworld similarities, which in turn determine truth-values.

By treating the relation between connection and conditionals as an indirect relation in this way, the theory is able to give a unified account of conditionals which explains the variations in their behaviour in different contexts.[20]

The proper response to this claim is flatly to deny[21] the claim to virtue, and flatly to assert that semifactuals are *not* parallel to counterfactuals. The counterfactual asserts that

> B, if A

while the semifactual asserts that

> B, even if A

It is the difference between

> B occurs *because* of the occurrence of A

and

> B occurs *independently* of the occurrence of A

The possible-world analysis of semifactuals proposed by Stalnaker simply denies this obvious distinction. On the other hand, the Humean account makes sense of it: counterfactuals *assert a connection* between antecedent and consequent while semifactuals *assert the independence* of antecedent and consequent.[22]

In fact, it would seem that Stalnaker blurs the distinction between a counterfactual in which the consequent *happens* to be true, and a semifactual. What justifies the assertion of a counterfactual is the law-asserted generality that permits the deduction of the consequent from the antecedent. The generality justifies asserting

> A → B

even if it turn out that A is false and B is true. The counterfactual asserts that A *would be* sufficient for B, and that can be so even when A does not occur and B does occur though brought about by some cause other than A.[23] Any counterfactual with a false antecedent and true consequent is indeed, as Stalnaker says, to be given an analysis exactly parallel to the analysis given to a counterfactual with false antecedent and false consequent. And if the Humean could not do this, then his account would be severely deficient in the virtues an analysis ought to display. But of course the Humean *can* account in parallel fashion for the two sorts of counterfactual. Stalnaker apparently wrongly infers that, since the Humean does not analyze semifactuals, all of which have a false antecedent and a true consequent, in a way parallel to how he analyzes counterfactuals with a false antecedent and false consequent, therefore he cannot analyze in a fashion parallel to the latter counterfactuals with a false antecedent and a true consequent.

It is sometimes said that one can deny a counterfactual

(s)        A → B

in two ways, one by asserting the contrary counterfactual

(s′)        A → ~B

and one by denying a connection between A and B by asserting the semifactual

(s″)        Even if A, B

(s′) denies (s) by denying there is a lawful connection between A and B, and it denies this by asserting a contrary connection between A and ~B. Thus, Goodman tells us that "In practice, full counterfactuals affirm, while semifactuals deny, that a certain connection obtains between antecedent and consequent."[24] On the other hand, it has also been suggested that (s″) is the denial, not of (s), but of (s′). Thus, Chisholm, for example, proposes to paraphrase "Even if you were to sleep all morning, you would be tired" as "It is false that if you were to sleep all morning, you would not be tired."[25] It is part of Stalnaker's point about semifactuals that both these construals are wrong.[26] It does not follow that his own account is correct — indeed, we have argued that it is not — nor that the Humean's is wrong. But his point is, it would seem, well-taken. For, what the semifactual asserts is the *independence* of its consequent from both A and ~A; it asserts that the laws $\mathscr{L}$ and particular facts $\mathscr{F}$ are such that B would occur whether A obtained or ~A obtained. And to *thus* assert the independence of the consequent from the antecedent is not quite the same as denying that there is a connection between antecedent and consequent; to so assert independence is compatible with allowing that *in certain contexts* the counterfactual (s) may be asserted. And if we assert (s), then that is tantamount to denying (s′). In this respect, Chisholm's paraphrase is closer to the truth than Goodman's. It would seem that Goodman has confused asserting independence with denying connection. But Chisholm's construal of (s″) is not quite correct either. He paraphrases it as a straightforward denial of (s′). But this is wrong. Asserting (s″), the independence of B from A, does not deny (s′) directly but rather justifies asserting (s) which in turn justifies denying (s′). However, to assert (s) requires there to be a connection between A and B. If (s″) is true and B is independent of A, then the sort of connection that is possible will, clearly, be rather unusual. In construing (s″) as a simple negation of (s′), Chisholm misses this oddity. It would seem that he, too, has confused asserting independence with denying connection. In this respect Stalnaker is quite correct in objecting to the way Goodman and Chisholm have construed semifactuals. But it does not follow, as we have said, that no adequate Humean account is possible.

Crucial to any account is the understanding of the distinction between asserting independence and denying connection, and in particular understanding that the former does *not* entail the latter. The relevant point can be brought out by considering a counterfactual that Stalnaker presents as an example that the possible-worlds analysis can handle but the Humean analysis supposedly cannot. We shall argue to the contrary that the Humean *can* adequately deal with the example, thereby showing that this argument, too, of Stalnaker's is a failure. And in showing how the Humean deals with it, we shall also exhibit how the Humean can hold that asserting that a connection obtains is compatible with asserting independence.

Stalnaker presents us with the following counterfactual:[27]

(*)     If the Chinese entered the Vietnam conflict, the United States will use nuclear weapons.

We are asked to suppose ourselves in a certain epistemic context, and then decide whether, in that context, we would decide that (*) was true or was false. The context is this: ". . . you firmly believe that the use of nuclear weapons by the United States in this war is inevitable because of the arrogance of power, the bellicosity of our president, rising pressure from congressional hawks, or other *domestic* causes. You have no opinion about future Chinese actions, but you do not think they will make much difference one way or another to nuclear escalation." In this context, we are told,

Clearly, you believe that . . . statement to be true even though you believe the antecedent and consequent to be logically and causally independent of each other. It seems that the presence of a "connection" is not a necessary condition for the truth of an if-then statement.[23]

On the other hand, of course, there is no reason why the statement should not be reckoned true if the possible-worlds analysis is accepted.

Now, in the context described one clearly law-asserts a causal regularity to the effect that

(t)         Whenever a nation is at war and its domestic conditions are so and so, then that nation will use nuclear weapons.

This regularity is clearly imperfect, and, when the conditions are actually specified to the best of our ability, still "gappy" and with "exceptions". We have noted that possibility of such complexities in Chapter I above, but they are irrelevant for present purposes, so we shall take the regularity Stalnaker clearly has in mind as exceptionless and unproblematic. This regularity is of the form

(t')        Whenever $F$ then $G$

This law-asserted regularity permits one to assert the *indicative reasoned prediction*

(**)        Since $Fa$, $Ga$

which is analyzed as the argument

> Whenver $F$ then $G$
>      $Fa$
> _____
> $\therefore Ga$

where the first premiss is law-asserted and is implicit, and the second premiss is also asserted. Now, if the generality (t) = (t') is true, then so is

(u)         Whenever a nation is at war and domestic conditions are so and so and some other nation enters the war, then that nation (the first) will use nuclear weapons

i.e.,

(u')        Whenever $F$ & $C$, then $G$

which is entailed by (t). But since this generality is law-assertible, and since '$Fa$' is *also* assertible, one can assert the *counterfactual conditional* (*)

> $Ca \rightarrow Ga$

which is analyzed as the argument

$$\text{Whenever } F \ \& \ C \text{ then } G$$
$$Fa$$
$$Ca$$
$$\overline{\phantom{xxxxxxxxxxxxxxxxxxxxxxxxxx}}$$
$$\therefore Ga$$

where the first *two* premisses are implicit, the first law-asserted and the second asserted, and the third premiss is not asserted. Hence, in the context Stalnaker supposes, since we can assert the indicative conditional (**), we can also assert the counterfactual conditional (*). Stalnaker is clearly correct in holding that in the context he describes we should assert (*). He is only wrong in holding that the Humean would disagree!

But if the Humean is committed to holding that there is a connection (u) holding between the antecedent and consequent of (*), he is *also* committed to holding that there is a law

(u″)      Whenever $F \ \& \ \sim C$, then $G$

*also* entailed by (t), which justifies asserting a second counterfactual conditional

(***)      $\sim Ca \rightarrow Ga$

And, of course, if we have both (*) and (***), then we also have

(†)      $Ga$

But, of course, we already have the reasoned prediction (**) that leads us to assert (†). Since we have the law (t) *and* the initial conditions *Fa*, we are in a position to assert (†). When we are, for that reason, in such a position, then we are *also* in a position to assert the counterfactual (*) [as Stalnaker points out] and not only that but *also* the counterfactual (***); and if we assert both those, or, what is the same, assert (†) independently of whether *a* is *C* or not, then can *also* assert the *semifactual*

($\overset{\dagger}{-}$)      Even if *Ca, Ga*

which states that the U.S. will use nuclear weapons *independently* of whether the Chinese will enter the war.

In short, suppose that the laws $\mathscr{L}$ and particular facts $\mathscr{F}$ permit a *reasoned prediction* of B. Then, for any A such that $\mathscr{L}$ and $\mathscr{F}$ do *not* entail ~A,[29] those laws entail a second law $\mathscr{L}'$ which, together with $\mathscr{F}$, justifies the assertion of the counterfactual

> A → B

But those same laws $\mathscr{L}$ also entail a third law $\mathscr{L}''$ which, together with $\mathscr{F}$, justifies asserting the counterfactual

> ~A → B

And to assert both these counterfactuals is to assert the independence of A from B, which is but another way of saying that the law $\mathscr{L}$ and facts $\mathscr{F}$ also justify asserting the semifactual

> Even if A, B

We see how the Humean can allow one to assert the independence of A and B, on the one hand, while allowing them, on the other hand, to be so connected as to justify asserting the counterfactual A → B.[30] We also see how the Humean can handle Stalnaker's proposed counter-example (*). In effect, Stalnaker makes the same error as Goodman and Chisholm, that of inferring that since A and B are independent there can be no connection between them sufficient to justify asserting the counterfactual

> A → B

We have seen that *in the epistemic context proposed* by Stalnaker, the Humean can agree with (*). We should also recognize that *in that context it would generally be odd to assert that counterfactual*. There is, of course, no conflict here, for, as Lewis has pointed out,[31] oddity is not falsity. Take the context of discourse to be defined by the proposed epistemic conditions and also by *the purpose of intelligent interference in a natural or social process*. In such a context, it is *misleading* to assert the counterfactual (s)

> A → B

since one can also assert the *stronger* claim that (s″)

> Even if A, B

If the context is that of *intelligent* interference, then one wants the *best* information available. Relative to the end of intelligent interference it is better to know (s″) than (s). Since the context demands the best information, it would less than candid and actually misleading if one asserted (s) rather than (s″). That is why it would be *odd* to assert the counterfactual when one could assert the corresponding semifactual. Indeed, perhaps it is this oddity that led Goodman to wrongly believe that to assert the semifactual is to deny the counterfactual.[32]

There are *two* respects in which the assertion of (s″) "Even if A, B" is *informationally better* than the assertion of (s) "A → B" *One.* (s″) gives more information about what particular facts are *actually occurring.* (s) makes a purely hypothetical assertion about the situation, while (s″) asserts categorically that B occurs. *Two.* The law (t′)

Whenever *F* then *G*

used to justify asserting (s″) is *less imperfect* than the law (u′)

Whenever *F* & *C* then *G*

used to justify asserting (s).[33] It is less imperfect because it has greater predictive power — and therefore (if our argument in Chapter I is successful) greater explanatory power. Since, other things being equal, it is better to act on the basis of less rather than more imperfect laws, it would be wrong in the context to give it to be understood that the best lawful knowledge one had was the more imperfect (u′). And, of course, one would misleadingly give it to be understood that one knew only the weaker (u′) if one asserted only the counterfactual (s) rather than the semifactual (s″).[34]

We may conclude that Stalnaker has not succeeded in presenting a counterexample to the the Humean position.

Let us now turn to another claim to superiority of the possible-worlds analysis, namely, the argument that it can account for the "fallacy of strengthening the antecedent" of counterfactuals, that is, that it can show why

$$\frac{A \rightarrow B}{\therefore (A \ \& \ C) \rightarrow B}$$

is invalid.[35] Clearly, our discussion of Stalnaker's proposed counter-example is relevant.

Consider the following two counterfactuals:[36]

(a)      If this match were struck, then it would light.

(b)      If this match had been soaked in water overnight *and* it were struck, then it would not light.

It is clear that we are (often) prepared to assert both of these. But if we are prepared to assert the second, then we are prepared to deny its contrary:

(c)      If this match had been soaked in water overnight *and* it were struck, then it would light

An inference from (a) to (c) is that of "strengthening the antecedent." We thus have a case where the premiss of this inference is asserted and its conclusion denied. It is therefore invalid, a fallacy.

Lewis' analysis of the situation is this. Assume that neither is vacuously true, that is, that there are accessible subspheres around the actual world in which the antecedents hold. Then Lewis says that if (a) is true that is because there is a subsphere S about our world such that: S contains at least one possible world in which the match (or, sorry, its counterpart) is struck, and the material conditional "This match is struck $\supset$ This match lights" holds throughout S. And if (b) is true there is a subsphere S′ such that: S′ contains a possible world in which the match is soaked overnight and struck, and the material conditional "This match is soaked in water overnight and is struck $\supset$ This match lights" holds throughout S′. For (a) and (b) both to be true, S′ must be a subsphere larger than, and therefore including, S; for, S′ has to contain possible worlds where the match is soaked, struck and lights, whereas S cannot contain any worlds in which the match is struck (soaked or not) and lights. But this is what we would expect: worlds where the match is soaked *and* struck are — supposing the match was never soaked nor struck — less like the actual world than are worlds where the match is struck but not soaked. So, we should, on Lewis' analysis, expect both (a) and (b) to be true, and therefore expect (a) to be true and (c) false. Clearly, then, on this analysis, it really is a fallacy to strengthen the antecedent.

Now, our discussion of Stalnaker's proposed counterexample shows that if we law-assert (t′)

Whenever $F$ then $G$

then we also can law-assert the more imperfect generalization (u')

Whenever $F$ & $C$ then $G$

It would seem, then, that wherever we can assert the counterfactual (**)

$Fa \rightarrow Ga$

then we can also assert

(****)    $(Fa$ & $Ca) \rightarrow Ga$

I.e., it would seem that, on the Humean analysis, strengthening the antecedent should *not* be a fallacy.

However, besides law-asserting (t'), we may also sometimes law-assert

(v')      Whenever $F$ then $\sim C$

This establishes that $F$ and $C$ are *not contenable*.[37] And in a context in which $F$ and $C$ are not contenable then *we cannot consistently assume the antecedent* of (****): where $F$ and $C$ are not contenable, the antecedent of (****) *cannot be assumed*, at least for contexts defined by the goal of intelligent interference, as we argued above. But if we cannot assume the antecedent of a conditional then we cannot assert the conditional itself, since the assertion of the latter involves the assumption of the former. It follows that, at least for discourse in contexts defined by the goal of intelligent interference, there are definite epistemic circumstances in which the Humean is prepared to assert (**) but not (****); that is, there are definite contexts in which the Humean is prepared to assert a counterfactual but not that counterfactual with a strengthened antecedent. It therefore turns out, after all, that Humean will not *in all cases* permit the inference from a counterfactual to that counterfactual with a strengthened antecedent. And if there are some cases where he will not permit that inference, then for him also it is a fallacy. Once again we find the supposed superiority of the possible-worlds analysis is illusory.

Bennett has proposed a variation on this theme as yet another challenge to the Humean. He points out that it could be true both that

(a')      If I walked on the ice, then it would remain firm

and also that

(b′)    If I walked on the ice and you walked on the ice, it would
        not remain firm

He then argues as follows: "One might try to argue that if (b′) is true,
then (a′) as it stands is not true, and that we accept it only because we
construe it as an ellipsis for 'If I *alone* walked on the ice, . . .' etc. But
we have to cope with the truth not just of (b′) but also of

(c′)    If I walked on the ice and I wore 60 lb. boots, it would not
        remain firm.

Must we then say that (a′) is really an ellipsis for 'If I alone walked on
the ice wearing normal footwear, . . .' etc.? But there are plenty more
where (b′) and (c′) came from, each requiring a further expansion of
(a′). Clearly, we shall get nowhere by trying to reconcile (a′) with (b′),
etc. through the plea that (a′) is an ellipsis."[38]
    Is there any way in which the Humean can accept that (a′) and (b′)
are both true, while not having to construe (a′) as an ellipsis for a
conditional with an indefinitely long antecedent? Bennett believes not,
and therefore reckons the possible-worlds analysis superior on this
account.[39] But in fact he is mistaken.
    There seems to be but *one* law involved in all of (a′), (b′) and (c′), to
wit,

($\lambda$)    For anything $x$, if $x$ is on a patch of ice, then that ice
        remains firm if and only if the weight of $x$ is no greater than
        $1.5 \times$ my weight.

This law itself may be deduced from other laws concerning the thickness of ice, and perhaps such other relevant factors as the history of
thawing and re-freezing that it has been through.[40] Note also that the
phrase 'my weight' is a *definite description* of a certain specific weight,
to wit, about 10 stone; it is the specific weight which is the relevant
factor, rather than the fact that it is *my* weight. The statement ($\lambda$)
makes this perfectly clearly, but some have been mistaken on this
simple point[41] so we had best be explicit about it.
    With the law ($\lambda$), the counterfactual (a′) can be analyzed in terms of

the argument

$(\lambda)$
My weight is less than 1.5 × my weight
I walk on this ice

∴ This ice remains firm

Similarly, (b′) can be analyzed in terms of

$(\lambda)$
My weight and yours together are not less than
1.5 × my weight
I and you walk on this ice

∴ This is does not remain firm.

(c′) receives a similar analysis. In short, the Humean can accept that both (a′) and (b′) are true without getting into the problem that Bennett raises.

Moreover, if (a′) and (b′) are both true, then (b′) denies the counterfactual derived from (a′) by strengthening to antecedent, to wit

If I and you walked on this ice, then it would remain firm.

And if (a′) is affirmed and the latter denied, then once again we have a case where the Humean does not permit the inference from a counterfactual to that counterfactual with a strengthened antecedent. Bennett's example shows that there are other grounds than that of non-contenable predicates for the Humean to reckon this inference a fallacy.

One is curious why Bennett overlooked the law $(\lambda)$. Perhaps it is because of the special role that the phrase 'my weight' plays in it. But we noted that that role is innocuous. Perhaps it is because $(\lambda)$, being imperfect, does not look to be a "real" law in the way that, say, Newton's laws are "real" laws. But imperfect laws are not the less laws for being imperfect. In any case, however, others besides Bennett have overlooked similar everyday law-asserted generalities, and when they have missed them they have thought themselves to have discovered counterexamples to the deductive-nomological model of explanation rather as Bennett has thought himself to have discovered a counterexample to the Humean analysis of counterfactuals.[42]

It is perhaps worth noting that, although with respect to *counter-*

*factuals* "strengthening the antecedent" is a fallacy, it is *not* so with respect to *laws*. That is, if

$(x_1)$      $(x)(Fx \supset Gx)$

can be law-asserted, then so can

$(x_2)$      $(x)[(Fx \ \& \ Cx) \supset Gx]$

which is the same law with a strengthened antecedent. Indeed, we used this relationship above[43] in order to explain why "A → B" follows trivially from "Even if ~A, B".

It might be objected that this inference, while valid for merely descriptive generalities, does not hold for laws. The following sort of example could be used to make this plausible. Consider the case where a pet dog regularly approaches its master when the latter comes into view. Let this regularity be described by $(x_1)$. But now let us add to the antecedent: let the master approach but in addition holding a whip; in that case the dog retreats. This process is to be described by

$(x_3)$      $(x)[(Fx \ \& \ Cx) \supset \ {\sim} Gx]$

rather than $(x_2)$. Given that there are things that are both *F* and *C*, $(x_3)$ and $(x_2)$ are contraries. Since $(x_3)$ is true, $(x_2)$ must be false. But $(x_2)$ is inferred from the true $(x_1)$ by "adding to the antecedent." False conclusions do not follow from true premisses, and so "adding to the antecedent" seems to be a fallacy here also.

That conclusion — which amounts to the claim that inductive connections can invalidate deductive connections! — would be hasty, however. For, it is not $(x_1)$ that describes what happens when the dog approaches its master, but rather

$(x_4)$      $(x)[(Fx \ \& \ {\sim} Cx) \supset Gx]$

or, what is the same,

Whenever *F* then *G* unless *C*.

At least, insofar as $(x_1)$ describes the process, it does so only imperfectly, omitting one crucial relevant variable. $(x_1)$ is, taken literally, false. So there is no question of fallaciously obtaining the false $(x_2)$ from the true $(x_1)$. $(x_1)$ is true only so far as it is recognized to be *conditioned*, that is, true conditionally upon the absence of *C*. And the point is, from $(x_4)$ we can validly deduce

$(x_5)$     $(x)[(Fx \mathbin{\&} \sim Cx \mathbin{\&} Cx) \supset Gx]$

by "adding to the antecedent." $(x_5)$ is true, so there is no problem of inferring a false conclusion from a true premiss.

On the other hand, we must also recognize that the antecedent of $(x_5)$ is *unrealizable*. It therefore cannot be used to support the assertion of counterfactuals in the context of discourse defined by the end of intelligent interference or potential interference in natural or social processes. But that fits with what we have just concluded about counterfactuals: for the latter, "adding to the antecedent" *is* a fallacy.

Let us now turn to another fallacy for which the possible-worlds analysis is supposed to the able to account while the Humean cannot: the "fallacy of transitivity."[44]

$$\begin{array}{c} A \to B \\ B \to C \\ \hline \therefore A \to C \end{array}$$

But it is easy enough to see that the invalidity of this is also easily accounted for by the Humean.

Bennett considers this example:[45]

> If there were snow on the valley-floor, I would be skiing along it
>
> If there were an avalanche just here, then there would be snow on the valley-floor
>
> If there were an avalanche, I would be skiing on the valley floor.

The first two can, clearly, be true, while the last is false. This is accounted for by Lewis if there are possible worlds in which the snow falls gently on the valley-floor (and I ski on it) which are more like our actual world than any possible world in which snow is deposited on the valley-floor by an avalanche. But the Humean can provide an adequate analysis also. Involved in the assertion of the first counterfactual will be the implicit assertion of a law to the effect that

$(m_1)$     Whenever there is snow in the valley, then I ski on it just in case there is no danger.

together with another implicit assertion that there is no danger. The exceptive clause "there is no danger" plays a role similar to that of the exceptive clause in (λ) [p. 210]. Failing to note the possibility of such an exceptive clause, Bennett again fails to see how the Humean can provide an analysis without an indefinitely expanded antecedent.[46] For the second counterfactual, the law is

($m_2$)    Whenever there is an avalanche in a valley, there is snow on the valley-floor

Besides these laws, we also have the law

($m_3$)    Whenever there is an avalanche, there is danger.

In the context of a law-assertion of ($m_3$), the law ($m_1$) justifies asserting the counterfactual

If there were an avalanche, then I would not be skiing on the valley-floor

But to assert this counterfactual is to deny the third of Bennett's counterfactuals. Thus, the Humean can, after all, explain why we should on occasion assert two counterfactuals A → B and B → C and yet deny A → C; for the Humean, too, transitivity is a fallacy

Lewis uses the following examples:[47]

If Otto had gone to the party, then Anna would have gone.

If Anna had gone, then Waldo would would have gone.

If Otto had gone, then Waldo would have gone.

As he explains it: "The fact is that Otto is Waldo's successful rival for Anna's affections. Waldo still tags around after Anna, but never runs the risk of meeting Otto. Otto was locked up at the time of the party, so that his going to it is a far-fetched supposition; but Anna almost did go. Then the premises are true and the conclusion false."[48] Lewis actually gives us here the generalizations upon which the Humean will rely to give an analysis of the three counterfactuals. Represent the three counterfactuals by

O → A
A → W
O → W

Then the assertion of the first can be analyzed as based on the assertion of a law to this effect:

Whenever O then A

The assertion of the second can be analyzed as based on the assertion of a law and a statement of individual fact:

Whenever A then W if and only if $\sim$O
$\sim$O

And the two laws justify the assertion of

$$O \rightarrow \sim W$$

which is the denial of the third of Lewis' counterfactuals. Once again the Humean has no trouble accounting for the facts that allegedly tell in favour of the possible-worlds analysis.

The final fallacy for which the possible-worlds analysis supposedly can account where the Humean cannot is that of contraposition,[49] the two invalid forms

$$\frac{A \rightarrow B}{\therefore \sim B \rightarrow \sim A} \qquad \frac{\sim B \rightarrow \sim A}{\therefore A \rightarrow B}$$

The example that Lewis gives is this:[50]

$$\frac{\text{If Boris had gone to the party, Olga will still have gone}}{\therefore \text{If Olga had not gone, Boris would still not have gone}}$$

He argues that this could have a true premiss and a false conclusion in the following circumstance: "Suppose that Boris wanted to go, but stayed away solely in order to avoid Olga, so the conclusion is false; but Olga would have gone all the more willingly if Boris had been there, so the premiss is true."[51] We are intended to take the premiss as a counterfactual rather than (what it sounds like) a semifactual asserting that the consequent is independent of the antecedent. Let us symbolize it as

(n$_1$)     $B \rightarrow O$

and the conclusion as

(n$_2$)     $\sim O \rightarrow \sim B$

It seems that Boris' presence is sufficient to attract Olga to a party. Thus, the assertion of $(n_1)$ can be analyzed as based on the assertion of a law to this effect:

$(n_3)$     Whenever B then O

This law, clearly, will also justify asserting the conclusion $(n_2)$. But Lewis tells us that it is another rule that Boris goes to parties if and only if Olga is absent. This is regularity is a law-assertible generalization to this effect:

$(n_4)$     Whenever ~O, and only then, B

This law will justify the assertion of

$(n_5)$     ~O → B

which is the denial of the conclusion $(n_2)$. thus, $(n_2)$ may or may not be inferred from $(n_1)$, depending upon which law is implicitly used to justify its assertion. If the law $(n_3)$ is implicit in both $(n_1)$ and $(n_2)$, then the assertion of the conclusion $(n_2)$ is justified if and only if the assertion of the premiss is justified. In this case, the inference from $(n_1)$ to $(n_2)$ *is* valid. Moreover, if we use $(n_4)$ to justify the denial of $(n_2)$, then we must also deny $(n_1)$, since the laws $(n_3)$ and $(n_4)$ are contraries. So in this case again the assertion of the premiss is justified if and only if the assertion of the conclusion is justified. Lewis construes the move from $(n_1)$ to $(n_2)$ as a move from a truth to a falsehood because he uses one law with respect to the premiss and a contrary law with respect to the conclusion. But this, clearly, is illegitimate. It is like arguing that, although *red* and *green* are contraries, since the inference

> *x* is red
> ―――――――
> ∴ *x* is not green

is valid, *tall* and *short* are not contraries, since the inference

> *x* is tall
> ―――――――
> ∴ *x* is not short

is shown to be invalid by its instance

> John is tall
> ―――――――
> John is not short

which has a true premiss — John is tall relative to Tom Thumb — but a false conclusion — John is short relative to Wilt Chamberlain. One must distinguish the sentence from what the sentence is used to assert. The latter varies from context to context since the *standard* of tallness appealed to is only implicit in each context. Once the standard is made explicit, and is the same for premiss and conclusion then *tall* and *short* *are* contraries. Similarly, in evaluating the validity of the inference from ($n_1$) to ($n_2$), one must make the relevant laws explicit, and when this is done, it becomes evident that the Humean has no reason for holding that Lewis' inference is invalid.

If Lewis' examples do not yield an instance of contraposition with a true premiss and a false conclusion, there are others that do. Stalnaker gives the following:

> If the U.S. halts the bombing, then North Vietnam will not agree to negotiate.
>
> If North Vietnam agrees to negotiate, then the U.S. will not have halted the bombing.

About these we are told that "A person would believe that [the first] statement is true if he thought that the North Vietnamese were determined to press for a complete withdrawal of U.S. troops. But he would surely deny [the second which is its] contrapositive. ... He would believe that a halt in the bombing, and much more, is required to bring the North Vietnamese to the negotiating table."[52] Let us symbolize the first by

($o_1$)     $H \rightarrow \sim N$

and the second by

($o_2$)     $N \rightarrow \sim H$

On the basis of knowledge about groups strongly determined to achieve various ends, it is evident we can law-assert an imperfect regularity to this effect:

($o_3$)     Whenever H then $\sim N$

The imperfection raises problems, but they are not relevant at this point so we may safely leave them for later discussion.[53] Given the law ($o_3$), we are justified in asserting $(o_1)$. If we accept that contraposition is valid, then we are also justified in asserting $(o_2)$

$$N \rightarrow \sim H.$$

With respect to the second of Stalnaker's counterfactuals, it is clear that we are meant to law-assert that in the context, something like halting the bombing is a necessary but not a sufficient condition for negotiations to occur, i.e., a law to this effect:

$(o_4)$       Whenever N then H

This justifies the assertion of the counterfactual

$(o_5)$      $N \rightarrow H$

which denies $(o_2)$. Thus, if we admit contraposition as valid, then we have a case where we must both affirm and deny $(o_2)$. And it would seem that the Humean must admit contraposition in this case, since if the law $(o_3)$ justifies asserting $(o_1)$ by the argument

$$\frac{\text{Whenever H then } \sim N}{\therefore \sim N}$$
$$\text{H}$$

then it equally well justifies asserting $(o_2)$ by the equally valid argument

$$\frac{\text{Whenever H then } \sim N}{\therefore \sim H}$$
$$\text{N}$$

It is evident that Stalnaker's example pushes the Humean harder than does that of Lewis.

However, if we law-assert both $(o_3)$ and $(o_4)$, then, via the inference pattern

$$\frac{\begin{array}{c} H \supset \sim N \\ N \supset H \end{array}}{\therefore \sim N}$$

we are committed to the nomic necessity of

$(o_6)$       $\sim N$

This means that, *given the generalizations we law-assert*, we *cannot consistently* assume the antecedent of ($o_2$). It follows that ($o_2$) cannot justifiably be asserted; it must be denied, at least in discourse in contexts defined by the goal of intelligent interference.

Once again, therefore, the Humean can account also as easily as does the possible-world analysis for a feature of the logic of counter-factuals that was supposed to argue for the latter and against the former.

Thus far we have replied to proposed counter-examples; it is time now to propose some of our own, to attempt to show that the law-deduction account of counterfactuals of the Humean in fact squares better than the possible-world account with ordinary usage. While this cannot, as we have insisted, be conclusive it nonetheless has a certain force, given the not unreasonable assumption that many ordinary counterfactuals are justifiably asserted in the context of intelligent interference.

Let us represent the counterfactual conditional

> If this were *F* then it would be *G*

by, as usual,

(1)      $Fa \rightarrow Ga$

where ' $\rightarrow$ ' represents "counterfactual implication".

Disposition predicates are defined in terms of counterfactual implication. Thus, for example,

(2)      *a* is soluble

is defined as

(3)      *a* is in water $\rightarrow$ *a* dissolves

"if *a* were in water then it would dissolve". Similarly, its contrary

(4)      *a* is insoluble

is defined as

(5)      *a* is in water $\rightarrow$ $\sim$(*a* dissolves)

"Solubility" and "insolubility" are contraries because, while at most one can be true of an object, both might be false. Thus, a lump of sugar is soluble and therefore not insoluble, while a match is insoluble and therefore not soluble. But the university, like the number four, is neither soluble nor insoluble: it is not the sort of thing that could be put in a sample of water, and could therefore be neither soluble nor insoluble.

Thus, (2) may be denied for two reasons, either because $a$ is insoluble or because $a$ is not the sort of thing that is either soluble or insoluble. Since (2) unpacks into the counterfactual conditional (3), the latter too may be denied on the basis of the same two reasons.

If we take (1) as representing counterfactual conditionals like (3) or (5), then it can be denied either because its contrary

(6)      $Fa \rightarrow \sim Ga$

is asserted, or because $a$ is of some sort $K$ such that it is a law that

(7)      $(x)(Kx \supset \sim Fx)$

Thus, (1) may be denied either because its contrary is true or because there is a law that establishes that neither it not its contrary is true. If (1) is denied because its contrary is true, then

(8)      $\sim(Fa \rightarrow Ga)$

is asserted, while if (1) is denied because there is a law establishing that neither it not its contrary are true, then both (8) and also

(9)      $\sim(Fa \rightarrow \sim Ga)$

are asserted.

Many attempt to analyze the semantics of counterfactuals like (1) in terms of "possible worlds". A possible world is an "$Fa$-world" just in case '$Fa$' holds in it. Then it is often held that a counterfactual like (1) is true if and only if:

(*)      '$Ga$' is true in the $Fa$-world that is most similar to the actual world.

But if (1) is to represent (3), then this is inadequate. For, in the $Fa$-world most similar to ours, either '$Ga$' is true in it or '$\sim Ga$' is. So, in that possible world, $a$ either dissolves or it does not. Hence, either (2) is true or (5) is true. That is, $a$ is either soluble or insoluble (though not both). However, we suggested that there are things which are neither soluble nor insoluble. Since the semantics of counterfactuals based on (*) does not allow for this, that semantics is not adequate for counterfactuals like (2) or (5) that unpack disposition concepts.

Some have proposed a possible world semantics for counterfactuals different from that given by (*) and which avoids the problem just raised for (*). Thus, as we know, Lewis proposes to analyze counterfactuals like (1) not as *strict* conditionals, as (*) proposes, but as *variably strict* conditionals. For Lewis, possible worlds are ordered into a series of concentric spheres by an accessibility relation of interworld similarity. A sphere S of possible worlds about the actual world contains those possible worlds that are similar to the actual world to at least a certain degree. Thus, the smaller the sphere, the more accessible it is from our worlds, or, what is the same, the more similar are the worlds in it to the actual world. A similarity relation will define a set $ of spheres about the actual world. Then, according to Lewis,[53a] a counterfactual like (1) "*Fa* → *Ga*" is true if and only if

> (**)        either (a) no *Fa*-world belongs to any sphere S in $, or (b) some sphere S in $ contains at least one *Fa*-world and in every such world '*Ga*' is true.

In case (a), where '*Fa*' is true in no possible world or at least in no sphere in the set $, the counterfactual is said to be "vacuously true", and the antecedent '*Fa*' is said to be "not entertainable" in our world as a counterfactual assumption. In case (b), '*Fa*' is an entertainable assumption, and within some sphere around our world that is large enough to reach at least one *Fa*-world — call such a sphere "*Fa*-permitting" — '*Ga*' is true at all *Fa*-worlds. As Lewis puts it, "a counterfactual is vacuously true if there is no antecedent-permitting sphere, non-vacuously true if there is some antecedent-permitting sphere in which the consequent holds at every antecedent would, and false otherwise".

Now consider our example in which neither (1) "*Fa* → *Ga*" nor (6) "*Fa* → ~*Ga*" is true. On the semantics (**), if either (1) or (6) is vacuously true, then both are. So, if neither are true, it must be the non-vacuous case. In the non-vacuous case, we have an *Fa*-permitting sphere. If '*Ga*' is true in every such *Fa*-world, then (1) is true, while if '*Ga*' is false in every such *Fa*-world, then (6) is true. So neither is true only in case that in an *Fa*-permitting sphere, for some *Fa*-worlds '*Ga*' is

true and for some *Fa*-worlds '*Ga*' is false. Thus, on the semantics (**),
the truth-conditions for the special case in which both (8) and (9)

$$\sim(Fa \rightarrow Ga)$$
$$\sim(Fa \rightarrow \sim Ga)$$

are true are that in some antecedent-permitting sphere, the consequent
'*Ga*' holds in some antecedent-worlds, and does not hold at others.

However, in any *Fa*-world, either the law (6) is false or else *a* is not
of kind *K*, nor of any other kind that entails that it is not *F*. But in our
example the ground for asserting that both (8) and (9) are true seems to
be precisely the truth in our world of such a law and such a kind
attribution. What occurs in a world in which either such a law or such a
kind-attribution does not obtain does not seem to be relevant.

The existence of such a kind and such a law entails that the antecedent of (1) and (6) is *not realizable* in our world. The existence of such
a law and such a kind can become relevant to the truth-conditions of
the counterfactuals if "not realizable" is equated with Lewis' "not
entertainable". Nor is this implausible. While it is unproblmatic that it is
possible that a match is soluble, it is somewhere "beyond the realm of
possibility" — if possible, it is an inaccessible possibility — that either
the university or the number four is soluble. If we identify "not realizable" with "not entertainable", then the kind and the law define the
limits of world-similarly: any world that violates them is either not a
possible world or, if it is, then it lies quite outside the system $ of
spheres of similarity; such a world is simply too dissimilar to our world
to be included in such a system.

In that case, however, both (1) "*Fa* $\rightarrow$ *Ga*" and (6) "*Fa* $\rightarrow$ $\sim$ *Ga*" will
be vacuously *true* rather than false. So, on the semantics (**), the law
and kind-attribution that provide the grounds for denying both (1) and
(6) can be made grounds determining the truth-conditions of such
counterfactuals only at the cost of making them grounds, not for
denying, but for asserting both (1) and (6).

It seems, then, that the possible world semantics has difficulty
accounting for the denial conditions of the counterfactuals that unpack
disposition concepts.

In contrast, the law-deduction account of counterfactuals proposed
by the Humean has no similar difficulties on these points.

According to this analysis, a counterfactual (1) "$Fa \rightarrow Ga$" can be asserted just in case that there is kind $K$ such that

$$Ka$$

and a law

$(\alpha) \qquad (x)[Kx \supset (Fx \supset Gx)]$

so that, in the context of this background knowledge, if the antecedent

$$Fa$$

of the counterfactual is *assumed* then the consequent

$$Ga$$

*follows deductively*. Upon this account, a counterfactual is, as we have been arguing, not so much a single sentence as a collapsed argument.

Since a counterfactual is not a single sentence, it is not true nor false in any literal way. Upon this account, to say that a counterfactual is true is to say no more than that it can justifiably be asserted. To say that it is false, is simply to deny that it can justifiably be asserted. Similarly, to negate a counterfactual is not to attach a connective to a single sentence. Negating a counterfactual is once again but a way of denying that it can be justifiably asserted.

Since there are on this account several assertion conditions for a counterfactual, there are several grounds for denying a counterfactual.

*One.* We may deny a counterfactual like (1) on grounds that its contrary is true. If we assert (6) "$Fa \rightarrow \sim Ga$" then we have a collapsed argument

$(\beta) \qquad (x)[Ha \supset (Fa \supset \sim Ga)]$
$\qquad\qquad Ha$
$\qquad\qquad Fa$

$$\overline{\qquad\qquad\qquad\qquad\qquad}$$

$\qquad$ So, $\sim Ga$

in which the first two premisses are asserted within our background knowledge while the third premiss is merely assumed. Assuming that there are $F$'s, the acceptability of this deduction precludes the acceptability of a deduction from '$Fa$' of '$Ga$'. For, the positive instances that confirm ($\beta$) would falsify any law like ($\alpha$) that could justify deducing

'*Ga*' from '*Fa*'. Thus, to deny a counterfactual on grounds that its contrary is true is to deny that there is a law that could sustain its assertion.

*Two.* We may deny a counterfactual like (1) "*Fa* → *Ga*" on grounds that *a* is of a kind that excludes its being *F*. In that case, it is part of our background knowledge that there is a kind *K* such that

$$Ka$$

and a law (7)

$$(x)(Kx \supset \sim Fx)$$

But this entails, again as part of our background knowledge, that

$$\sim Fa$$

which means that in the context determined by this background knowledge one cannot consistently assume the antecedent of the counterfactual — nor of its contrary. Thus, to deny a counterfactual on grounds that the individual that it mentions is not the sort of thing for which the antecedent could hold is to assert that background knowledge entails that the antecedent cannot consistently be assumed.

In short, the "collapsed argument" analysis of counterfactuals of the Humean, in contrast to the analysis in terms of a possible world semantics, can account quite nicely for the two ways of denying the counterfactuals that unpack dispositions. This argues that the "collapsed argument" analysis fits more accurately the ordinary uses of counterfactuals than does the alternative analysis.


## (ii)  *Causal Discourse*

Another argument that has been proposed as telling in favour of Lewis' analysis of counterfactuals and against the Humean or any "regularity" account of counterfactuals, is that the former yields a more adequate account of causation than the latter.[54] Specifically, of course, it is said to be an advantage to any analysis that it allows there to be causation even where there are no regularities;[55] or no known regularities.[56] Lewis' allows this, and the Humean's doesn't. So the former is to be preferred. As I have argued elsewhere, claims that there are causes where there are no regularities and that we can know causes without knowing regularities, are unsound.[56a] The point of the present remarks is to show that Lewis' analysis can be reckoned virtuous in these respects only if these more general claims are sound. The issues in this context have not so much to do with Lewis' analysis itself as with these more general claims. Our final response, then, to Lewis can be found only in these other discussions.

As we saw above, Lewis begins with the idea of *counterfactual dependence*.[57] If $A_1$, $A_2$, ... , and $C_1$, $C_2$, ... are two families of propositions in each of which no two are compossible, then the C's depend counterfactually on the A's just in case each counterfactual $A_i \rightarrow C_i$ is true. If the $A_i$ and the $C_i$ are propositions about particular events, then the events $C_i$ *depend causally* on the events $A_i$ just in case the C's counterfactually depend on the A's.

Lewis also defines causality for single events rather than families.[58] If C and E are two distinct events, then E *depends causally* on C if and only if the family $\{C, \sim C\}$ depends counterfactually on the family $\{E, \sim E\}$, that is, just in case that whether E occurs or not depends on whether C occurs or not. In this case the dependence is a matter of two counterfactuals

$$C \rightarrow E$$
$$\sim C \rightarrow \sim E$$

being true. And there are two cases. *One.* If C and E do not occur, then, upon Lewis' analysis of counterfactuals, the second of the two counterfactuals is automatically true, and E depends causally on C if and only if $C \rightarrow E$ is true, i.e., if and only if E would have occurred if C had occurred. *Two.* If C and E do occur, the first of the two counterfactuals is automatically true, and E depends causally on C if and only if $\sim C \rightarrow \sim E$ is true, i.e., if and only if, if C had not been then E would not have been.

*Causation* is not the same as causal dependence.[59] Causal dependence implies causation but not conversely. For, causation is transitive but causal dependence need not be. To see this, note first that causation is a relation among two events C and E only if C and E are actual. If C and E are actual, and E depends causally on C, then C causes E. That is, C causes E if C and E are actual and if $\sim C \rightarrow \sim E$ is true. But if those conditions are sufficient for causation, they are not necessary. Consider three actual events A, B, and C such that C depends causally on B ($\sim B \rightarrow \sim C$ is true), and B depends causally on A ($\sim A \rightarrow \sim B$ is true). In this case it is clear that A causes C. But, since $\rightarrow$ is not transitive the two relations of causal dependence do not guarantee that $\sim A \rightarrow \sim C$ is true. Thus, we can have it true that C causes E even if C is not causally dependent on A, that is, even if $\sim[\sim A \rightarrow \sim C]$. Causation and causal dependence coincide for two *contiguous* events, but for sequences they do not. Nonetheless, causal dependence is what is crucial for defining causation. A *causal chain* is a sequence of actual events A, B, C, ... such that B depends causally on A, C on B, etc. Lewis then says that C causes E if and only if there is a causal chain leading from C to E.

E is *nomically dependent* on C just in case that there are laws $\mathcal{L}$ and particular facts $\mathcal{F}$ such that $\mathcal{L}$ and $\mathcal{F}$ entail C $\supset$ E but $\mathcal{F}$ alone does not. As we saw, Lewis' account has the claimed desirable consequence that E may be causally dependent upon C and be caused by C even if E is not nomically dependent upon C.[60] Now, in Section (IV) above, we have already challenged the claim that there can, in the contexts we are considering, be counterfactuals which involve no laws. Given that argument, the separation of causal dependence from nomic dependence cannot be achieved. Nonetheless, Lewis gives independent arguments for the separation, and it is these we must consider.

Lewis provides a sense in which nomic dependence is, in certain cases, *reversible*.[61] The special cases involve nomic dependence between one family $A_1$, $A_2$, ... , of non-composible events, and a second family $C_1$, $C_2$, ..., of non-composible events. Each $C_i$ depends nomically on the corresponding $A_i$ by virtue of $\mathcal{L}$ and $\mathcal{F}$. Now let A be the disjunction $A_1 \vee A_2 \vee$ ... ; then, to say that an event A occurs is to give a *determinable* description of that event, asserting that *there is* one of the *determinate* descriptions $A_1$, $A_2$, ..., which holds of it (but not more than one, since no two of the family are composible), without however asserting specifically which one of these determinate descriptions applies. We can now form a further family A & $C_1$, A & $C_2$, ... It is easy enough to show that, given that no two of the $C_i$ are composible, then no two of the A & $C_i$ are composible.[62] It is also easy enough to show that if $\mathcal{L}$ and $\mathcal{F}$ entail, for each $i$, $A_i \supset C_i$, and no two $A_i$ are composible and no two $C_i$ are composible, then $\mathcal{L}$ and $\mathcal{F}$ entail, for each $i$, (A & $C_i$) $\supset A_i$.[63] Since $A_i$ and (A & $C_i$) form two families of non-composible events, and since $\mathcal{L}$ and $\mathcal{F}$ entail, for each $i$, (A & $C_i$) $\supset A_i$, it follows that the $A_i$ are nomically dependent on the (A & $C_i$) by virtue of the very same facts $\mathcal{L}$ and $\mathcal{F}$ that render the $C_i$ nomically dependent on the $A_i$. In this sense, nomic dependence is *reversible*.

But if nomic dependence is thus reversible, Lewis' counterfactual dependence is not.

Make the assumptions of counterfactual dependence: $A_i$, $A_2$ ... and $C_1$, $C_2$, ... are two families of non-compossible events; and $A_i \rightarrow C_i$ is true for each $i$. We ask: if the $C_i$ depend counterfactually on the $A_i$, do the $A_i$ depend counterfactually on the $(A \& C_i)$? that is, do we have $(A \& C_i) \rightarrow A_i$ true for each $i$? Now, no two $(A \& C_i)$ are compossible. So our question is this: given that no two $A_i$ are compossible and no two $(A \& C_i)$ are compossible, does this *guarantee* that $A_i$ holds in the nearest $(A \& C_i)$-world or is $(A \& C_i)$ compatible with $\sim A_i$? The answer is that $(A \& C_i)$ is consistent with $\sim A_i$.[64] Since they are consistent it may well turn out that the interworld similarity test will judge as the nearest $(A \& C_i)$-world a possible world in which $A_i$ does *not* hold. In that case $(A \& C_i) \rightarrow A_i$ will be false and $A_i$ will not be counterfactually dependent on $(A \& C_i)$. Thus, counterfactual dependence is not reversible in the way nomic dependence is. As a model of a case where counterfactual dependence fails, Lewis gives this:[65] @ is our actual world, the dots are other worlds, and distance on the page represents similarity "distance."

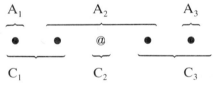

The counterfactuals $A_1 \rightarrow C_1$, $A_2 \rightarrow C_2$ and $A_3 \rightarrow C_3$ are true in the actual world, so the $C_i$ depend on the $A_i$. But we do not have the reverse dependence of the $A_i$ on the $(A \& C_i)$. For, the reverse dependence obtains only if $(A \& C_1) \rightarrow A_1$, $(A \& C_2) \rightarrow A_2$ and $(A \& C_3) \rightarrow A_3$ all hold, but we have, instead of the latter two, $(A \& C_2) \rightarrow A_1$ and $(A \& C_3) \rightarrow A_1$. In fact, given the non-compossibility of the $A_i$, these latter entail the denials of the counterfactuals that would have to be true if the reverse counterfactual dependence were to hold.

Lewis has thus got himself a distinction. However, it is one thing to have a distinction, and quite another for that distinction to explicate a difference in the logic of counterfactuals and causation. In order to show that it does this, Lewis presents an example the details of which, he argues, his analysis enables us to understand.

The barometer reading depends counterfactually on the pressure — that is as clear-cut as counterfactuals ever get — but does the pressure depend counterfactually on the reading? If the reading had been higher, would the pressure have been higher? Or would the barometer have been malfunctioning? The second sounds better: a higher reading would have been an incorrect reading. To be sure, there are actual laws and circumstances that imply and explain the actual accuracy of the barometer, but these are no more sacred than the actual laws and circumstances that explain the actual pressure. Less sacred, in fact. When something must give way to permit a higher reading, we find it less of a departure from actuality to hold the pressure fixed and sacrifice the accuracy, rather than vice versa. It is not hard to see why. The barometer, being more localized and more delicate than the weather, is more vulnerable to slight departures from actuality.[66]

Now, there *is* an asymmetry in the example. If Lewis has correctly located it, then his analysis does account for it. It is not entirely clear, however, whether Lewis has in fact correctly located it.

We have a family $h$ of barometer heights $h_i$ which form a set in which no two are compossible, and a family $p$ of atmospheric pressures $p_i$ which also form a set in which no two are compossible. We have a law that establishes a correlation between these:

$\mathcal{L}$        When a barometer is working, and only then, the $h$ that is present = the $p$ that is present.

(This is not tautological, since what it is for a barometer to be in good working order can be defined in terms of its own mechanism, and not merely as any state such that $h = p$.) We also have the assumption of individual fact

$\mathcal{F}$        This barometer is working

It is evident that the $h_i$ are nomically dependent on the $p_i$. If $P$ is the determinable characteristic that the situation has some pressure or other, i.e., if $P = p_1 \lor p_2 \lor \ldots$, then $(P \& h_i)$ is a situation in which a barometer with height $h_i$ exists where an atmospheric pressure is being exerted on it. Then, as Lewis says, the $p_i$ are nomically dependent on the $(P \& h_i)$. However, Lewis also says, though the $p_i$ are nomically dependent on the $(P \& h_i)$, it does not follow that the $p_i$ are counterfactually dependent on $(P \& h_i)$. That is, we may have a case where the nearest $(P \& h_i)$-world is one where $p_i$ does not hold. In such a possible world, one or both of $\mathscr{L}$ and $\mathscr{F}$ will be false. However, the law can still hold in *our* world, thereby giving the nomic dependence in the absence of counterfactual dependence.

Lewis makes his point by denying that, where $\mathscr{L}$ and $\mathscr{F}$ are true, the counterfactual

(c)     If this barometer were $(P \& h_i)$, then the pressure would be $p_i$.

must also be true. Because this counterfactual could be false, nomic dependence does not guarantee counterfactual dependence. Lewis' analysis explains why (c) is false — *if it is*. But, of course, if it is not then that tells against, if not Lewis' analysis of counterfactuals, then at least his analysis of counterfactual dependence, and his claim to have captured the asymmetry clearly present in his example. And one must, I think, *flatly assert* [67] that the counterfactual (c) is *true*: if I walked into the room, and the height of the barometer were to be higher than it actually is, then, given $\mathscr{L}$ and $\mathscr{F}$, that *would* imply that the pressure is higher than it is now.[68]

Lewis' reply to this is that if the height were greater than it is, then that would imply that the barometer is broken. But, of course, it wouldn't imply that at all. It would imply that only if the pressure were not also different from what it is.

It *is* true that if the height were now, in an instant, to *change* from $h_i$ to a quite distinct value $h_j$, then it would be reasonable to explain this *change* by a failure in the barometer. For, *normally*, atmospheric pressure does *not* change abruptly, in an instant, from $p_i$ to a quite distinct value $p_j$. Normally, then, an abrupt change in $h$ is not accompanied by an abrupt change in $p$. That is, normally, when there is a change from $h_i$ to $h_j$, there is not also a change from $p_i$ to $p_j$, so that the upshot is a situation in which the $h$ that is present $\neq$ the $p$ that is present. $\mathscr{L}$ now permits us to infer that $\mathscr{F}$ is false, that the barometer is not in working order.

We have this argument

> Atmospheric pressure does not change in an instant.
> The height in this barometer changes in an instant
> _____
> $\therefore$ This barometer is broken

The first premiss is law-asserted, and, since the normal happens more

often than not, we are normally justified in asserting the second premiss. This justifies asserting the counterfactual

(c′)     If the height in this barometer were to change in an instant,
        then the barometer would be broken.

This counterfactual must be distinguished from

(c″)     If the height in this barometer were different from what it is,
        then the pressure would be different from what it is.

and from

(c‴)     If the height in this barometer were different from what it is,
        then the barometer would be broken.

Lewis denies (c″), or, what amounts to the same, (c), because he asserts
(c‴). This is reasonable, because the consequent of (c‴) denies one of
the implicit assumptions that must be asserted if (c″) is asserted. The
difficulty is elsewhere: Lewis has no grounds for asserting (c‴). It
would seem likely that Lewis illegitimately mistakes (c‴) for (c′). Given
that (c′) is justified, such a mistake would account for why Lewis thinks
he can assert (c‴).

Lewis does advance as a reason for asserting (c‴)[69] the further
counterfactual

(c*)     If the reading of this barometer were other than it is, then it
        would be incorrect.

This counterfactual is justified by the definition

A barometer reads correctly if and only if $h = p$

and the assumption of particular fact

(a)      Atmospheric pressure is $p_i$

Together, these justify

For any $j$, such that $h_j \neq h_i$, if this barometer reads $h_j$, then
it reads incorrectly.

Assuming further that

This barometer reads $h_i$

we obtain (c*). Thus, (c*) is justified only under the assumption (a) that the pressure is not other than it is. In this context we assert the second premiss of this argument:

$\mathscr{L}$

(a)
The height in this barometer $\neq h_i$

∴ This barometer is broken

And this argument justifies asserting (c‴).

However, Lewis cannot use (c*) as a reason for asserting (c‴) and denying (c″), or, what is the same, (c). For, asserting (c*) or (c‴) requires, given $\mathscr{L}$, the assertion of (a), and the assertion of (a), in the context of assuming the antecedent of (c‴), implies the denial of the consequent of the latter, the therefore the denial of (c‴) itself. (c*) makes no other assumption than that the atmospheric pressure is not other than it is while the barometer reading is as it is. It is therefore the same assertion as (c‴) and not something independent that provides grounds for asserting (c‴). If we are prepared flatly to assert that (c) is true, then that is flatly to deny that (c‴) is true, and equally flatly to deny that (c*) is true.

Given that we law-assert $\mathscr{L}$, and assume that the barometer reading is other than it is, then we must give up *either* the assumption (a) that the atmospheric pressure is other than it is *or* the assumption $\mathscr{F}$ that the barometer is in good working order. In denying (c), Lewis is in effect saying that we must *always* give up $\mathscr{F}$. And when we announce that we are prepared flatly to deny (c‴) and flatly to assert (c), then we are saying that it is at least sometimes correct to assert $\mathscr{F}$ and allow that the pressure is other than it is. Moreover, it *is* sometimes correct to assert that the barometer is in working order and therefore that if it read other than it does, the atmospheric pressure would be other than it is. So we are right and Lewis is wrong.

Lewis' grounds for always holding the atmospheric pressure constant consists in the claim that a world with the same atmospheric pressure as ours but with a broken barometer will always be more similar to ours than a world with a working barometer and different atmospheric pressure. Perhaps so. But he cannot cite this as a reason for affirming (c‴) and denying (c). For the issue is whether his analysis of counterfactuals makes sense of the barometer case. The decision about (c) must be made independently of the analysis. It would seem that he *is* wrong about (c). The correct conclusion is that his analysis, with its appeal to interworld similarities, is mistaken. On the other hand, the Humean analysis does explain why we are sometimes justified in asserting (c). This counts in its favour.

But the Humean must account for the asymmetry that appears in the barometer example. He can do this by noting that $\mathscr{L}$, and the lawful knowledge we have of changes in atmospheric pressure justify asserting (c′)

> If the height in this barometer were to change in an instant, then the barometer would be broken

and denying

> If the height in this barometer were to change in an instant, then the atmospheric pressure would have changed.

This means that we *cannot* effect a change in atmospheric pressure by altering the height of the column of mercury in the barometer. On the other hand, any change in atmospheric pressure will effect a change in the height of the barometer. Thus, from the viewpoint of intelligent interference, the laws that govern the behaviour of atmospheres and barometers determine that there is an asymmetry in the situation: a change in pressure can effect a change in the barometer reading, but a change in the reading will not effect a change in the pressure. As we say, although pressure and reading are *correlated* with each other, where correlation is always symmetrical, *causation* in contrast is asymmetrical with the pressure causing the reading but not conversely. Much must be said on this topic,[70] but these remarks and those of Chapter I, Sec. I I think suffice to show that the Humean can account for the presence of an asymmetricality between pressure and reading in the barometer case, in spite of the nomic dependence of the one on the other being symmetrical, or, as Lewis says, reversible.

The barometer example is similar to another issue that Lewis discusses and claims that his account of causation can handle but which cannot be handled by any account that analyzes the causal relation in terms of laws. I refer to the problem Lewis' solution to which we discussed above, namely, the problem of effects.[71]

Assume that C causes E but that E does not cause C. Assume further that the laws $\mathscr{L}$ and individual facts $\mathscr{F}$ are such that C could not have failed to cause E, i.e., that $\mathscr{L}$, $\mathscr{F}$ and C jointly entail E. This makes E nomically dependent on C. In that case the counterfactual

(d)        If E had not occurred then C would not have occurred

seems to follow. This amounts to the problem of contraposition, since (d)

$$\sim E \rightarrow \sim C$$

is the contrapositive of the counterfactual

$$C \rightarrow E$$

justified by $\mathscr{L}$ and $\mathscr{F}$. Now, as we saw, there are cases where the Humean will deny that if one is justified in asserting a counterfactual then one is justified in asserting its contrapositive. Nonetheless, there remains many contexts in which the inference remains sound. So, for the Humean, there will be many cases in which, under the given assumptions about C, E, $\mathscr{L}$ and $\mathscr{F}$, he is justified by them in asserting (d). But if he asserts (d), then he asserts the causal dependence of E on C, and therefore the causation of C by E, contrary to the assumption that E does not cause C. Thus, if we explain causation in terms of nomic dependence, then one confronts this *problem of effects*. Lewis avoids the problem by basing the assertibility of counterfactuals on judgments of interworld similarity rather than on laws.

Lewis simply flatly denies that the counterfactual (d) follows:[72] holding C fixed, he asserts, and giving up either the laws $\mathscr{L}$ or the circumstances $\mathscr{F}$, always or almost always yields a world more similar to ours that a world without C but the same laws and circumstances. For, holding $\mathscr{L}$ and $\mathscr{F}$ constant but changing C will, assuming the laws are deterministic, require a change in conditions immediately prior to C, and then immediately prior to that, and, in fact, in the whole preceding history.[73] This will involve a massive change in individual fact that will make any world without C but with the same laws as ours less similar to ours than a world with the same history as ours up to C and with a minor miracle or minor change in the laws.[74]

We have earlier raised problems concerning judgments of interworld similarity, so it is not clear that Lewis' solution to the problem of effects goes through. But even if it does, that does not necessarily tell in favour Lewis' analysis and against the Humean's. It does so only if the Humean is unable to solve the problem of effects.

This we can do by challenging Lewis' claim that the counterfactual (d)

> If E had not occurred then C would not have occurred

entails that E is causally dependent upon C. For the Humean, (d) will indeed obtain. For him, *counterfactual dependence is determined by laws. But counterfactual dependence does not entail causal dependence, since not all laws are causal laws.*[75] For example, as in the barometer example, the laws may determine that both

$$C \rightarrow E$$

and

$$\sim E \rightarrow \sim C$$

and also that, while C causes E, E does *not* cause C, being merely correlated with it. *Distinguishing between causal laws and correlations enables the Humean to solve the problem of effects.* Of course, it also challenges Lewis' analysis of causation in terms of counterfactual dependence as opposed to nomic dependence. But that should not surprise us, since we are in the process of arguing that there is no counterfactual dependence apart from nomic dependence. That being so, the problem of effects can be solved only by invoking different sorts of nomic dependence, rather than by invoking a form of dependence other than nomic dependence.

### (iii)  *Causal Pre-emption*

Another part of Lewis' case for the possible-worlds analysis of counter-factuals consists of the claim that his analysis can, where the law-deduction analysis cannot, account for what Lewis calls cases of "causal pre-emption." We must examine his argument. It will turn out that the argument fails, and that its unsoundness consists in imposing a simplistic view of causal dependence and causal laws upon the Humean or anyone who holds that the consequent of a counterfactual must be law-deductible from the antecedent. In fact, Lewis imposes this view because he takes it for granted that the law-deduction theorist accepts Lewis' explication of causal dependence in terms of counterfactual dependence.[76] Now, we have just argued that the Humean can defend himself against the charge that Lewis can handle certain features of causal discourse which he (the Humean) cannot, provided that he (the Humean) can provide a basis for distinguishing causal relations as a subclass of lawful or nomic relations. We have suggested in Chapter I, Section I, that the Humean can indeed draw this distinction.[76a] But be that as it may, none of this is to argue *directly* that Lewis' analysis of causation in terms of counterfactuals is actually mistaken. In fact, all that we did — though in imitation of Lewis' own practice — was flatly assert that Lewis was wrong about the assertibility of a crucial counter-factual.[77] Our present discussion will support that assertion: it will no longer be so flat. For, the upshot of our discussion of the case of causal pre-emption will yield a direct argument against Lewis' analysis of causation: it will show that Lewis' is just not a plausible account of causal discourse. Thus, not only does the case of pre-emption not support Lewis against the Humean but actually counts against him and for the latter.

Lewis gives the following description of causal pre-emption.[79] We have three events $C_1$, $C_2$ and $E$. These are all actual. $C_1$ causes $E$. $C_2$ does not cause $E$. But $C_2$ would have caused $E$ if $C_1$ had been absent. Hence, $C_2$ is a potential alternate cause of $E$, but is pre-empted by $C_1$. $C_1$ and $C_2$ over-determine $E$, but do so asymmetrically. The problem is: which analysis of causation can better account for this asymme-tricality?

Lewis gives no example, but seems to have in mind something like the following example.

Smith shoots his well-aimed gun at Jones who is riding by on his bicycle. This would normally cause Jones' death. But just at that moment Brown comes along in his car driving recklessly. The car comes between Smith and Jones and is struck by the bullet. Hence Smith's shooting the gun does not cause Jones' death; it is prevented from doing so by Brown's recklessly driving the car. However, Brown's car proceeds to violently knock Jones off his bicycle, which in turn causes the latter's death. So Brown's recklessly driving the car causes Jones' death by causing him to be thrown from his bicycle. Jones' death is over-determined; but the capacity of Smith's action to cause Jones' death has been pre-empted by Brown's action.

Or here is another example.

Some carelessly lit rubbish causes a shed to catch fire, and this in turn causes the destruction by fire of the building to which it is attached. At the same time the fire in the rubbish causes a circuit breaker to trip, cutting off electricity to the wires in the shed, which in turn prevents a short circuit in the wiring from developing sufficient heat to cause the same conflagration in the main building, something which it would otherwise have done.

Clearly, the laws that explain these processes are complex indeed.

Lewis proposes that he is able to capture the asymmetry of the determination of $E$ by $C_1$ and $C_2$ by deploying his notions of *counterfactual dependence* and *causal chain*. He argues that if these notions are explicated as he proposes in terms of *counterfactuals*, then the possible-worlds account of the latter shows why the asymmetry obtains while the Humean or law-deduction account of counterfactuals is unable to show that there is any asymmetry in the case. So Lewis concludes that his account of counterfactuals and of causation ought to be adopted, that of the Humean rejected.

An event $E$ is *counterfactually dependent* on $C$ just in case $C \rightarrow E$ and $\sim C \rightarrow \sim E$ *both* hold.[79] If $C$ causes $E$ but $E$ is not counterfactually dependent on $C$, then it does so by means of a *causal chain*.[80] $C$ causes $E$ just in case *either* $C$ and $E$ are both actual and $E$ is counterfactually dependent on $C$, or there is a chain of events $C$, $D_1$ and $D_2$ and $D_2 \ldots$, $E$ such as $D_1$ is counterfactually dependent on $C$, $D_2$ is counterfactually dependent on $D_1$, and so on, with, finally $E$ counterfactually dependent on its predecessor. Since the events are all actual, the counterfactuals $C \rightarrow D_1$, $D_1 \rightarrow D_2$, $\ldots$ are all trivially true, on Lewis' account. The counterfactual dependence is thus given by the counterfactuals $\sim C \rightarrow \sim D_1$, $\sim D_2 \rightarrow \sim D_1, \ldots$ etc.[81]

Now, in the pre-emption case, if $\sim C_1$ obtains, then $C_2$ would still cause $E$. Thus, $E$ would occur even if $C_1$ doesn't. Thus, the counterfactual $\sim C_1 \rightarrow \sim E$ does *not* hold. In that case, $E$ is not counterfactually dependent on $C_1$. Similarly $\sim C_2 \rightarrow \sim E$ does *not* hold and so $E$ is not counterfactually dependent on $C_2$ either. But by hypothesis, $C_1$ *does* cause $E$. If $E$ is not counterfactually dependent on $C_1$ but $C_1$ nonetheless causes $E$, then it can do so only if there is a causal chain that leads from $C_1$ to $E$. For the sake of simplicity assume it is a chain of one link, to wit, $D$. In that case $E$ is counterfactually dependent on $D$, and $D$ is counterfactually dependent on $C_1$; that is, the two counterfactuals $\sim D \rightarrow \sim E$ and $\sim C_1 \rightarrow \sim D$ hold. Counterfactual dependence is not transitive since $\rightarrow$ is not transitive. But causation *is* transitive: $C_1$ causes $D$ and $D$ causes $E$, so $C_1$ causes $E$. $E$ is counterfactually dependent on neither $C_1$ nor $C_2$; yet $C_1$ causes $E$. Thus, although $C_1$

and $C_2$ over-determine $E$, there remains an asymmetry in the relations between $C_1$ and $E$, on the one hand, and $C_2$ and $E$ on the other.

In order to see how Lewis uses this against the law-deduction account of counterfactuals, it is necessary to note what the latter is committed to if it accepts Lewis' analysis of causation in terms of counterfactual dependence.

To say that $E$ depends counterfactually on $C$, both $C \to E$ and $\sim C \to \sim E$ must hold.[82] In cases of causation, $C$ and $E$ are both actual, so, for Lewis, $C \to E$ is trivial. For the law-deduction account, however, $C \to E$ will not be a trivial matter: on this account there must be a connection between $C$ and $E$ such that $E$ is law-deducible from $C$. The other counterfactual, $\sim C \to \sim E$ is not trivial, even on Lewis' account. For the law-deduction account, for this counterfactual to hold $\sim E$ must be law-deductible from $\sim C$, or, *what is the same*, given the nature of deductive connections, $C$ must be law-deductible from $E$. Hence, for the Humean, Lewis' explanation of counterfactual dependence amounts to saying that $E$ depends counterfactually on $C$ just in case that the laws $\mathscr{L}$ and individual facts $\mathscr{F}$ entail that $C$ is *necessary and sufficient for $E$*.

In the schema of causal pre-emption, we have the causal chain $C_1$, $D$, $E$ in which each link is counterfactually dependent on its predecessor. Hence, Lewis' account of the latter, if accepted by the law-deduction theorist, commits that theorist to the position that, given the individual facts $\mathscr{F}$, the laws $\mathscr{L}$ permit the law-assertion of both

$(a_1)$     $C_1 \equiv D$ [= Anything is a $C_1$ if and only if it is a $D$]

$(a_2)$     $D \equiv E$ [= Anything is a $D$ if and only if it is an $E$]

$(a_1)$ immediately justifies the assertion of the counterfactual

(b)     $\sim D \to \sim C_1$

as well as the counterfactual

(c)     $C_1 \to D$

Lewis now reasons as follows, raising a possible objection to his analysis. Upon the latter, $E$ is supposed to depend counterfactually on $D$. But it would seem that it does not. For, "if $D$ had been absent then $C_1$ would have been absent [by (b)] and $C_2$, no longer pre-empted, would have caused $E$."[83] That is, $E$ would occur even if $C_1$ didn't, and, since $D$ is counterfactually dependent on $C_1$, i.e., since $\sim C_1 \to \sim D$ holds, $E$ would also occur even if $D$ didn't. So $E$ does *not* depend counterfactually on $D$, and therefore Lewis has not only not been able to explain the asymmetry of the causal over-determination but has also failed to explain how $C_1$ is the cause of $E$.

To this possible objection, Lewis points out that it depends upon accepting the counterfactual (b). This he rejects. "We may reply by denying the claim that if $D$ had been absent then $C_1$ would have been absent. That is the very same sort of spurious reverse dependence of cause on effect that we have . . . rejected in simpler cases. I rather claim that if $D$ had been absent, $C_1$ would somehow have failed to cause $D$. But $C_1$ would still have been there to interfere with $C_2$, so $E$ would not have occurred."[84]

This ought perhaps to be elaborated a bit. On Lewis' possible-worlds account of counterfactuals, to assert (b) is to assert

(b*)      In the nearest $(\sim D)$-world, $\sim C_1$ obtains

The objection he is considering relies upon the point that

If $\sim C_1$ were to be the case, then $C_2$ would cause $E$

which is the counterfactual

(c)        $\sim C_1 \to [C_2$ causes $E]$

Upon the possible-worlds account, to assert (c) is to assert

(c*)      In the nearest $(\sim C)$-world, $C_2$ and $E$ both hold, and so do the two counterfactuals
$C_2 \to E$ and $\sim C_2 \to \sim E$.

(For simplicity we consider only the simplist case in which $C_2$ causes $E$, that in which there are no intermediate links in the chain.) The commitment to (c*) is clear: if $C_2$ causes $E$ in the nearest $(\sim C_1)$-world, then $C_2$ and $E$ must both be actual *in that world*, and $E$ must depend counterfactually on $C_2$ *in that world*, i.e. the two counterfactuals $C_2 \to E$ and $\sim C_2 \to \sim E$ must be true *in that world*. The first of these will trivially be true in that world. For $\sim C_2 \to \sim E$ to be true in the nearest $(\sim C_1)$-world, then we must have

In the $(\sim C_2)$-world that is closest to the $(\sim C_1)$-world that is nearest to our world, $E$ holds.

Now, by the counterfactual dependence of $D$ on $C_1$, we have

(d)        $\sim C_1 \rightarrow \sim D$

and to assert this is to assert

(d*)      In the nearest $(\sim C_1)$-world, $\sim D$ holds

Moreover, since $C_2$ does hold in our world, a $(\sim C_1)$-world in which $C_2$ obtains is nearer to our world than one in which $C_2$ does not obtain. Hence, the nearest $(\sim C_1)$-world is also a $C_2$ world. But this $(\sim C_1 \ \& \ C_2)$-world will, by (d*), also be a world in which $\sim D$ holds, and, by (c*), also a world in which $E$ holds. However, if $\sim D$ and $E$ hold in that world, then $\sim D \supset \sim E$ does *not* hold. This, if the $(\sim C_1 \ \& \ C_2)$-world is *also* the $(\sim D)$-world that is nearest to ours, then $\sim D \supset \sim E$ will not hold in that nearest $(\sim D)$-world, and the counterfactual

(e)        $\sim D \rightarrow \sim E$

will be false in our world. But this is so. For, if we accept (b) = (b*), then the nearest $(\sim D)$-world is a $(\sim C_1)$-world, and this will be nearer to ours if $C_2$ holds in it than if it doesn't. So the nearest $(\sim D)$-world is also a $(\sim C_1 \ \& \ C_2)$-world, which means, by (c) = (c*), that $E$ holds in it. So (e) is false.

But if $E$ is to depend counterfactually on $D$, then (e) must be true. Hence, if the above argument that (e) is false is sound, then $E$ does not depend counterfactually on $D$, and there is no causal chain leading from $C_1$ via $D$ to $E$. And so it would seem that Lewis' account of causation cannot deal adequately with the case of causal pre-emption.

Lewis' way out is to deny the soundness of the objection. As we saw, he simply denies the counterfactual (b). Which is to say that he denies that the nearest $(\sim D)$-world is the nearest $(\sim C_1)$-world.

Lewis does assert

$C \rightarrow D$

When he denies (b) he is denying the contrapositive of this. Thus, Lewis' move in his discussion of causal pre-emption is simply part of his more general case that contraposition is invalid for $\rightarrow$.[85]

If (b) is false, then the nearest $(\sim D)$-world is one is which $C_1$ holds. In that case, $(a_1)$ will not be law-assertible in that world. This is part of Lewis' more general point that the criterion of inter-world similarity that he uses does *not* require that the worlds most similar to ours share the laws that hold in our world.

But as Lewis sees it, the law-deduction theorist is committed to just such a scale of inter-world similarity. For the latter, a world in which our laws were violated could never be more similar to ours than one in which our laws obtain. So for the law-deduction theorist, the nearest $(\sim D)$-world must be one in which $(a_1)$ is law-assertible. The law-deduction theorist is therefore committed to accepting (b). Lewis' move is not open to him. It would seem, therefore, that the law-deduction theorist cannot, where Lewis can, account for the counterfactual dependence of $E$ on $D$ and thereby account for $C_1$ causing $E$ through the causal chain $C_1, D, E$.

Now, it is certainly true that if the Humean law-asserts $(a_2)$, i.e., $D \equiv E$, then he is thereby committed to asserting the counterfactual (e), $\sim D \rightarrow \sim E$, which Lewis denies. However, if Lewis' analysis of causation commits the Humean to law-asserting $(a_2)$, it also commits him to law-asserting $(a_1)$, i.e., $C \equiv D$. It is also the case that $E$ is supposed to occur even if $C_1$ does not; for, if $C_1$ does not occur then $C_2$ will cause $E$ to occur. So assume that $C_1$ does not occur, i.e., that $\sim C_1$. In that case we have, by $(a_1)$, $\sim D$, and in turn, by $(a_2)$, $\sim E$. However, $E$ *does* occur; it is caused by $C_2$. What this establishes is that $(a_1)$–$(a_2)$ do *not* adequately represent the logical complexities of the laws that govern contexts of causal pre-emption. On the other hand, Lewis' account of causation in terms of counterfactuals commits the law-deduction theorist to accepting $(a_1)$ and $(a_2)$. And this is to say in turn that law-deduction theorist or the Humean ought not to accept Lewis' counterfactual analysis of causation. Indeed, *given our argument that counterfactuals ought not to be asserted unless the consequent is law-deducible from the antecedent, it follows that Lewis' case for his counterfactual analysis of causation is unsound. That is, given the soundness of our case for the law-deduction requirement, it follows that Lewis' analysis of causation is simply inadequate as an account of causal discourse.*

The difficulty that Lewis supposes he has raised for the Humean disappears once we recognize the inadequacy of his counterfactual analysis of causation. The analysis of causation is used to impose a simplistic model of the laws to be used in the analysis of the pre-emption case. And the difficulty he creates for the Humean with the causal asymmetry of the over-determination in the pre-emption case exists only if that simplistic model is accepted. But why accept that model? Indeed, as we have just suggested, what the difficulty argues for is not the rejection of the Humean account of counterfactuals but rather the rejection of Lewis' counterfactual analysis of causation!

Another way in which the inadequacy of Lewis' analysis can be seen is this. The following two counterfactuals *correctly* describe the pre-emption case:

(f$_1$)        $C_1 \rightarrow E$

(f$_2$)        $\sim C_1 \rightarrow E$

If the Humean accepts Lewis' counterfactual analysis of causation, then that requires him to law-assert (a$_1$) and (a$_2$). But in that case he is committed to asserting $\sim C_1 \rightarrow \sim E$, and therefore to denying (f$_2$). But the latter *is* justifiably assertible in the pre-emption case. Once again, the appropriate response of Humean to this difficulty is to reject the counterfactual analysis of causation.

Now, if we assert (f$_1$) and (f$_2$), then the laws must be such that we can deduce that $E$ will occur independently of whether $C_1$ occurs. This is, of course, so in the case where causal pre-emption occurs. This means that we can assert the semi-factual

(g)        Even if $C_1$, $E$ [86]

which asserts that $E$ is independent of $C_1$.[87] It would seem that this invalidates (f$_1$), which *does* correctly describe the pre-emption case. However, we must recall our previous argument,[88] that the independence by $E$ of $C_1$ is compatible with $E$ also being dependent on $C_1$ in the way asserted by (f$_1$): asserting (g) does not require one to deny (f$_1$).

In fact, if we can assert (g), then (f$_1$) follows trivially. Suppose that we assert (g) on the basis of the law

(h)        All $F$ and $E$

From (h) it follows trivially that

(h′)        All $F$ & $C_1$ are $E$

This "piggy-back" law[89] justifies the assertion of (f$_1$). Similarly, the other trivial consequence of (h), namely,

(h″)        All $F$ & $\sim C_1$ are $E$

justifies the assertion of (f$_2$).

This is disturbing, however. There is, in the pre-emption case, surely something more than trivial about $(f_1)$ and $(f_2)$. For, as we argued, if $(f_1)$ and $(f_2)$ were merely trivial consequences of (g), then it would be *odd* to assert the former in a context in which the latter was asserted.[90] If we asserted (g) because we had a law like (h) and knew the process to be *F*, then (g) would make a *stronger* assertion than either $(f_1)$ or $(f_2)$, since the assertion of (g) would be based on the law (h) rather than on the *less imperfect* laws (h') and (h") on which the assertions of $(f_1)$ and $(f_2)$ would be based. The point is that, in the case of the pre-emption process, it does *not* seem *odd* to assert $(f_1)$ and $(f_2)$ as well as (g). It is incumbent upon the Humean to show why this is so. Naturally he must do so by appealing to the laws involved.

In the pre-emption case, (g) is *not* asserted on the basis of a law like (h). Rather, the relevant law is more like this

(i)        Whenever either $C_1$ or $C_2$ then $E$

This is a first approximation for a lawful description of the causal process in the pre-emption case, since it omits the role of *D*, and leaves it open whether $(C_1$ or $C_2)$ is necessary as well as sufficient for *E*. That is, (i) is imperfect relative to the full description of the pre-emption process. But imperfect laws are still laws. And (i) suffices for our present purpose, that of understanding the non-triviality of the counter-factuals $(f_1)$ and $(f_2)$

Given (g), the Humean would analyze $(f_1)$ in terms of the collapsed argument

$$\text{(i)}$$
$$C_1$$
$$\overline{\quad\quad}$$
$$\therefore E$$

where the first premiss is law-asserted and the second premiss is merely supposed. Similarly, the counterfactual $(f_2)$ is analyzed in terms of the collapsed argument

$$\text{(i)}$$
$$C_2$$
$$\sim C_1$$
$$\overline{\quad\quad}$$
$$\therefore E$$

where the first premiss is law-asserted, the second premiss is asserted, and the third premiss is merely supposed. What justifes (g) is not a law like (h), but, once again, the law (i) and the fact that the pre-emption process Lewis describes is one in which $C_2$ does hold. The argument

$$(i)$$
$$\frac{C_2}{\therefore E}$$

where the first premiss is law-asserted and the second is asserted, justifies the assertion of the semi-factual (g). *In each case the same law is used to justify the assertion.* In particular, it is not the case that a more imperfect law is used to justify the assertion of $(f_1)$ and $(f_2)$ than is used to justify the assertion of (g). Because the *same law* is used to justify each, the assertion of (g) is *not* stronger than the assertion of $(f_1)$ and $(f_2)$. And so, in the pre-emption case, it is not odd, as ordinarily it would be, to assert the counterfactuals $(f_1)$ and $(f_2)$ when one asserts the semi-factual (g). The difference is, of course, the special laws that apply to the pre-emption case.

We may conclude, then, that, contrary to Lewis, the Humean can adequately deal with processes in which causal pre-emption occurs; and moreover, we can also conclude that such cases provide strong grounds for rejecting Lewis' counterfactual analysis of causation.


### (iv)  *Which Belief-Contravening Assumptions Are Reasonable?*

The Humean provides no algorithm for reducing subjunctive conditionals to the complex assertions which their uses express. Given that the Humean's analysis has the implicit content varying from context to context, the best that he can do is provide some rough guidelines. They are rough in two respects. So far as concerns the interpretation of the other fellow's assertions, one has available only what he explicitly says and what one can assume he might reasonably have included among the implicit parts of this assertion. Often what he is has implicitly included is clear enough. But perhaps just as often it isn't clear. In such a case all one can do is recognize that fact, and *ask him* to make explicit what he implicitly includes in his assertion. Moreover, for both the other fellow and oneself, what one might reasonably include among the implicit or explicit parts of a counterfactual assertion will in fact vary from context to context, depending upon such things as what law-assertions are reasonable on the basis of the evidence available, and other such things. It is this latter sort of context-dependence in particular that rules out any algorithm. Nonetheless, such context-dependence does not, as we have suggested, exclude rough guidelines.

These guidelines tell us, in effect, when it is reasonable to collapse the argument

$$\text{All } F \text{ and } G$$
$$Fa$$
$$\overline{\phantom{All F and}}$$
$$\therefore Ga$$

into the counterfactual

$$Fa \to Ga$$

or expand the counterfactual into the argument. Clearly, what is reasonable will be determined by the context in which the counterfactual discourse occurs. Specifically, of course, we have argued that where the counterfactual is used in a context defined by the goal of intelligent interference or potential interference then any assertion of a counterfactual *ought* to be expandable into an argument which includes among its (essential) premises a law-asserted generalization.[91]

Moreover, as we have also argued, this same context-defining goal of the discourse places constraints on what can be assumed in the antecedent of the counterfactual.[92] Specifically, we argued that, in that context, the antecedent, including both its explicit and implicit parts, must be logically and lawfully realizable, that is, it must be logically self-consistent, and consistent with (known) laws of nature. (So far as the conditions for subjective assertibility are concerned, the antecedent must be consistent with those generalizations that, on the best evidence available, we may take to be law-assertible. It may turn out — given the openness of induction — that an antecedent is consistent with accepted laws but unrealizable due to laws that we have not yet discovered. In that case, of course, we would be subjectively justified but objectively unjustified in asserting a counterfactual with the falsely-believed-to-be-realizable antecedent.)

For many purposes, the above guidelines for what one can take as an antecedent of a counterfactual conditional are more than sufficient. Yet they can be elaborated, and it will pay to do so. Another who has reflected carefully on these guidelines is Goodman, in his *Fact, Fiction, and Forecast.* Yet Goodman does not, in the same way as ourselves, relate the guidelines he proposes to the cognitive interests that define the context of the counterfactual discourse. As a consequence, the guidelines he proposes have at times an air of adhocery about them: their rationale is never quite clear, save as descriptions of how we normally use counterfactuals; and the latter is not enough since one wants to know not just how such statements *are* used (the descriptive issue) but how they *ought* to be used (the normative issue).

Goodman considers a counterfactual

(1)     A → C

to be assertible just in case that there is a law-assertion L, and assertions of particular fact S such that

(2)     L & A & S ∴ C

is valid. He also wishes to account for the fact that one can assert (1) only if one denies

(3)     A → ~C

He notes that in asserting the counterfactual (1), we are not saying that it holds *if* S obtains, but that we are, more strongly, asserting S, committing ourselves to the truth of these further relevant conditions.[93]

The first condition that he imposes on the S,[94] is that it be compatible with A, both logically compatible and also non-logically compatible in the sense that the conjunction A & S not violate a non-logical law, i.e., a law of nature. We have seen already the rationale for this condition if the context of discourse is defined by the goal of intelligent interference or potential interference.

Goodman next lays down the rule[95] that S must be (both logically and non-logically) compatible with both C and ~C. For, if S is not compatible with ~C, then it by itself implies C. Since S is given as true, it follows that we are justified in asserting the semi-factual

(4)     C, even if ~A

Goodman believes that to assert this semi-factual is to deny (1). On this ground, he wants to lay down the condition that S be compatible with ~C. But, as we saw,[96] Goodman is wrong in believing that to assert (4) is to deny (1). However, as we also saw,[97] in the context defined by the goal of intelligent interference or potential interference, where one can assert (4) there one *ought not* to assert (1) since there is, *for purposes of (reasoned) prediction and explanation,* another assertion that *could be made that is informationally stronger.* The assertion of (1) can be justified, given the laws and evidence available, but so can an informationally stronger statement, and since the use of the latter is better *relative to the cognitive interest defining the context of discourse,* it follows that to assert (1) would be inappropriate.[98] Thus, Goodman's rule that S be compatible with $\sim$ C has a sound rationale.

As for the rule that S be compatible with C, if it were not, that is, if S entailed $\sim$C, then we would be in a position to assert the semi-factual

(5)        $\sim$C, even if A

If we are justified in asserting this, then we are also justified in asserting the weaker (3) — though it would be, for reasons just indicated, inappropriate to assert (3) when one can assert (5). And if we can assert (5), then we are also justified in denying (1). Hence, if S is incompatible with C, then (1) as analyzed into (2) ought to be denied. Thus, there is also a sound rationale for this rule that Goodman gives, that any S used to analyze (1) into (2) must be compatible with C.

Furthermore, if we let S be $\sim$(A & $\sim$C), then A & S alone entails C, without L; and if we let S' be $\sim$(A & C) then A & S' alone entails $\sim$C, without any law.[99] Goodman works with the examples [100]

> If Jones were in Carolina, he would be in South Carolina
>
> If Jones were in Carolina, he would be in North Carolina.

These may be represented by

($a_1$)     $(A_1 \lor A_2) \rightarrow A_1$

($a_2$)     $(A_1 \lor A_2) \rightarrow A_2$

Since

($a_3$)      $A_1 \equiv \sim A_2$

holds, if we have ($a_2$) then we also have

($a_4$)     $(A_1 \lor A_2) \rightarrow \sim A_1$

which denies ($a_1$). Since ($a_1$) and ($a_2$) are counterfactuals, both $\sim A_1$ and $\sim A_2$ are true. If we let S be $\sim A_2$, then we obtain the consequent of ($a_1$) from an A & S; and if we let S' be ~$A_1$, then we obtain the consequent of ($a_2$) from an A & S'. Again, the inference goes through without appeal to any L. In these cases, we have by suitable choice of S and S' both consistent with A, ended justifiably both asserting a counterfactual and denying it. Thus, it is not sufficient that we let the S in (2) satisfy the condition of being consistent with both C and ~C: we must also impose certain restrictions on the relation of S and A. Goodman suggests[101] that we impose the condition that ~A does not lead by law to either S or ~ S.

However, this misses an important point about these cases, or at least important relative to the context of intelligent interference or potential interference. It is this: *in these examples the consequent is inferred from the antecedent and other statements of individual fact alone without any lawful generalization being essential to the deduction.* Now, we have already argued that for the contexts of discourse that we are considering, it is a necessary condition for the counterfactual to be assertible that the consequent be deducible from the antecedent and other statements of individual fact *and that a law be essential to the deduction.* Since, in the examples, no law is essential to the deduction, this suggests an alternative condition for meeting the problem Goodman confronts, and, moreover, one that receives its rationale not just from the elimination of troublesome cases but from cognitive interests that define the context of the discourse. This is the condition that A & S *be an instantiation of the antecedent of a law that is essential to the deduction of* C. Note that this is a positive criterion, unlike Goodman's negative; this one does not merely exclude, but excludes by telling us just which S's can be included.

The idea is still not clear enough, however. For, the notion of an antecedent is sufficiently unclear to render the criterion unworkable. Thus, if it permits

($i_1$)     $(x)[(Fx \ \& \ Gx) \supset Hx]$
          $Fa$
          $Ga$
          $\overline{\phantom{(x)[(Fx \ \& \ Gx)}}$
          $\therefore Ha$

to justify

          $Fa \rightarrow Ha$

it does not permit

(i$_2$)        $(x)[\sim Hx \supset (\sim Fx \vee \sim Gx)]$
              $Fa$
              $Ga$
              _____

              $\therefore Ha$

to justify that counterfactual, even though the two law-statements are logically equivalent.[102] It is not so much the idea of "antecedent" that is important, but rather the idea of *a potential basis for a reasoned prediction.*

What we have seen is that in discourse defined by the end of intelligent interference or potential interference, what is important is that *if A were true* then the L and S be such that A & S on the basis of L *would yield a reasoned prediction of* C. In this sense, both (i$_1$) and (i$_2$) yield a reasoned prediction of C on the basis of A & S. The point can be put this way: In the deductive-nomological account of explanation, any explanation is a potential prediction; indeed, it is a necessary and sufficient condition for an inference based on a law and individual facts to constitute an explanation of a fact C is that the inference satisfy the *principle of predictability,*[103] that the inference could have been a prediction had the individual facts been known prior to the coming to know of the facts that make the consequent true. Similarly, in a context of discourse defined by the end of intelligent interference or potential interference, a necessary and sufficient condition for an inference based on a law L and individual facts S to justify the assertion of the counterfactual A → C is that *the inference satisfy the same principle of predictability, that if A as well as S were known to be true then* A & S & L *would yield a reasoned prediction of* C.

Goodman is concerned[104] that his conditions on A and S admit, concerning a normal match *m* on my desk that has never been struck, both the acceptable counterfactual

(5)        If *m* had been scratched, then *m* would have lighted

and also the unacceptable counterfactual

(6)        If *m* had been scratched, *m* would not have been dry

since the true statement "*m* did not light" could be taken as (a clause in) S, and this and A would, by the law that justifies (5) yield the consequent of (6).

The relevant law is something like this:

> (L$_1$)    Whenever a match is scratched, then it lights if and only if there is oxygen present and the match is dry.[105]

Then (5) is to be analyzed into the argument

> (5′)    (L$_1$)
> Oxygen is present
> Match $m$ is dry
> Match $m$ is scratched
> ———————————————
> ∴ $m$ lights

The first premiss of (5′) is law-assertible. The second and third premisses are *normally* true of matches found, like $m$, on a person's desk; the *statistical likelihood* of their being true is *very high*. This probability affords us good grounds for asserting the second and third premisses are true. Let us, following Sellars, call such premisses that normally are true "standing conditions."[106] That certain conditions can be treated as standing conditions is an *objective fact*.[107] Moreover, by definition of 'standing condition', the probability that the standing conditions *actually* hold is very high. This probability establishes that in contexts such as that of matches on a person's desk, one is *subjectively justified* in asserting that those premisses obtain. It does not follow that we are always *objectively justified* in asserting that the standing conditions hold. For, what is statistically very likely may also not obtain. However, be the latter as it may, it is clear that if we are subjectively justified in asserting (5) = (5′), then we are, since dryness is a standing condition, justified in asserting the semi-factual

> Even if $m$ were scratched, it would be dry

and therefore the weaker counterfactual

> If $m$ were scratched, then it would be dry

which is the denial of (6). It would seem, then, that if we assert (5) = (5′), then we are in a position to deny (6).

Yet his point remains about the S he chooses: "$m$ did not light." This we know to be true. We can therefore employ the argument

(6')       ($L_1$)
           Oxygen is present
           Match $m$ does not light
           Match $m$ is scratched
           ————————————————
           $\therefore m$ is not dry

to justify the assertion of (6). The first premiss is law-asserted; we are justified in asserting that the standing condition mentioned in the second premiss holds; and we know the third premiss to be true. So we seem to be required both to deny (6) and to assert it.

How can we exclude (6) = (6')? Clearly, the problematic point is the third premiss of (6'). What grounds have we for preventing its inclusion as a clause in the S which, together with ($L_1$), justifies the assertion of the counterfactual (6)? Clearly, some further condition is needed.

To solve this problem, Goodman suggests that we impose the requirement that if one is to be justified in asserting the counterfactual $A \to C$, where C is law-derivable from A by means of auxiliary assumption S, then A and S must be *cotenable*.[108] To say that A and S are cotenable and that A & S is self-cotenable is to say that it is not the case that S would not be true if A were, i.e., just in case that

$$\sim(A \to \sim S)$$

holds.[109] Now, we have seen already that the requirement of cotenability does indeed find a rationale in any context of discourse defined by the goal of intelligent interference or potential interference. For, that goal imposes the conditions that the antecedent of any counterfactual be realizable, and if A & S is not self-cotenable then it is not realizable.

Let us take

$$A \to C$$

to be the counterfactual (5) = (5'). Let the auxiliary S needed for (5) be T. In that case, C is law-derivable from A & T. The counterfactual (6) becomes

$$A \to \sim T$$

and its auxiliary S is C. The antecedent and the auxiliary hypothesis must be cotenable. In the case of (6), this means that

$$\sim(A \rightarrow \sim\sim C)$$

i.e.,

$$\sim(A \rightarrow C)$$

holds. But given (5), it does not hold. Hence, if (5) is justifiably assertible, then the antecedent of A is not cotenable with the other facts that are needed for the law-derivation of its consequent. Moreover, as we have argued, given the law $(L_1)$ and the standing conditions needed to assert (5), we are in fact *normally justified* in asserting that counterfactual. It follows that normally we are justified in denying (6) by virtue of the uncotenability of its antecedent with the facts that must be included as premisses for the law-derivation of its consequent.

The question of *realizability* which we used to justify the rule of cotenability is a difficult one. It cannot mean just logical consistency; otherwise we would not need the additional condition of cotenability. On the other hand, it cannot mean technical realizability; otherwise it would exclude perfectly good counterfactuals like

If Mars were a million miles closer to the sun, then . . .

Cotenability provides a reasonable sense for 'realizability' in this context: it is realizability in the sense of being consistent not just logically but with existing individual facts and with the laws of nature. For, A and S are cotenable just in case existing individual facts and the laws are not such that A's occurrence would cause S not to be; or, more accurately, just in case the individual facts and laws do not entail that A and S cannot jointly obtain. Thus, cotenability is a step towards providing an *explication* of the pre-analytic notion of realizability with which we have been working. But further analysis will be needed before the pre-analytic notion has been fully clarified.

Goodman has raised an important objection to introducing the concept of cotenability into the discussion of the assertibility-conditions for counterfactuals. "In order to determine the truth of a given counterfactual it means that we have to determine, among other things, whether there is a suitable S that is cotenable with A and meets certain further requirements. But to determine whether or not a given S is cotenable with A, we have to determine whether or not the counter-factual 'If A were true, then S would not be true' is itself true. But this means determining whether or not there is a suitable $S_1$, cotenable with A, that leads to ~S and so on. Thus we find ourselves involved in an infinite regressus or a circle, for cotenability is defined in terms of counterfactuals, yet the meaning of counterfactuals is defined in terms of cotenability. In other words to establish any counterfactual, it seems that we first have to determine the truth of another. If so, we can never explain a counterfactual except in terms of others, so that the problem of counterfactuals must remain unsolved." [110] But it does not seem as desperate as Goodman makes out. In order to determine whether a counterfactual is, so far as we can tell, worthy of assertion — that is, in order to determine whether it is subjectively worthy of assertion, we have to take into account the generalizations $\mathscr{L}$ that we law-assert and the individual facts $\mathscr{F}$ that we accept. If A → C is to be worthy of assertion, then there must be an L in $\mathscr{L}$ and an S in $\mathscr{F}$ such that L & S & A ∴ C is valid and such that S is cotenable with A. Hence, if A → C is to be worthy of assertion then there must not be an $L_1$ in $\mathscr{L}$ and an $S_1$ in $f$ such that $L_1$ & $S_1$ & A ∴ ~S is valid and such that $S_1$ and A are cotenable. But then we must go on and do the same for $S_1$ and A. And so on. The regress looms. Note, however, that in each case A occurs. Moreover, all the S, $S_1$, ... are actual facts to which A's obtaining would be causally or lawfully relevant. The regress can be stopped provided that we can find some way of delimiting the list of actual facts to which A's obtaining would be lawfully relevant.

Consider first the case where the particular fact C is actual. Let A be the fact that $c$ is $G$. If we are correct, then to explain C ideally one must find (1) a system $\mathscr{S}$ of which $c$ is a part, (2) a set $\mathscr{V}$ of relevant variables exemplified in $\mathscr{S}$, (3) a process law for $\mathscr{S}$ and $\mathscr{V}$, and (4) a complete set of boundary conditions for $\mathscr{S}$ (closure is a special case of boundary conditions); and this knowledge will explain why $c$ is $G$ provided that from it, and a knowledge of the state of the system at any one time, one can deduce that $c$ is $G$. The crucial point is that since the set of relevant variables is complete, we need know nothing more about what goes on *inside* $\mathscr{S}$ in order to explain C; and since the boundary conditions are complete, we need to know nothing more about what goes on *outside* $\mathscr{S}$ in order to explain C. Thus, given that we can explain C by process knowledge then we need to take nothing else into account in order to explain C. Moreover, the process law *completely describes* the interactions among the relevant variables; it omits nothing about how one variable determines other variables or how it in connection with other variables determines still other variables or how other variables determine it.

Now let A not be asserted but merely supposed to be true. That is, for our purposes, take it to be the antecedent of a counterfactual. Since A is a statement of particular fact, let it be the statement that $c$ is $F$. In order to see whether $C = Gc$ follows law-deductively from this A, then, ideally, the deduction would be by way of process knowledge of the sort just described. In the process knowledge we have a complete set of relevant variables, a process law, and a complete set of boundary conditions. The relevant individuals $a_1$, $a_2$, ... , and the relevant particular facts $\mathscr{F}(a_i)$ about them are just those such that

(*)      $\mathscr{F}(a_i)$ and $Fc$

describe the state of a system to which the process knowledge applies. If (*) and the process knowledge jointly entail $Gc$ then the counterfactual $Fc \rightarrow Gc$ can justifiably be asserted. It may turn out, however, that the process law describing the interaction of $a$'s being $F$ with existing facts may establish that (*) does *not* describe a possible state of a system containing $c$, and $a_1$, $a_2$, .... . In other words, it may determine that (*) is not self-cotenable. But if (*) is self-cotenable and if (*) and the process knowledge entails $Gc$, then $\mathscr{F}(a_i)$ will include facts S such that C is law-derivable from A & S. But it may well include more facts. The more, however, will be *all* the relevant facts and *only* the relevant facts. It will, in short, include *all* the fact $S_1$, $S_2$, ... etc. that are relevant to determining whether (*) is self-cotenable. As for whether (*) is cotenable with other facts in the world, that is, whether there are laws leading from (*) to the denial of some other existing fact, then the answer is that there are no such laws, and therefore that (*) *is* cotenable with all other existing facts. This cotenability or the fact that there are no relevant laws follows immediately from the nature of process knowledge, and, more specifically, the fact that, given a complete set of boundary conditions nothing more about what happens outside the system need be known in order to determine what happens inside the system. Thus, if we have process knowledge, the determination of whether A and S are cotenable does *not* involve an infinite regress.

We are not, of course, everywhere in a position to analyze counterfactuals in terms of process knowledge. The point of the above remarks is to make clear two things. The first is that Goodman has raised a problem but has given us no reason to suppose that the inclusion of the cotenability requirement in the analysis of the conditions for justifiably asserting a counterfactual leads to a vicious or interminable regress. The second is to indicate how, in the less-than-ideal case which is the norm, one is to proceed in evaluating whether the cotenability condition is fulfilled: one brings to bear to the assumption A and the relevant facts S the best, i.e., least imperfect, lawful knowledge that one has with respect to the set of relevant variables, the set of boundary conditions, and the set of laws describing the interactions among the variables exemplified in the situations described by A and S. Our knowledge may in fact be "gappy" enough that there is no clear termination. Such is life — at least for those of us who are not omniscient. In such cases all we can do is recognize the source of the problem, and live with it until we have an opportunity — if ever — of improving our knowledge of the appropriate laws.

Consider, however, the conjunction (*) of statements of actual fact $\mathscr{F}(a_i)$ and the assumption $Fc$. If $Fc$ is the assumption of a *counterfactual*, then $Fc$ is false and $\sim Fc$ is true. But evidently, one cannot include $\sim Fc$ among the actual facts $\mathscr{F}(a_i)$. Indeed, in imposing the cotenability requirement, we have implicity excluded $\sim Fc$ from among the actual facts that can be introduced for the law-deduction of C from A; for, after all, $Fc$ is not even consistent with $\sim Fc$, let alone cotenable with it.

Thus, in considering the conditions that actually obtain that are causally relevant to the A—C relation, we delete the condition ~A from the list. We must delete not only ~A, but such other statements as are inconsistent with A. For, if ~A is not deleted, an inconsistency arises in our set of beliefs plus assumptions. A is deleted to eliminate the inconsistency. Moreover, the inconsistency between A and ~A is only the simplist sort. Any inconsistency that arises when A is added must be eliminated, but there may well be ambiguities (unlike the case of ~A) as to how the inconsistency can be eliminated. What we have to do is spell these matters out in more detail, in accordance with such conditions as we have laid down, in particular the principle of predictability and requirement of cotenability. Moreover, it will pay to do this in a way that ties in with the "possible worlds" way of speaking.

For the latter, we must, of course, use an *explicated* sense of 'possible world'. Carnap has provided the relevant idea in his concept of "state description."[111] A state description is a conjunction such that each conjunct is either an atomic sentence or the negation of an atomic sentence, and such that, for every atomic sentence, either that sentence or its negation (and not both) appears as a conjunct. The notion of 'every' here is relative to the language used. For our purposes, we need not suppose that we are working with an "ideal language" capable of yielding a complete description of the actual world, but only, as we have just seen, a language adequate to describing the portion of the world that is *relevant* to the A—C process. For brevity's sake, however, let us refer to this portion of the actual world as "the actual world".

We have, then, a set of state descriptions for the world. These state descriptions are mutually exclusive — at least one must be true — where the relevant senses of 'can' and 'must' are the strict logical, i.e., syntactical, and therefore explicated, senses.

Let $\mathscr{F}$ be the true state description, the one that describes the actual world. The other state descriptions describe other worlds that are possible relative to the actual world; or rather "describe" those worlds, that is, describe in an unproblematic and explicated sense and not in the literal but problematic sense that Lewis uses.

Let $\mathscr{L}$ be the process law that describes and explains events and processes in the actual world. ($\mathscr{L}$ entails all more imperfect laws.)

Let A and C be statements of individual fact that are antecedent and consequent respectively of the counterfactual A → C concerning which we are inquiring whether its assertion is justified or not.

This question is to be answered, to put it roughly, in terms of whether A, $\mathscr{F}$ and $\mathscr{L}$ entail C. But this answer is rough because it is not $\mathscr{F}$ itself, but some other state description or partial state description that is to be used in deciding whether C follows from A and $\mathscr{L}$. It is the principle of predictability and the requirement of cotenability that tell us how to restrict $\mathscr{F}$.

The principle of predictability tells us that the conjunction of A and the relevant individual facts must be such as, with $\mathscr{L}$, would be able, in the appropriate epistemic context, to provide a reasoned prediction of C. Hence, the principle of predictability directs us to use as the auxiliary statements of individual fact in our derivation a partial state description derived from $\mathscr{F}$ by deleting C from $\mathscr{F}$, or ~C, whichever occurs in $\mathscr{F}$, and, of course, anything else in $\mathscr{F}$ that could decide, without reference to $\mathscr{L}$, whether C or ~C holds in $\mathscr{F}$.

Specifically, the principle of predictability directs us to delete from

$\mathscr{F}$ conjuncts so as to obtain the *partial* state description $\mathscr{F}'$ of the actual world, where $\mathscr{F}'$ is the maximal set of conjuncts such that, if ~C is conjoined the result is not inconsistent and if C is conjoined the result is not inconsistent.

[If there are several such maximal sets of conjuncts, let $\mathscr{F}'$ be their disjunction.]

$\mathscr{F}'$ provides a partial description of a possible world; it is a description up to a certain point, a *node*, if you wish, from which two branches emerge, one which leads to the $\mathscr{F}'$-world being a C-world and one which leads to the $\mathscr{F}'$-world being a ~C-world. $\mathscr{F}'$ is, if you wish, a history of the world up to the point at which the truth-value of "C" remains to be determined. Except that the notion 'history' suggests a temporal relation between $\mathscr{F}'$ and the occurrence or non-occurrence of C. Sometimes it is, as in the case of the match $m$ that Goodman told us about: $m$'s lighting comes at a later point in the history of the match than the supposed co-occurrence of the scratching and the standing conditions. However, the crucial relation is not temporal but epistemic: Goodman's problem about concerning the match $m$ can be made to arise even when all the events are supposed to be simultaneous, as the following example shows.[112] Consider a bolt $b$, which at time $t$ is black and iron. The normal counterfactual

(i)     If the temperature of $b$ had been 650° at time $t$, then $b$ would have been red at $t$

is established by taking as our auxiliary statement S of relevant conditions, the sentence "$b$ was iron at $t$", together with some other clauses. But the deviant counterfactual

(ii)    If the temperature of $b$ had been 650° at $t$, then $b$ would not have been iron at $t$

will also be admitted if we take as the auxiliary statement "$b$ was black at $t$", together with some other clauses. This example shows that Cooley is mistaken in his suggestion that Goodman's worries concerning match $m$ can be overcome simply by noting the temporal order of the events of scratching and lighting. A counterfactual A → C is justifiably assertible using auxiliary individual facts to law-deduce C from A only if there is, as it were, an epistemic screen between those auxiliary facts and C. The temporal order can provide such a screen, but it is the idea of the screen which is basic and not that of the temporal order. This idea of an epistemic screen between the auxiliary individual facts and C finds its clarification in the idea we have just developed of a partial world-description that is *nodal* with respect to C. To limit the idea of a nodal description to *histories* that branch into different possibilities at nodal times, as some do,[113] is to commit the same error as Cooley.

Now, not only is $\mathscr{F}'$ a partial description of a possible world, it is a partial description of the actual world. In effect, the idea behind the asserting of A $\rightarrow$ C is that, given $\mathscr{L}$, then as it were *adding* A to the world as thus described is sufficient to *determine* this world to be C.

This places restrictions on both A and $\mathscr{F}'$. The relevant sort of *determination* is *determination by laws*: the laws $\mathscr{L}$ must be *essential* to a deduction of C from A & $\mathscr{F}'$. This is, of course, a condition imposed by the principle of predictability. It means that neither A alone nor $\mathscr{F}'$ alone nor the conjunction A & $\mathscr{F}'$ alone can entail C. If these conditions are not fulfilled then the counterfactual A $\rightarrow$ C cannot be justifiably asserted.

However, we can justifiably assert the counterfactual A $\rightarrow$ C when ~A obtains in the actual world. If it does so obtain, then it necessarily also holds in any $\mathscr{F}'$-world. Hence, if by "adding A to $\mathscr{F}'$" we mean only "conjoining A to $\mathscr{F}'$" then A & $\mathscr{F}'$ would in general be inconsistent and would by itself entail C. The relevant auxiliary facts for the law-deduction of C from A cannot simply be those of the appropriate nodal description $\mathscr{F}'$. *We have not only to suppose that* A *is true but also suppose that sufficient changes have been made in* $\mathscr{F}'$ *that the latter can accomodate the truth of* A, *i.e., changes sufficient that* A *can consistently be conjoined to it.* This means that we must revise $\mathscr{F}'$ by deleting ~A if it occurs in it, and, indeed, a set of conjuncts sufficiently large that, when they are deleted from $\mathscr{F}'$, ~A is not entailed by the result.

Specifically, we must delete from $\mathscr{F}'$ conjuncts so as to obtain the partial state description $\mathscr{F}''$ which is the maximal set of conjuncts such that if A is added as a further conjunct the result is not inconsistent.

[If there are several such maximal sets of conjuncts, then let $\mathscr{F}''$ be their disjunction.]

[If $\mathscr{F}'$ is a disjunction, then, since $p \lor q$ entails $r$ if and only if $p$ entails $r$ and also $q$ entails $r$, it follows that ~A holds in $\mathscr{F}'$ if and only if it holds in each disjunct. Let $\mathscr{F}'$ be the disjunction $D_1' \lor D_2' \lor \ldots$ of conjunctions $D_i'$. For each disjunct $D_i'$ form the sentence $D_i''$ which is the maximal set of conjuncts of $D_i'$ such that A can be consistently conjoined, or, if there are several such maximal sets, then it is the disjunction of these. Then $\mathscr{F}''$ is the disjunction $D_1'' \vee D_2'' \vee \ldots .$]

Now form the partial state description $\mathscr{F}''$ & A.

[If $\mathscr{F}''$ is a disjunction, then, since $((p \vee q)$ & $r)$ is logically equivalent to $((p$ & $r) \vee (q$ & $r))$, it follows that conjoining A to $\mathscr{F}''$ is logically equivalent to conjoining A to each disjunct of $\mathscr{F}''$.]

We still cannot take $\mathscr{F}''$ to be the partial state description necessary for the law-deduction of C from A, however. For, we can have a reasoned prediction of C from A when conjoined to $\mathscr{F}''$ only if the state of affairs A & $\mathscr{F}''$ is *realizable*. That is, we must once more revise our notions in order to meet the *requirement of cotenability*.

Specifically, A must be self-cotenable, in the sense that A is consistent with $\mathscr{L}$. If it is not, then A $\to$ C cannot be justifiably asserted.

Moreover, A must be cotenable with $\mathscr{F}''$ in the sense that A & $\mathscr{L}$ must not entail $\sim \mathscr{F}''$, or, what is the same, $\mathscr{F}''$ & $\mathscr{L}$ must not entail $\sim$ A. If $\mathscr{F}''$ does not satisfy this condition, we must delete conjuncts until we have a partial state description that does.

[If $\mathscr{F}''$ is a disjunction, and A is not cotenable with $\mathscr{F}''$, then, since if A entails $\sim(p \vee q)$ then it entails both $\sim p$ and $\sim q$, it follows that A is not cotenable with any disjunct of $\mathscr{F}''$.]

Specifically, if A is not cotenable with $\mathscr{F}''$ then we must delete from $\mathscr{F}''$ conjuncts so as to form the partial state description $\mathscr{F}^+$ which is the maximal set of conjuncts such that, when $\mathscr{L}$ is assumed, A can be consistently conjoined. Note, by the way, that this may involve deleting *every* conjunct of $\mathscr{F}''$.

[If there are several such maximal sets, then let $\mathscr{F}^+$ be their disjunction.]

Now, to $\mathscr{F}^+$ conjoin the negations of all the conjuncts that were deleted from $\mathscr{F}''$ in order to obtain $\mathscr{F}^+$. Let the result be $\mathscr{F}^*$.

[If $\mathscr{F}''$ is a disjunction, then delete from each disjunct sufficient conjuncts that A is cotenable with the result; then conjoin to this result (or results, if there is more than one such sufficient set) the negations of the deleted conjuncts; then $\mathscr{F}^*$ is the disjunction of the resulting sentences.]

Now form the partial state description $(\mathscr{F}^*$ & A$)$.

A can, consistently with the laws $\mathscr{L}$, be added to an $\mathscr{F}^+$-world and therefore to an $\mathscr{F}^*$-world, where, in general, it cannot, consistently with $\mathscr{L}$, be added to an $\mathscr{F}'$-world. That is, an $(\mathscr{F}^+$ & A$)$-world and an $(\mathscr{F}^*$ & A$)$-world are *both logically and physically realizable* where, in general, $(\mathscr{F}'$ & A$)$-world is not.

The distance between an $\mathscr{F}'$-world and an ($\mathscr{F}^*$ & A)-world is a measure of what must at a *minimum* be changed in the actual world at the nodal point in order for it to be *lawfully (physically) possible* for A to obtain in it.

The difference between $\mathscr{F}^+$ and $\mathscr{F}^*$ is what must be assumed as contrary to fact along with A, when one asserts the counterfactual A → C. Or rather, it is minimum contrary-to-fact assumption that must be made.

An $\mathscr{F}'$-world is nodal with respect to C. We must impose the same condition on our (A & $\mathscr{F}^*$)-world. That is, if (A & $\mathscr{F}^*$) *by itself* entails C, then the counterfactual A → C cannot justifiably be asserted — that is, of course, in a context of discourse defined by our cognitive interest in intelligent interference or potential interference.

We thus obtain: If an (A & $\mathscr{F}^*$)-world is nodal with respect to C and if $\mathscr{L}$ guarantees deductively that C holds in an (A & $\mathscr{F}^*$)-world, then the counterfactual A → C is justifiably assertible.

We now see the connection between the Goodman-type criterion that we have developed as part of a Humean analysis of counterfactuals and Lewis' criterion in terms of similarity of possible worlds. An (A & $\mathscr{F}^*$)-world is obtained from the actual world by modifying the latter in the *minimal* way that is consistent with the laws $\mathscr{L}$ and with A's obtaining in such a world. Thus, if the nearest A-world must be one with the process law of the actual world, then no A-world is closer than an (A & $\mathscr{F}^*$)-world. Moreover, if the nearest (A & $\mathscr{F}^*$)-world must be one with the same laws as the actual world, then no (A & $\mathscr{F}^*$)-world could be nearer than one in which C holds. Thus, if we assume — as Lewis does not — that our criterion of interworld similarity is such that every world that differs in laws from our world is less similar to ours than any world that has the same laws as ours then, if the laws permit us to justifiably assert A → C, then the nearest A-world is also one in which C holds, so that A → C is also justifiably assertible upon Lewis' criterion. Moreover, if the laws $\mathscr{L}$ are deterministic, in the sense of entailing either that C occurs in an (A & $\mathscr{F}^*$)-world or that ~C occurs in it, then, if A → C is true on Lewis' criterion, then it is justifiably assertible by ours. For, A → C will be true if C holds in the nearest A-world; which means that C will hold in the nearest (A & $\mathscr{F}^*$)-world, since that's the nearest A-world; but if C holds, then it has been determined to do so by virtue of $\mathscr{L}$; so that C is law-derivable from (A & $\mathscr{F}^*$) and A → C therefore justifiably assertible. Hence, *if we assume that similarity of laws outweighs all other dissimilarities in* *interworld comparisons, and if we assume the world is deterministic in the mentioned sense, then the Humean analysis of counterfactuals that we have developed and Lewis' possible-worlds analysis counterfactuals, judge exactly the same counterfactuals to be justifiably assertible.*

The conditions we have proposed, and in particular the requirement of cotenability, tell us where the relevant A-world must differ from the actual world: it must differ at those places with which the supposition of A is not cotenable. Whatever else remains may be used for the law-deduction of C from A, provided it is compatible with the principle of predictability. The requirement of cotenability imposes a *minimum* requirement on the differences, leaving the remainder untouched, and it is for this reason that our account yields results so close to those yielded by Lewis' account in terms of *greatest similarity.*

It should be remembered, too, that the conditions that we have imposed are *not ad hoc*, introduced merely as devices sufficient to give an account that more or less fits ordinary usage. Rather, our conditions are *normatively justified: in any context of discourse defined by our cognitive interest in intelligent interference or potential interference, they are conditions that ought to be fulfilled by the assertion of a counterfactual.*

Rescher[114] has proposed an analysis of counterfactuals that, while an alternative, is in certain respects is very similar to ours.

He considers the triad of beliefs:

(a)    Julius Caesar is a person.
(b)    Julius Caesar has no tail.
(c)    All lions have tails.

We now make the contrary-to-fact assumption that

(d)    Julius Caesar was a lion.

The law (c) and the assumption (d) together entail

(e)    Julius Caesar has a tail.

This justifies the assertion of the counterfactual

(7)    If Julius Caesar were a lion, then he would have a tail.

Now, (e) and (b) contradict each other. Unless one or the other is eliminated from the context then an inconsistency results. *We* resolve this problem by means of the principle of predictability that directs us not to assert (b) in any context in which (7) is asserted. As we just put it, the description of the world that we are allowed to use in the context must be partial, and, specifically, *nodal* with respect to the consequent. Rescher proposes another rule: *when one introduces an assumption into one's beliefs that leads to a contradiction, then in the context where the assumption is made delete beliefs so as to restore consistency* [115] *and so as to leave law-beliefs unchanged.*[116] To restore consistency one must eliminate either (b) or (c), and Rescher's rule directs us to delete (b) rather than (c).

Rescher argues[117] that his rule shows why one can assert (7), while one must deny

> If Julius Caesar were a lion, then he would be a tail-less lion [since Julius Caesar had no tail].

or, what amounts to the same,[118]

(8)      If Julius Caesar were a lion, then he would have been tail-less.

which is the denial of (7). (8), he argues, cannot be asserted because his rule directs that its consequent (b) cannot be introduced into a context in which its antecedent (d) is assumed. However, in the belief situation assumed by Rescher *there is no law* that would permit the deduction of the consequent of (8) from its antecedent. So (8) cannot be asserted in any case, and there is no need to have an additional rule like Rescher's to eliminate it.

However, where we believe (a), (b), and (c), we generally also have the law-belief

(f)      No person has a tail.

and assumption (d) and beliefs (a) and (f) entail (b), which seems to justify the assertion of (8). The assumption (d) and law (c) demand the introduction of (e); this contradicts (b); Rescher's rule therefore directs us to remove this inconsistency by deleting (b). Now the assumption (d), belief (a) and law (f) restore the inconsistency by entailing (b). Rescher's rule directs us to remove this inconsistency. To do so we must block the entailment by deleting one of the premises. We keep the assumption, of course, and Rescher's rule directs us to keep the law (f), so that we must delete the belief (a).

Rescher's rule is not the only one that can block the assertion of (8), however. Our own requirement of cotenability does the job as well. For, besides laws (c) and (f), we also have the law

(g)      Nothing is both a lion and a person.

This law justifies the assertion of the counterfactual

(9)      If Julius Caesar were a lion then he would not be a person.

which establishes that its consequent (a) is not cotenable with the antecedent of (8). It therefore cannot be used as an auxiliary premiss in order to derive the consequent of (8) from its antecedent. It is, however, essential if (f) is to justify the asserion of (8). It follows that, by virtue of the unfulfillability of the requirement of cotenability, one cannot justifiably assert (8).

Now, Rescher's rule, that one preserve laws as true when contrary-to-fact assumptions are added to our sets of accepted beliefs, is one that we have in effect accepted. For, adopting Goodman's cotenability requirement amounts to saying that when an assumption is added to our assertions and the result is inconsistent with accepted laws and individual facts, then we retain the laws and delete, i.e., reject for inferential uses in that context, the individual facts that are, given the laws, not cotenable with the assumption. Rescher's rule, then, can be given a sound rationale, at least so far as concerns discourse in contexts defined by our cognitive interest in intelligent interference or potential interference. On the other hand, Rescher himself proposes no such rationale. Or rather, his only rationale consists of an attempt to, on the one hand, preserve consistency when belief-contravening assumptions are added, and, on the other hand, to provide a criterion that separates those ordinary counterfactuals that seem okay from those that do not.[119] However, in order to do the latter, Rescher is forced to add further rules. Thus, for example, the rule that one retain laws does not *require* of counterfactuals that there be a lawful connection between antecedent and consequent. To be sure, in the context in which he looked at (8), his rule succeeded in eliminating this "unnatural" counter-factual. But there is no guarantee this will always work. Indeed, it suffices to add an assumption that is causally and lawfully irrelevant to the consequent and to the other facts accepted in the context, e.g., we could obtain

> If I'm a monkey's uncle, then the moon is made of green cheese.

Such counterfactuals do, of course, have a rhetorical use. Yet they are clearly paradoxical when compared even to (8). Clearly, the paradox arises precisely from the fact that there is no connection between antecedent and consequent. Rescher's tack is to dismiss them as paradoxical[120] and simply impose the *further* rule that for a counterfactual to be justifiably assertible there must be a "genuine connection" between antecedent and consequent.[121] This succeeds in solving the specific problem, but *philosophically* it leaves much to be desired. That is, there is a deeper issue or problem that he leaves unexamined, to wit, whether there is any connection between the two rules he proposes, and whether there is any reason why people *ought* to conform to these rules. That they *do* on the whole conform to these rules does seem to be the case; but the question is, are they doing as they ought? and are they doing it because they ought to? Rescher nowhere answers such questions, nor even asks them.

The failure to develop a systematic philosophic treatment of counterfactuals leads Rescher into at least one error — or, at least, something that is, from our perspective, an error. As we have argued, Rescher's two rules that we have just examined can be given a sound rationale in any context of discourse defined by the goal of intelligent interference or potential interference. But we have also argued that in such a context thee is no reason why one should not admit as justifiably assertible counterfactuals with antecedents that *happen* to be true. Of course, considerations based on the principle of predictability or the requirement of cotenability will require that this fact not be used in any law-deductions in the context in which the counterfactual is asserted. That, however, is another matter. The point is that this does not preclude one being justified in the assertion of a subjunctive conditional the antecedent of which turns out to be true. Rescher, however, *without any rationale* introduces the categorical requirement that the antecedent of a counterfactual must be false.[122] Since most subjunctive conditionals do in fact have false antecedents, this brings his account into line with a large part of *normal usage*. But as a result he dismisses as *just wrong* certain deviant cases. A deeper analysis, such as that which we have developed, reveals that these cases, though deviant, are *not wrong*. At this point, at least, Lewis and Stalnaker are closer to the correct account than is Rescher, since they allow counterfactuals with true antecedents to be true and therefore to be justifiably assertible. Only, as we have argued, they give the wrong reasons for doing so.

It would seem, then, that Rescher's account of counterfactuals has both its sound points and its unsound ones, and that where his position is sound, it is already incorporated into ours.

There is another point that Rescher makes that is also worth looking at.

He considers Goodman's match $m$, and uses the law

($L_2$)     Whenever a match is scratched, then if there is oxygen present and the match is dry, then the match lights.[123]

(Note that this is not the law ($L_1$) that we used above[123a] in our own discussion of Goodman's example.) We now make the assumption

(A)     Match $m$ is struck

and add this to our beliefs

($B_1$)     Match $m$ is dry
($B_2$)     Oxygen is present where match $m$ is located
($B_3$)     Match $m$ has not been struck
($B_4$)     Match $m$ is not lit.

The result is an inconsistency. We must give up some beliefs in order to restore consistency and do so in a way that retains ($L_2$). This means we must give up ($B_3$) and *one of* ($B_1$), ($B_2$) and ($B_4$). If we retain ($B_1$) and ($B_2$) and give up ($B_4$), the result is the counterfactual (5):

If $m$ had been scratched, then $m$ would have lighted.

If we retain ($B_4$) and ($B_2$) and give up ($B_1$), the result is the counterfactual (6):

If $m$ had been scratched, then $m$ would not have been dry.

If we retain ($B_4$) and ($B_1$), the result is the counterfactual

(10)     If $m$ had been scratched, then $m$ would not have been in the presence of oxygen.

Rescher suggests that (5) alone of these three is "plausible", that the other two are implausible.[124] In this he is, no doubt, correct. Yet one wants to know just what 'plausibility' amounts to. Rescher raises no such question, construes 'implausible' as 'deviant' and sets about finding a criterion that will exclude them as unworthy of assertion. Certainly, there is nothing in his rules, as thus far stated, that justifies (5) over the other two. Nor, — which is just as important to us — do the rules we have developed give us any grounds for preferring (5) to the others. If Rescher is faced with a problem, then so are we. But what we have to do is not merely exclude the two deviant counterfactuals but find a *rationale* for excluding them. However, in a way with which we are by now familiar, that rationale may establish that their deviance is not due to their unassertibility but to something that leaves them assertible but "odd".

Rescher proposes as a corollary as it were to his rule for retaining laws when belief-contravening assumptions are made that we ought also to retain the auxiliary assumptions which assure its applicability.[125] Then, he claims, when we make assumption (A), the law $(L_1)$ applies only if we retain $(B_1)$ and $(B_2)$. His subsidiary rule therefore apparently suggests that we can assert (5) but not the other two, since the other two require us to give up one or the other of $(B_1)$ and $(B_2)$.

Evidently if this worked, it would be as available to us as it is to Rescher. Unfortunately, it does not do even the job Rescher asks of it. It is hardly likely, then, that we shall be able to find a rationale for it. The problem is that if $(B_1)$ and $(B_2)$ can be retained as boundary conditions to get (5), so can $(B_2)$ and $(B_4)$ be retained as boundary conditions to get (6), and $(B_1)$ and $(B_4)$ can be retained to get (10). In each case what is taken as a boundary-value fact is different, and there seems nothing that prevents them from being so taken.[126]

Certainly, there is nothing abut $(L_2)$ that picks out $(B_1)$ and $(B_2)$ as special. For, $(L_2)$ is of the form

$(L_2)$ $\quad (x)[Sx \supset ((Ox \;\&\; Dx) \supset Lx)]$

which is logically equivalent to

$(L_2')$ $\quad (x)[Sx \supset ((\sim Lx \;\&\; Ox) \supset \sim Dx)$

and to

$(L_2'')$ $\quad (x)[Sx \supset ((\sim Lx \;\&\; Dx) \supset \sim Dx)$

There is nothing about $(L_2)$, therefore, that picks out $D$ (the match is dry) and $O$ (oxygen is present) as special boundary-value conditions, where $L$ (the match lights) is not.[127] Rescher suggests that, while there is no logical difference between $(L_2)$, on the one hand, and $(L_2')$ and $(L_2'')$, on the other, in the sense that they are *deductively* equivalent, nonetheless there is a *nomological* difference: $(L_2)$ is not *inductively* equivalent to $(L_2')$ or $(L_2'')$.[128] Now, it is arguably true that statements that are logically equivalent, and therefore interchangeable in contexts within the domain of deductive logic, are often *not* interchangeable in contexts within the domain of inductive logic. Certainly, many have argued with considerable plausibility that data that yield a confirming instance of one hypothesis need not yield a confirming instance of a logically equivalent hypothesis. One has only to think of Goodman[129] and Popper.[130] For example, data that yield a confirming instance of

   (r$_1$)     All A are B

might not yield a confirming instance of its contrapositive

   (r$_2$)     All non-B are non-A.

But if we are prepared to law-assert (r$_1$) then *that* fact justifies our using (r$_2$) to explain and predict. For if we use either of the predictive arguments

| (r$_1$) | (r$_2$) |
|---|---|
| This is A | This is non-B |
| ∴ This is B | ∴ This is non-A |

then we could make *exactly the same predictions* if we replaced (r$_1$) by its logical equivalent (r$_2$). The data we mentioned therefore *directly* support the law-assertion of (r$_1$) and *indirectly* support the law-assertion of (r$_2$). In this way the epistemic route from data to an hypothesis may well be different from the epistemic route from the same data to a logical equivalent of that hypothesis. We may well agree with Rescher when he makes this claim.[131] Nonetheless, as we have just said, once we have data that support the law-assertion of hypothesis, then the fact that we law-assert that hypothesis justifies our law-asserting any of its logical equivalents: *if we are justified in using an hypothesis to explain and predict, then we are justified in using any of its logical equivalents to explain and predict the same facts.* Moreover, in any context defined by the goal of intelligent interference or potential interference, what concerns us are counterfactuals which could have yielded predictions, i.e., reasoned predictions based on laws. It follows that whatever the inductive differences relative to available data that there are between an hypothesis and some of its logical equivalents, those differences make no difference for what we are about. Rescher may well be right, therefore, that such differences exist, but he is wrong in invoking them for the purposes he does: they will not distinguish laws from their logical equivalents so far as concerns their use in predictions, nor, therefore, so far as concerns their use in supporting the assertion of counterfactuals.[132]

On the other hand, if we use, instead of $(L_2)$, the law $(L_1)$

> Whenever a match is scratched, then it lights if and only if there is oxygen present and the match is dry

which if of the form

$$(x)[Sx \supset (Lx \equiv (Ox \& Dx))]$$

then we do have a difference in logical role between $L$, on the one hand, and $O$ and $D$, on the other. $(L_1)$ supports the same counter-factuals as $(L_2)$, since it entails the latter. The converse entailment does not hold, however. The difference is that $(L_2)$ asserts $O$ and $D$ are sufficient for $L$ when $S$, whereas $(L_1)$ asserts that they are sufficient *and also* necessary. This perhaps provides us with a way out of Rescher's problem concerning the "plausibility" of the counterfactual (5) as opposed to the implausibility of (6) and (10). The principle of pre-dictability that we have proposed tells us that, if we assert (5) then we cannot also in that context assert (6) or (10), since in order to assert the latter it is necessary to assert $(B_4)$ which the principle of predictability has instructed us to delete when we assert (5). Similarly, if we assert (6) then we cannot, in that context, assert (5) or (10); and if we assert (10) then we cannot, in that context, assert (5) or (6). But this does not provide any grounds for reckoning (5) more "plausible" than (6) or (10).[133] So perhaps the different roles that $O$ and $D$ play relative to $L$ in $(L_1)$ can provide us with a reason. Now, if our aim is simply to come up with a criterion that will rule out the deviant cases without asking whether this *ought* to be the criterion we use, then this might be satisfactory. At least until the next counter-example is proposed. Yet for us, this will not do. The question would remain for us, what is it about these different logical roles of the relevant concepts in $(L_1)$ that renders "plausible" one counterfactual that the law supports and renders deviant the other two that it also supports.

But surely the difference is fairly obvious. The counterfactual (5)

> If $m$ had been scratched, then $m$ would have lighted

corresponds to a *causal relation* whereas the counterfactuals (6)

> If *m* had been scratched, then *m* would not have been dry

and (10)

> If *m* had been scratched, then *m* could not have been in the
> presence of oxygen

correspond to *lawful but non-causal* relations. Striking a match where the match is dry and oxygen is present *causes* the match to light: whereas striking a match where the match does not light but oxygen is present *does not cause* the match to be not dry though *as a matter of lawful fact* the latter must be so in the circumstances: nor does striking a match which does not light though it is dry cause the absence of oxygen though as a matter of lawful fact the latter must be so in the circumstances.

Matches have a definite function in the world of man: they are used to light fires, pipes, etc. That is, the flame they can be made to produce can be used to do these things. Normally, then, we have the lighting of a match as a proximate goal that is part of a larger task. The flame can often be produced if a match can be made to strike. That, however, is something achievable by human action. We have, then a receipe for producing a regularly desired end: strike the match and it will light. Still, the end is not always achieved. What we would like to know, then, are the conditions that are necessary and sufficient for the desired end to occur once we have made the means to happen. These conditions are what we above called "standing conditions". The receipe is a useful one provided that these standing conditions *normally* obtain — as they do in the match case. The law ($L_1$), in contrast to Rescher's ($L_2$), encapsulates the lawful knowledge that relates means, end and standing conditions. That is why laws of that form are more common in causal contexts than laws of the weaker form to which Rescher appeals. Yet what is crucial are not the different logical roles that $L$, on the one hand, and $O$ and $D$, on the other, play in ($L_1$), but rather the different *causal roles* that $S$, $L$, $O$ and $D$ play, of which ($L_1$) itself captures only a small, *though essential*, part. By bringing $S$ about, $L$ can be made to happen, provided $O$ and $D$ do, as is normal, also obtain. But by bringing $S$ about, not-$D$ is not made to happen, even when $O$ and not-$L$ obtain; nor can not-$O$ be made to happen by bringing about $S$, even though $D$ and not-$L$ obtain. Thus, by virtue of what are normal human interests, and the causal (means-ends) relations that hold under normal conditions, (5) will be by far the most commonly asserted counterfactual of the three (5), (6), (10). And, of course, when it is asserted neither of the other two can be. For this reason, then, (5) sounds "plausible" to the ear and the other two sound deviant. Yet deviance for this reasons does not render them non-assertible; oddity, as Lewis has emphasized, is one thing, non-assertibility another.

In similar fashion, where we above[134] used the principle of predicability to establish about the bolt $b$ that is black and iron at $t$, that if we assert

(i)      If the temperature of $b$ had been 650° at time $t$, then $b$ would have been red at $t$

then we cannot also in that same context assert

(ii)     If the temperature of $b$ had been 650° at $t$, then $b$ would not have been iron at $t$

On the other hand, we provided no grounds that would exclude asserting (ii) in every context. Nonetheless, (i) is normal and (ii) is clearly deviant. We can now see why this is so: (i) corresponds to a causal relation, while (ii) corresponds to lawful but non-causal relation.

It follows that if we can make good on the distinction between causation and correlation then we have solved Rescher's problem about match $m$ and bolt $b$. And of course, if our argument of Chapter I, Sec. I is sound, then the Humean can indeed make good this distinction.

We should perhaps add, however, that Rescher's instincts were sound when he indicated that when we assert counterfactuals about what would be the case were the match $m$ struck, what we normally hold constant are the standing conditions. Where Rescher went wrong was in identifying these as conditions for applying a law rather than as conditions that must obtain if the implementation of a means is to achieve its customary end. Modified to fit this revision Rescher's rule may be said to stand, and it receives a rationale in *many* contexts defined by our cognitive interests in intelligent interference or potential interference. The rule concerning the revision of belief-sets when belief contravening assumptions are added would be this: (1) modify the belief-set, deleting beliefs sufficient to restore consistency, while (2) retaining all laws, and, (3) retaining so far as is consistent with the previous requirements, standing conditions that must obtain if antecedent and consequent are to stand in their normal means-end relation-ship.[135] We have seen that conditions (1) and (2) receive a rationale in contexts defined by the goal of intelligent interference or potential interference. *If the context is narrowed by specifying that the antecedent-consequent relation must not be merely lawful, but causal as well*, then condition (3) receives its rationale. Rescher doesn't make clear why (3) should be weaker than (2). We must now see why, however: it is because (3) is justified for a narrower range of contexts of discourse than is (2).

We must now recognize, however, that (2) is not to be adhered to rigidly either. Rescher, in proposing (2), is standing strongly opposed to David Lewis' account, who, as we know, is prepared to argue that a counterfactual might be justifiably asserted even if, in a possible world where the antecedent and consequent are both true, one of the laws that holds in our world would be violated. Up to now we have stood more or less alongside Rescher in this disagreement with Lewis, but now I want to argue that we should go part way towards Lewis' position. To be more specific: On the account of counterfactuals that we have proposed, when one asserts a counterfactual, then one assumes the antecedent, asserts the relevant laws, and deduces the consequent from there, and, moreover, one also assumes the negation of what is, given the laws, not cotenable with the assumption of the antecedent. This, in effect, coincides with Rescher's rule that, when one makes a belief-contravening assumption, one restores consistency to the belief set by retaining the laws and deleting from one's belief set such beliefs about individual facts that will be minimally sufficient to restore consistency. Since the requirement of cotenability requires us to make a *minimal* change from the actual world when we make our belief-contravening assumption, then, provided that the actual and possible worlds are deterministic and provided that our scale of interworld similarly ranks as more similar to each other worlds that share laws than worlds that don't, *then* our account and Lewis' pick out the same counterfactuals as worthy of assertion. Lewis, however, rejects this standard of interworld similarity, and holds (though without much argument) that worlds that differ in laws might be more similar than worlds that share laws: a world in which there is a minor violation, i.e., miracle, of a generaliza-tion that is a law in our world, might be more similar to ours than any world in which all the laws of the actual world also obtained. If we put Lewis' point into our framework, then what it amounts to is this: in some contexts in which a counterfactual is asserted, then rather than also assuming the denial of individual facts not cotenable with it, one may instead retain those facts and assume the denial of the law that renders them non-cotenable with the antecedent of the counterfactual. Or in Rescher's terms, what Lewis is saying is this: when one makes a belief-contravening assumption, rather than retaining the laws and deleting individual facts in order to restore consistency, one may instead restore consistency by retaining those facts and deleting a law. *Is there any context in which it is reasonable to do this*? If the answer to this question is affirmative, then this will show that the intuitions on the basis of which Lewis constructs his possible-worlds account of counterfactuals are intuitions that can be shown to be justified in terms of (what we have argued is) the more justified account of counterfactuals that we have been developing.

In Rescher's way of putting these things, when we make belief-contravening assumptions we restore consistency by retaining law-assertions and assertions of standing conditions, and if we must choose between these, then we give up the latter before the former. We have just seen how this can be given a rationale, provided the context of the counterfactual discourse is defined by the goal of intelligent inter-ference. *What must be recognized is that in such contexts of discourse, one is sometimes justified in retaining an individual fact and giving up a law.* Note, however, that even if this is admitted, and even if we go this far with Lewis in admitting that worlds with "miracles" can be more similar to ours than worlds which share our laws, one can *not* draw the further conclusion, which Lewis wants to defend, that a counterfactual may be worthy of assertion even if the consequent is not law-deducible from the antecedent. That is, even we allow that it is sometimes reasonable to delete certain law beliefs, it does not follow that the laws connecting antecedent to consequent are among those that may be deleted. In *this* respect, whatever insights we are prepared to grant to Lewis' intuitions, we are *not* prepared to make concessions on the central point of disagreement.

The relevant point may be made simply. Some laws hold unconditionally, others hold only conditionally. (The latter will, of course, be imperfect, but nonetheless laws for all that.) In contexts of counterfactual discourse, we are often justified in retaining an individual fact and giving up a law where the law that we give up holds only conditionally. Normally that will require one to give up further individual facts, to wit, those the obtaining of which is sufficient for the obtaining of the generality to be given up. To make this more concrete, consider once again the barometer case, only this time we must specify the relevant laws in greater detail.

Let '$Wxt$' indicate that barometer $x$ is in working order at $t$; let '$Rxt$' indicate that $x$ gives reading $R$ at $t$; and let '$Pxt$' indicate that $x$ is situated in a pressure $P$ at $t$. We have the following law about barometers:

(M$_1$)    $(x)(t)[Wxt \equiv (Rxt \equiv Pxt)]$

Moreover, if we take $a$ to be an individual barometer, then we have the following law about $a$:

(M$_2$)    $(t)[Wat \equiv (Rat \equiv Pat)]$

Now, it is a standing condition with respect to $a$ that it is in working order:

(M$_3$)    $(t)(Wat)$

Hence the following *conditioned* law also holds of $a$:

(M$_4$)    $(t)(Rat \equiv Pat)$

Assume we are at time $t_0$. The law (M$_4$) justifies the assertion of the counterfactual

(N$_1$)    $\sim Rat_0 \rightarrow \sim Pat_0$

which says that if the barometer were to have a different reading then the pressure would be different. But, as Lewis points out on the basis of his intuitions,[136] we might also be inclined to say that if the barometer were to read differently the pressure would *not* be different. In that case we would be making the belief-contravening assumption

(O$_1$)    $\sim Rat_0$

while retaining the belief

(O$_2$)    $Pat_0$

However, given (M$_4$), (O$_2$) is *not* cotenable with (O$_1$). If we follow Rescher's rule, of always retaining the laws, then we cannot both suppose (O$_1$) and retain (O$_2$). Thus, Rescher's rule would exclude Lewis' intuitions. On the other hand, if Lewis' intuitions are sound, then we must give up the law-belief ($M_4$). However, this is a *conditioned* law. If we give it up, we can still retain the *unconditioned* law ($M_1$), provided that we give up the belief that the normal standing condition ($M_3$) obtains, that is, the condition that must obtain if the conditioned law ($M_4$) is to hold. And in fact, the unconditioned law ($M_1$), and the fact ($O_2$), jointly justify the assertion of the counterfactual Lewis is inclined to assert, to wit,

$$(N_2) \qquad \sim Rat_0 \rightarrow \sim Wat_0$$

which says that if the barometer were to have a different reading then it would be broken.

Lewis, of course, supports the validity of his intuition that ($N_2$) is often assertible by appeal to his possible-worlds analysis of counterfactuals, or, more specifically, to his scale of interworld similarity: an ($O_1$)-world in which ($O_2$) holds is more similar to ours than ($O_1$)-world in which the law ($M_4$) of our world holds. But does his intuition have validity within the framework we are proposing?

We can, perhaps, get at an answer to this question by looking at some other aspects of Lewis' intuitions, and perhaps at the same time recognize that the underlying concept that guides his thinking as he develops his analysis is not so very far from ours.

Lewis relates the intuitive notion of inter-world similarity that he employs to an equally intuitive notion of comparative possibility. "It is *more possible*," we are told, "for a dog to talk than for a stone to talk, since some worlds with talking dogs are more like our world than is any world with talking stones. It is more possible for a stone to talk than for eighteen to be a prime number, however, since stones do talk at some worlds far from ours, but presumably eighteen is not a prime number at any world at all, no matter how remote."[137] Recall, too, that "Possibility is true at some *accessible* world . . .",[138] so that the more possible a world is the more accessible it is from our, the actual, world. Mondadori and Morton tell us that if one "physical process . . . is more possible" than another, then it would "*make fewer demands* on the way the world actually is . . .".[139] Stalnaker tells us that the possible world selected for testing the truth of a counterfactual must "differ minimally from the actual world. This implies, first, that there are no differences between the actual world and the selected world except those that are required, implicitly or explicitly, by the antecedent. Further, it means that among the alternative ways of *making the required changes*, one must choose one that does the least violence to the correct description and explanation of the actual world."[140] The metaphors that recur are suggestive: the more possible world is more accessible because it requires that fewer changes be made in the actual world, requiring the least violence be done to the latter; the number of points of difference between the possible world and the actual world is a *measure of what must be done to change* the actual world into the possible world, or, what is the same, to transform the possible world into the actual world; and the *nearest* relevant world, one that is *most accessible*, is the world that is accessible with *the least amount of doing*; the relevant world that *differs minimally* from ours is the one to which we can accede through *the least effort*. That is, the metaphors seem to suggest that a possible world is accessible through *doing something* and that the more similar it is then *the less needs to be done*. Thus, the nearest relevant world is the most possible in the sense that it is the world that is *most easily brought about*, the world the achievement of which requires *the least interference* in the processes of the actual world.[141] Or, in short, *the metaphors that apparently guide Lewis and others in constructing their possible-world analyses reveal that they, too, tend to locate counterfactuals in an area of discourse defined by the aim of intelligent interference.*

Lewis wishes to say that (sometimes) a possible world with an individual fact that violates a generality that is a law in our world is more similar to ours than another possible world which has the same laws but different individual facts. What this means, we are suggesting, is that the former world is more easily achieved through interference. But this *is* often so. Let $W_a$ be a possible world in which the barometer reads other than it does, $(O_1)$, in which the pressure remains the same, $(O_2)$, and in which the law $(M_4)$ that describes the barometer in our world does not hold. Let $W_b$ be a possible world in which the barometer reading is different, and $(M_4)$ obtains, so that the pressure must be different. *It is evident that the world $W_a$ is more easily achieved than world $W_b$. Since it is often possible to change the conditions that guarantee that certain imperfect laws hold, antecedent-worlds in which such laws do not hold are often more accessible, or more possible, than antecedent-worlds in which those laws do hold; which is to say that the former are more similar to our world than the latter.*

If we take interworld similiarity to be a (crude) measure of the amount of interference that is necessary to making the antecedent of a counterfactual actual, then Lewis is correct in his judgment that certain worlds which differ from ours in laws are more similar to our world than others which share our laws. Moreover, Rescher is wrong when he lays down the rule that when we make a belief-contravening assumption we ought never to restore consistency by deleting a law. To the contrary, we ought sometimes to give up a law-belief. The rule is: if, in the context in which a belief-contravening assumption is made, interference can more easily bring about a state of affairs in which the assumption is true and the law false than it can bring about a state of affairs in which the assumption and the law are both true and some other individual fact is made false, then in order to restore consistency delete the law. Or to put it in our terms: if an individual fact is, by virtue of a certain law, not cotenable with the assumption of the antecedent of a subjunctive conditional, then we should assume the negation of the law rather than the negation of the individual fact just in case that it is easier to achieve the former by interference than it is the latter.

We should note, of course, that whether or not certain states of affairs *can* be achieved by interference, or *can be more easily achieved*, is itself a matter of what laws obtain: whether by doing something one can achieve something else depends upon the laws describing the interactions of human beings with other natural (including social) events.[141b] Thus, to make only the most obvious point, a law can be "changed" only if it is a conditioned rather than an unconditioned regularity, and whether a law is conditioned, and what those conditions are, depends upon the laws, including (where relevant) the less imperfect laws that hold in the area.

All this clearly has a rationale in the framework we have developed.

This framework consists in locating counterfactuals in the context of discourse defined by the end of intelligent interference or potential interference in natural (including social) processes. It is this end that determines that the consequent of a counterfactual ought to be law-deducible from its antecedent. For, where one's end is *intelligent* interference or potential interference, then one is concerned with the lawful relations that hold among sorts of events, i.e., the relations that determine whether interference is even possible (lawfully possibly) and if it is, what exactly it can achieve. This provides a generic constraint on the counterfactuals that we may justifiably assert. But the discourse may be defined not merely by a generic interest in the possibilities of interference, but by a more specific interest in what interference, were it to occur, could actually achieve. This more specific constraint on counterfactuals requires, as we saw just above in discussing Rescher, that the connection between antecedent and consequent be causal. What we have now seen is that this constraint may be made more specific yet, to pick out the mode of interference that could most easily bring about the consequent. If this further constraint is put on the context of discourse, then it might well turn out, as Lewis says, that we are justified in giving up a law (i.e., assuming its denial) and letting a minor "miracle" stand. Only, we also see that this talk of "miracles" is thoroughly misleading.

There is another, not unrelated, context for counterfactuals that also justifies the giving up of *conditioned* laws, and assuming their negations when one assumes the antecedent of the counterfactual. This is a context we have mentioned before, namely, that in which responsibility is ascribed. If we can justifiably assert that

(*)     If Jones had done A, then X would not have happened

then we have *prima facie* grounds for holding Jones responsible for X's having happened. And, of course, if Jones is responsible for X then certain things will (likely) also happen to Jones, e.g., he may be charged and convicted of manslaughter due to negligence. However, there are ways of excusing Jones from his responsibility. One way is (as we suggested above, in our discussion of semi-factuals) by arguing that X would have happened anyway, even if Jones had done A. There is a second way in which he might be excused that is of interest in the present context. This is by arguing that Jones *could not have done* A, i.e., that that state of affairs was *not accessible* through his choice. There are two possibilities concerning this "could not have done". First, there is the case of external compulsion: someone else has forced Jones not to do A, to do something else. In this case we transfer responsibility from Jones to the person who was coercing him. Secondly, we might argue that Jones could not have done A because it was contrary to his character. Now, a person's character is constituted by various regular patterns of behaviour, regular ways of responding to various stimuli (both external and internal). But these regular patterns of response are all learned; that is, they are *conditioned* regularities. If one is *strict* in his ascriptions of responsibility, then one will hold Jones responsible for X even though he could have avoided X only by doing A, something which was out of character. Such a one will assume the antecedent of (*) in ascribing responsibility to Jones and will assume along with that that certain regularities in Jones character do not hold; that is, such a one will allow, if you wish, a small miracle of character, an uncountable free choice. On the other hand, if one is *liberal* in his ascriptions of responsibility, then one will assume the antecedent of (*) but will not give up the regularities of Jones character; but that will render some particular fact uncotenable with the antecedent and one will have also to assume its negation, either the condition that evokes in Jones the response of not doing A or the condition the presence of which in the past determines the present regularity in Jones' character; and whichever such fact or facts is assumed will have the responsibility transferred to it.[142]

There are, therefore, contexts in which one is justified in including as part of one's assumption, when one asserts a counterfactual, the negation of a law.[142a] In addition there are also contexts in which one assumes a law in its own right. In such cases one must also assume the negations of any individual facts inconsistent with the law. (One may, of course, often be able to do this in a variety of ways. In that case all that one need assume is that *there is* one or another of these alternatives that is fulfilled.)

The most important such context is that of *research*.[143] Here one often reasons prior to performing experiments as follows: "If this law were true, then if we did so and so, then such and such would be the upshot." We then do so and so, and if such and such *is* the upshot then we have evidence that confirms as true the generalization we have in our reasoning merely supposed to be true; or such and such is *not* the upshot and we have falsified the law we had supposed to be true. Such reasoning is common in science; as Popper, Kuhn, and otheers[144] have emphasized, the scientist is not a *mere* observer: to use Bacon's terms, he *interrogates* nature,[145] asks whether an hypothesis that he has proposed fits or does not fit the facts. Popper thinks that this is incompatible with inductivism or positivism. His reason is that according to the inductivist or the positivist one cannot accept an hypothesis prior to the availability of confirming data. But in order to defend the positivist or inductivist from this charge of being unable to account for experimental practice in science, all one must do is distinguish, as we did above in Chapter I, Section I, between acceptance$_1$ and acceptance$_2$, that is, between accepting as an hypothesis worthy of testing and accepting as confirmed,[146] or true so far as available data testify. Be that as it may, however, the present point is that in contexts of interrogating nature, one may assume laws, and these assumptions are often, as we come to discover, contrary-to-fact.

In this sort of reasoning in the context of research, we would *assume* a generalization like

> All A are B

to be lawful, and then *assert* the counterfactual

> This is A → this is B

For this counterfactual, the connection between antecedent and consequent is provided not by a generalization that is law-asserted but only by a generalization that is merely assumed to be a law. In our terms, the legitimacy of this is clear enough. We have located counterfactuals in a context of discourse defined by the end of intelligent interference and potential interference. This end also defines the context of research. For what research aims at is the discovery of *laws*, that is, the matter-of-fact generalities that determine whether interference is possible and, if it is, then what it can achieve. Since research aims to discover lawful connections that are not yet known, and since one must reason about the alternative possibilities if one is to intelligently interrogate nature, it follows that in such contexts one is fully justified in asserting counterfactuals in which the connection between antecedent and consequent is not asserted but merely assumed.

In Lewis' terms, to assert

> This is A → this is B

when we assume the generality

> All A are B

is to pick out as the A-world most similar to ours that one in which the generality holds. Since the generality may well not hold in our world, this means that the criterion of interworld similarity that is being used does *not* impose the condition that similar worlds must share the same laws. Lewis' approach provides nothing by way of a rationale that makes sense of such a shift in criteria of similarity. Nor is there anything in his vague talk of similarity, or in Stalnaker's talk of a "selection function", that shows why the relaxation of the shared-law criterion is, in the context of research at least, intimately related to the similarity criterion through an abiding concern with the lawful connections among the shorts of events mentioned in the antecedent and the consequent. In contrast, our account provides just such a rationale.

The laws that we assume in experimental contexts are most often at the specific level. We accept as true a generic theory, and then assume only such specific-level hypotheses as are compatible with this generic theory.

However, there are contexts in which one may assume laws that are contrary to the deepest, i.e., most generic, theory. For, there are scientific contexts in which even the deepest theory is called into question, and a researcher begins to assume a contrary theory is true, and search for data that would confirm or disconfirm this hypothesis. One has only to think of Einstein re-thinking the kinematical basis of physics. Such episodes are relatively rare in the history of science — not without reason, Kuhn refers to such periods as "revolutionary" [147] — but they do occur.

All this is fairly clear. For our purposes the important point is that we may well *assume* to be true generalities that are contrary-to-fact, and these generalities need not be only those that hold conditionally, but also unconditioned generalities, *including those at the deepest level*.[147a] Now, it is only conditioned regularities that are in any sense open to change by human interference. Thus, Lewis is wrong when he implies, as he does, that inter-world similarity, and therefore the assertibility conditions of counterfactuals, is to be given in terms of a notion of accessibility that requires the most accessible = most similar world to be that which is most easily attained through human effort.[147b] Rather, as we pointed out in our discussion of Rosenberg above,[148] the criterion of inter-world similarity — or, what amounts to the same, the assertibility conditions of counterfactuals — varies from context to context: there are many scales of inter-world similarity. Or, to put it less misleadingly: there are different criteria for what is to be assumed as part of the antecedent of counterfactuals, and for which short of laws may be used to law-deduce the consequent. Which criterion is to be used varies from context to context, and the relevant factor determining which criterion to use is the purpose that defines the context of discourse in which the counterfactuals are used.

In general, then, we may conclude that our approach to counterfactuals by reference to the ends that define the context of discourse in which they occur can provide a set of standards for their assertibility that establishes a rationale for their use which is largely absent from a possible-worlds analysis of the sort provided by Lewis and Stalnaker. In particular, we have shown how Lewis' intuitions about counterfactuals, when suitably qualified, do in fact have a rationale. But that rationale is provided by our own account and not by the possible-worlds account. Thus, insofar as Lewis bases much of his case for the possible-worlds analysis on an appeal to his intuitions, we see that, once again, his case does not stand up to critical scrutiny.

### (v)  *Counterfactuals and Inductive Support*

The Humean analysis of the use of counterfactuals reduces the problem of the latter to two other problems, namely, (1) the problem of induction or of inductive support for laws, and (2) the problem of the contexts in which it is and is not reasonable to make various suppositions. The latter is relevant in two ways: (2a) it determines that counterfactuals are assertible only if the consequent is law-deducible from the antecedent; and (2b) it determines what other beliefs one must give up when one makes the supposition that the antecedent is true. Given (2a), then the criteria of inductive support mentioned in (1) determine which generalities may be used to support counterfactual assertions, i.e., be among the general premises which are implicitly asserted when a counterfactual is asserted.[149] It is not the purpose of the present essay to solve the problem of induction, to fully lay out and justify rules for inductive support. That there are good inductive reasons, good grounds justifying law-assertibility, we are taking for granted rather than arguing for. To be sure, in our discussion of scientific method in Chapter I, above, we have *illustrated* certain *patterns* of inductive support that are especially relevant to what we are about.[150] But we have not attempted to justify them against possible sceptical attacks.[151] On the other hand, the Humean account of counterfactuals *does* presuppose that the issue (1) be addressed. To that extent, our account is incomplete.[152] Moreover, it is a point at which an opponent can attempt to find counter-examples. That is, he can attempt to find two counterfactuals in which the connection supporting the one has less inductive support than the connection supporting the other but where the former is assertible and the latter is not. Not surprisingly, we find Lewis proposing such a counter-example, and to defend the Humean we must show that it does not, after all, constitute a counter-example. And fortunately, in order to show that we do not have to solve the problem of induction; we need no idea of what constitutes good inductive support than that which we have already provided.

It will pay to quote Lewis' discussion in full:

I am a moderate Warrenite. I think it quite probable that Oswald killed Kennedy, that he was working alone, and that there was no second killer waiting. But I think it slightly probable that Oswald was innocent, and that someone else killed Kennedy. I think it overwhelmingly probable that one or the other of these two hypotheses is true; and negligibly probable, for instance, that Kennedy was not killed at all. Then what happens when I . . . [add the belief-contravening assumption] 'Oswald did not kill Kennedy' to my stock of beliefs as if it were an item of new knowledge? Clearly I continue to believe that Kennedy was killed (perhaps not with quite as much certainly as before, but still very strongly indeed), and give up the belief that there was no killer but Oswald. That is: when my most probable hypothesis is ruled out, most of the probability goes to what was my next-most-probable hypothesis. According to the [Humean] . . . , this means that 'Kennedy was killed' is cotenable for me with the supposition that Oswald did not kill him, and 'No one but Oswald killed Kennedy' is not. I should therefore assert such counterfactuals as

(a)        If Oswald had not killed Kennedy, someone else would have.

rather than

(b)        If Oswald had not killed him, Kennedy would not have been killed.

But this is just backward from the truth. Actually I assert the second and deny the first. Further, I regard 'No one but Oswald killed Kennedy' as cotenable with the supposition that Oswald did not kill him, and I regard 'Kennedy was killed' as not cotenable with that supposition. The reason is plain. According to my actual system of beliefs — beliefs that have not really been revised under the impact of the belief-contravening supposition — probably I inhabit one of the worlds where Oswald did kill Kennedy, working alone, with no other killer waiting. These worlds (except for some negligible probability according to my beliefs) are worlds where the second counterfactual is true and the first is false. That is because they are worlds to which worlds with no killing are closer than worlds with a different killer. Therefore the second counterfactual is probably true, according to my beliefs, and the first is probably false. Therefore I assert the second and deny the first. To summarize: there is no reason at all why my most probable antecedent-worlds should be the same as the antecedent-worlds closest to my most probable worlds. The [Humean account] gives me the character of the former worlds, but the assertibility of counterfactuals depends on the character of the latter worlds.[153]

We have the following propositions:

(1)        O killed K
(2)        Only O killed K

(3)      O worked alone
(4)      O was innocent
(5)      Someone other than O killed K
(6)      K was not killed

Let ($\gamma$) be the conjunction: (1) & (2) & (3)
Let ($\delta$) be the conjunction: (4) & (5)
Let ($\alpha$) be the disjunction: ($\gamma$) $\vee$ ($\delta$)
Let ($\beta$) be the proposition: (6)
Let ($\varepsilon$) be the disjunction: ($\beta$) $\vee$ ($\delta$)

If ($\alpha$) is true, then ($\beta$) is false, and conversely, since $\sim$ (6) follows from both disjuncts of ($\alpha$). Hence, if we accept ($\alpha$) we must reject ($\beta$) and conversely. Similarly, if K was killed then either O did it or O didn't, i.e., either O did it or O was innocent, so that if ($\alpha$) is true, then either ($\gamma$) is true or ($\delta$) is true but not both. Thus, if we accept ($\alpha$), and reject ($\gamma$) then we must accept ($\delta$). *At least, so Lewis holds.*

For Lewis, ($\beta$) is negligibly probable. He therefore rejects ($\beta$) and accepts ($\alpha$), which is overwhelmingly probable. Lewis thus believes that (6) is false. Moreover, ($\gamma$) is for Lewis highly probable, so he believes that this is true, and therefore believes that ($\delta$) is false. Within this framework of probabilities, Lewis now considers what would happen if he made the supposition that

O did not kill K

i.e., the supposition that (5) is false. This is a belief-contravening assumption, since it contradicts the belief (1), and therefore the belief ($\gamma$). To restore consistency we must reject ($\gamma$), and therefore accept ($\varepsilon$). But if we reject ($\gamma$), and ($\alpha$) is overwhelmingly probable, then we must hold that the remaining disjunct of ($\alpha$), namely ($\delta$), has become overwhelmingly probable. And if we accept the disjunction ($\varepsilon$), and accept that its one disjunction ($\delta$) is overwhelmingly probable and that its other, and mutually exclusive disjunct ($\beta$) is negligible, then one must accept that ($\delta$) is true. Thus, when one adds the belief contravening assumption that (5) is false, one revises one's beliefs to now accept ($\delta$).

This in turn permits one to assert the counterfactual (a):

If O did not kill K, then someone other than O would have killed K that is,

$$(4) \rightarrow (5)$$

And this in turn requires one to reject (b)

If O had not killed K, then K would not have been killed

that is,

$$(4) \rightarrow (6)$$

which is entailed by (b) since (5) entails not-(6)

But, says Lewis, it is not (a) that one ought to assert, but rather (b). And therefore the Humean account is in error.

Lewis' reason for asserting (b) rather than (a) is that any antecedent-world (i.e., any possible in which O did not kill K) in which K was not killed is more like our (or what is most probably our) world than any world in which someone else killed K.

Now, the first point to be made is that *this* does *not* constitute an argument against the Humean. For, Lewis finds (b) assertible and (a) non-assertible on the basis of his own account of counterfactuals. And this is simply to beg the question in favour of his own account, and against the Humean's! Lewis can use his example to argue against the Humean *only if it is intuitively clear, on theory-independent grounds, that* (b) *is, and* (a) *is not, worthy of assertion.* And in fact, it is clear that intuition provides no sure guide at all that (b) is more worthy of assertion than (a). Indeed, what is interesting about (a) and (b) is that both have an element of plausibility, even though the assertion of one requires the rejection of the other. This means, in turn, that the more adequate analysis will be the one that accounts for this ambivalence about (a) and (b), rather than one that comes down, as Lewis' does, on the side of one of these two counterfactuals. Be that as it may, however, Lewis' case is unpersuasive so long as it rests merely upon the assertion of the correctness of his own position.

Moreover, it is not at all evident that, even on Lewis' own analysis, (b) is more worthy of assertion than (a). Consider an antecedent-world, that is, one in which (4)

O did not kill K

holds. (b) is more worthy of assertion than (a) only if such a world in which (5) holds, and therefore in which K is killed, is *less like* our world than is an antecedent-world in which (5) holds, i.e., in which K is not killed. Surely, however, any possible world in which K is not killed has to differ far more from the actual world than one in which K is killed, not by O but by, say, some other Dallas assasin. Surely, the consequences of Kennedy not being killed — serving out his term, succeeding himself, Johnson not becoming president, and so on — are sufficient to make any such world vastly more dissimilar to ours than any world in which Kennedy is killed by some assasin coincidently succeeding where Oswald failed, or by a second Oswald who took the real Oswald's place at the crucial moments.[154] And in that case, by Lewis' own analysis, it is (a) that ought to be asserted and not (b)! That is, it would seem that, contrary to what Lewis himself apparently believes, his analysis actually agrees with the position that he attributes to the Humean.

However, Lewis does have a way out of this criticism. For, he argues that "we put more weight on earlier similarities of fact than on later ones."[155] Lewis is not sure of the reason for this: "Perhaps," he says, "it is just brute fact"[156] that we so weigh earlier similarities to count more than later ones. However, he also suggests that

... perhaps my standards are less discriminatory than they seem. For some reason — something to do with the *de facto* or nomological asymmetries of time that prevail at *i* if *i* is a world something like ours — it seems to take less of a miracle to give us an antecedent-world exactly like *i* in the past than it does to give us one exactly like *i* in the future. For the first, all we need is one little miraculous shove, applied ... at the right moment. For the second, we need much more.[157]

This point is surely well taken, at least insofar as we gloss this as we argued it should be glossed, in terms of the rule that that world is most accessible, most similar to ours, the bringing about of which requires the least interference in the processes of the actual world.[158] It would have been far easier — it would require a "smaller shove", it would be a "smaller miracle" — if Oswald's finger had slipped, causing the shots to miss, and leading to Kennedy's survival, than it would have been to arrange for a large group — a second Oswald, or back-up man, but in any case a conspiracy involving others, or alternatively, a second assassin or conspiracy doing the deed months later or even minutes earlier.[159] Moreover, as Lewis points out, there is an asymmetry between cause and effect: we bring about the cause in order to bring about the effect, and not conversely. So, if we were contemplating interference, we would aim to achieve the antecedent of the conditional in order to bring about the consequent. What counts in determining accessibility is the difficulty in bringing about the antecedent because, *given the laws connecting antecedent and consequent,* in bringing about the antecedent one has *thereby* brought about the consequent. It is worth noting that it is Lewis himself who introduces as relevant to inter-world similarity judgments the idea of nomological asymmetries and thereby the idea of nomological connections; it is not we ourselves who did this, trying somehow to impose a Humean view on Lewis. And what is clear is that *Lewis' emphasis upon weighing earlier similarities of fact more than later ones can be given a rationale in terms of the account of counterfactuals that we have proposed that locates them in a context of discourse determined by the aim of potential interference.*[160]

If this is so, then it would seem that the Humean, as well as Lewis, ought to agree that it is the counterfactual (b) rather than (a) that ought to be asserted. What, then, are we to say about Lewis' argument that the Humean is committed to asserting (a)?

The relevant point can be best brought out, perhaps, by addressing another problem that we have already mentioned, namely, the plausibility of *both* (a) *and* (b) — in spite of the fact that they are apparently not jointly assertible. This, we suggested, tells against any analysis which forces one to come down on one side or the other without accounting for their being both plausible. Indeed, Lewis argues that the Humean asserts (a), and if he is correct, that counts against the Humean; Lewis himself commits himself to (b), and that therefore counts against his analysis; and we have just argued that the Humean might well accept Lewis' point and assert (b), which would, again, count against him. Is no one correct?

What must be noted is that (a) and (b) are by no means as clear in their import as Lewis suggests. In fact, to take only (a), it is ambiguous: (a) might be used to assert

(a′)     If O did not kill K, then someone other than O would have, not right then but eventually, since the disaffection of certain groups was sufficient to ensure it.

or to assert

(a″)     If O did not kill K, then someone other than O would have done it there, since there were others involved in the conspiracy.

or to assert

(a''')     If O did not kill K, then someone other than O must have
          done it, since K was in fact killed at that time in Dallas.

Nor are these the only possibilities. Moreover, similar ambiguities
appear in (b). And of course, what is the crucial point, one might be
justified in asserting one of these, say (a'), while denying others, say (a'')
and (a'''). Thus, if we consider just (a) by itself, we are tempted to say
*both* that it is plausible *and* that it tends to conflict with (b) *and* that
(b) is plausible. It is because there are these ambiguities in (a) and (b)
that no analysis is adequate that judges just one and not the other to be
assertible *without qualification*.

     It is clear that Lewis' account does not force him to take either (a) or
(b) as unambiguous. Once the ambiguity is noted, he is free to qualify
as appropriate his committment to asserting (b) and rejecting (a).
Certainly, noting the ambiguity in (a) and (b) in no way entails rejecting
his criterion of inter-world similarity that involves weighing earlier
similarities more than later ones — a criterion that we also saw could be
given a rationale within our Humean framework.

     However, the disambiguating of (a) and (b) does tell against Lewis to
a certain extent. If the Humean is correct, then to assert (a) is to assert
*implicitly* certain laws $\mathscr{L}$ and individual facts $\mathscr{F}$ that permit the law-
deduction of the consequent from the antecedent. On this account, (a)
can be ambiguous because what is implicitly asserted may vary from
context to context. It is for this reason that the Humean is unable
to provide a neat algorithm for reducing subjunctive conditionals to
complex assertions. The ambiguity in what a subjunctive conditional is
used to assert can be eliminated by making explicit what is only
implicit. Clearly, disambiguating (a) into (a'), (a''), (a'''), etc., consists of
making explicit what must be implicit in any context in which it is used
to make a clear assertion. Counterfactuals thus carry with them implicit
content, that varies from context to context. This testifies somewhat to
the claims of the Humean. Lewis does suggest that what context
resolves is not what is implicitly asserted, but rather the vagueness of
inter-world similarity in a way favourable to the truth of one counterfactual rather than another.[161] One who asserts (a) = (a') is weighing
inter-world similarities somewhat differently than one who asserts (a) =
(a''). No doubt his position can thus be defended. Nonetheless, the

Humean's account seems more plausible since the facts that are made explicit in, e.g., (a'), (a''), (a'''), are facts that are relevant to the inferrability of the consequent of (a) from its antecedent.

Nonetheless, one must admit that the ambiguity of (a) and (b) tells only moderately in favour of the Humean: Lewis' account can be made to fit the data. The more important point is, however, that when Lewis has the Humean assert (a) he makes the whole discussion turn only on the probabilities of various individual facts (1)—(6), and completely ignores the laws and individual facts that are needed, if the Humean is correct, for the *law-deduction of antecedent from consequent*. What the ambiguity in (a) consists in, according to the Humean, is *different ways of law-deducing* the consequent of (a) from its antecedent. Lewis tries to force the Humean to assert (a) *without once introducing considerations of how the law-deduction is to proceed*.

Lewis construes the Humean as committed to picking out the most probable of those worlds in which the antecedent is true and then noting whether the consequent is true in that possible world.[162] But this is simply wrong.[163] In the *first* place, this imports too much of Lewis' own analysis into the Humean's. It makes the truth, or assertibility, of a counterfactual depend upon antecedent and consequent holding in some possible world. But for the Humean what is crucial to assertibility is law-deducibility. Lewis' construal of the Humean *simply ignores* this, and, therefore, is no version at all of what the Humean is committed to. In the *second* place, it is perfectly open to the Humean to adopt some criterion other than probability for picking out the antecedents of assertible counterfactuals. Thus, one can use the criterion of cotenability that we, following Goodman, have proposed, and which, moreover, can be given a rationale in terms of certain ends defining contexts of discourse in which counterfactuals are used. Interestingly enough, Lewis proposes no rationale at all for the probabilistic criterion that he proposes that the Humean must adopt. He himself adopts a criterion of inter-world similarity the main import of which can, we have argued, be given a rationale which shows it to be acceptable to the Humean. There seems no reason, then, why the latter should not agree with Lewis in rejecting assertibility conditions for counterfactuals that are based on discovering which is the most probable antecedent-world.

We may conclude, therefore, with Lewis, that the probability considerations of the Warrenite do indeed not establish the assertibility of (a) rather than (b). But when we so agree, we are agreeing *as Humeans*.

Lewis' counterexample may be a counter-example to something, but it is certainly no counterexample to a Humean analysis of counterfactuals.

## IX. CONCLUSION

We have now examined in detail the challenges to the Humean account of counterfactual conditionals that arise from the analysis of such conditionals in terms of a possible-worlds semantics. We have found that this battery of arguments and counter-examples fails to establish the unsoundness of the Humean analysis. Or at least, we have succeeded in replying to these challenges if a number of promissory notes that we have issued can be cashed in.

Thus, in the first place, we have used a notion of greater and lesser content in laws, i.e., the distinction between process knowledge, which is the ideal, and imperfect knowledge, which falls short of that ideal. We have assumed this distinction in the present Chapter. This clearly presupposes the defence of this distinction that we sketched in Chapter I.[1]

Secondly, we have introduced a distinction between scientific explanations, causal explanations, and correlations. Causal explanation is a species of scientific explanation; correlation is another; process, we might add, is still a third. But these distinctions and claims also need to be defended. That defence also was sketched in Chapter I. We have further claimed that causation involves an asymmetry between cause and effect that is not present in general with correlations or other sorts of lawful connection. We have, however, done little more than hint at an account of what constitutes this asymmetry. In this case the material is too complex to be dealt with in this essay, and the reader must be referred elsewhere for a detailed discussion.[2]

Thirdly, we have relied upon Lewis' idea that it can be "odd" to say something even though it is not absolutely wrong. We have discussed this point in Chapter II, but this notion, too, requires further discussion, that cannot be given here. Again, the reader must be referred elsewhere for a more detailed treatment.[3]

Fourth, and finally, in our discussion of "adding to the antecedent" in the case of counterfactuals, we noted something similar in the case of laws. Specifically, we noted that some laws ride "piggyback" upon other laws. This raises certain obvious questions. The distinction between perfect and imperfect knowledge goes some way towards answering these questions. But detailed discussion is clearly beyond the scope of the present essay. Yet once again, therefore, the reader must be referred elsewhere.[4]

One should also mention, perhaps, at this point, that other analyses of conditionals exist, which are important, but which we have not addressed. One might mention, for example, the proposal by E. Adams[5] to analyze predictive conditionals of the form "If A then C" as assertible if and only if the agent's conditional subjective probability for C, given A, is high. But if we have not dealt with these other cases, we have, I think, developed the tools to do so. Thus, for the case of Adam's analysis, if we reasonably suppose that an agent's subjective probability is justified just in case that it is based on a (reasonably accepted) statistical law or actuarial assessment;[6] then, if it is based on the former there is no control over individual events, as we saw in Chapter I, Sec. I, and therefore, given a cognitive interest in control, the conditional, contrary to Adams, is *not* assertible; whereas if the subjective probability is based on an actuarial assessment, then the conditional is assertible in the context defined by a cognitive interest in control, not by virtue of the statistics, or at least not by them alone, but because of the (imperfect) knowledge that provides the connection between A and C — even though the law (this is why the actuarial statistics are relevant) permits only *ex post facto* explanation and no prediction. In short, what I am predicting — on the basis of past experience — is that the Humean themes have been articulated in sufficient detail to permit the proponents of the analysis to quietly accept as in fact non-problematic all purported counterexamples. This claim is, of course, one that is made in knowledge of the Humean limits.[7]

What all this shows, I think, is that so far as the Humean is concerned, the major problems concerning counterfactuals are not significantly different from the major problems confronting the Humean account of explanation, that is, the deductive-nomological model. The crucial point is that the issues raised lead directly into the discussion of issues in the philosophy of science. Recent discussions of counterfactual conditionals have created the impression that the problems such conditionals generate are to be solved by logicians fiddling about with modal logics and possible worlds. The conclusion that we have reached is that this approach is thoroughly wrong-headed, and that the problems of how to analyse counterfactuals ought to be transferred from the logicians and the set theorists to the philosophers of science. If the present essay has established only this much, then even that would go some way towards justifying the major thesis of this essay, namely, that the Humean analysis is, and the possible worlds analysis is not, the correct account of counterfactual conditionals.

# NOTES

## CHAPTER I

*Section I*

[1] Thomas Taylor, 'Preliminary Dissertation in the Platonic Doctrine of Ideas,' in his translation of *The Philosophical and Mathematical Commentaries of Proclus*, Vol. I, p. xxxii.

[2] Cf. Wilson, 'The Lockean Revolution in the Theory of Science.'

[3] This is the task of Chapter II, below.

[4] B. Russell, 'Mysticism and Logic,' p. 26.

[5] Spinoza, *Ethics*, Bk. IV, Prop. LXII.

[6] A. Bain, *Mind and Body*, pp. 120—2.

[7] Cf. Wilson, 'Hume's Cognitive Stoicism' and 'Hume's Defence of Science.'

[8] Neurath, 'Empirical Sociology,' p. 326; his italics.

[9] *Ibid.*

[10] Cf. Veblen, 'The Place of Science in Modern Civilization.'

[11] Russell, 'The Place of Science in a Liberal Education,' p. 39.

[11a] Neurath, 'Empirical Sociology,' p. 329.

[11b] Galileo, *Dialogues concerning Two New Sciences*, p. 286ff. Cf. Hempel, 'Aspects of Scientific Explanation,' pp. 364—5. Also Kuhn, 'Mathematical vs. Experimental Traditions in the Development of Physical Science,' p. 55ff.

[12] Cf. Veblen, 'The Place of Science in Modern Civilization.'

[13] "Observable fact" is to be understood as "the concepts are such as could appear in the empiricist's language."

[14] For an excellent discussion of the objectivity issue, cf. M. Mandelbaum, *The Anatomy of Historical Knowledge*, Chapters Six and Seven. The securing of objectivity presupposes, of course, the logical or ontological separability of knower and known.

[14a] Cf. the discussion on objectivity in M. Brodbeck, *Readings in the Philosophy of the Social Sciences*, pp. 79—138.

[15] Thus, scientists, as experts in the scientific method, can claim no special sanction from expertise for any social activist activities they undertake. Experts they may be, but their social goals are to be evaluated in the same way as everyone else's: on moral grounds. Cf. G. Bergmann, 'Ideology.'

[16] This is, of course, simply the old and philosophically familiar act/object distinction, or what is the same, the (onto)logical separability of knower and known.

[16a] Cf. Wilson, "Hume's Defence of Science."

[17] C. A. Hooker, 'Philosophy and Meta-Philosophy of Science: Empiricism, Popperianism, and Realism,' pp. 203—4.

[17a] Cf. P. Frank, 'The Variety of Reasons for the Acceptance of Scientific Theories,' pp. 22—4.

[18] G. Ryle has argued ('"If", "So", and "Because"') that the predictive assertion ought not to be analyzed as an enthymeme, but this argument does not stand up to scrutiny; cf. N. H. Colburn, 'Logic and Professor Ryle'; and Wilson, *Explanation, Causation and Deduction*, Ch. 3, Sec. 5.

[19] Cf. J. L. Mackie, 'Counterfactuals and Causal Laws'; N. Rescher, *Hypothetical Reasoning*.

[19a] Cf. F. Kaufmann, *The Methodology of the Social Sciences*, Ch. IV.

[20] Hempel ('Deductive-Nomological vs. Statistical Explanation') has suggested empiricists should supplement the deductive-nomological model of explanation by an "inductive-statistical" model based on laws like (S), where the idea is that if a statistical law like (S) confers a high probability upon an (individual) event, then the latter has thereby been explained. The above discussion amounts to an argument against introducing Hempel's inductive-statistical model into the empiricist concept of explanation. Here, we follow such positivists as R. von Mises; cf. his *Positivism*, pp. 172—3, 180—1. For some reason F. Suppe identifies acceptance of the inductive-statistical model as one of the essential theses of positivism (cf. his 'Afterword-1977' to his *The Structure of Scientific Theories*, p. 619ff); the cases of von Mises, Frank, Bergmann and others, show how wrong is such an identification. N. Rescher, *Scientific Explanation*, develops Hempel's ideas. For criticism of the Hempel-Rescher position, cf. M. Brodbeck, 'Explanation, Prediction and "Imperfect" Knowledge', and F. Wilson, 'Review of Rescher's *Scientific Explanation*.' Salmon has presented an account of "statistical" explanation which treats it as a special case of (A) rather than (S) and as involving what we shall later call "imperfect laws"; cf. W. Salmon, *Statistical Explanation and Statistical Relevance*.

[20a] Cf. J. S. Mill, *A System of Logic*, Bk. III, Ch. XII, Sec. 1.

[21] Hempel and Oppenheim, 'Studies in the Logic of Explanation,' Sec. 4.

[22] *Ibid.*, p. 15.

[23] *Ibid.*, pp. 14—5.

[24] Cf. A. Grünbaum, 'Causality and the Science of Human Behaviour'; L. Addis, *The Logic of Society*.

[24a] Cf. also Wilson, *Explanation, Causation, and Deduction*, Sections (3.2) and (3.3), for an extended discussion of some formalist criticisms of the deductive-nomological model of explanation that are largely based upon a simple failure to take this hermeneutical point seriously.

[25] Hempel and Oppenheim, 'Studies in the Logic of Explanation,' p. 9.

[26] *Ibid.*

[27] *Ibid.*, p. 12.

[28] *Ibid.*, p. 11; this idea is treated at length in Wilson, *Explanation, Causation and Deduction*.

[28a] Cf. Wilson, *Explanation, Causation, and Deduction*, Chapter III, where the importance of the Principle of Predictability in defending the deductive-nomological model of explanation is developed in detail.

[29] Hempel and Oppenheim, 'Studies in the Logic of Explanation,' p. 11.

[30] *Ibid.*, pp. 10—11.

[31] For a discussion of the analytic/synthetic distinction of the early positivists, cf. Wilson, 'The Notion of Logical Necessity in Carnap's Later Philosophy,' Chapters I and II.

[32]  Cf. F. Kaufmann, *The Methodology of the Social Sciences*, p. 69.

[33]  Scriven, 'Explanations, Predictions, and Laws,' p. 177; Scriven's views are discussed in detail in Wilson, *Explanation, Causation and Deduction*.

[34]  Cf. Hempel, 'Aspects of Scientific Explanation,' p. 367.

[35]  Scriven, 'Explanations, Predictions, and Laws,' p. 182ff; 'Truisms as the Grounds for Historical Explanations,' p. 486.

[36]  Cf. Wilson, 'Goudge's Contribution to the Philosophy of Science'; Brodbeck, 'Explanation, Prediction, and "Imperfect" Knowledge'; Wilson, *Explanation, Causation and Deduction*.

[37]  Scriven, 'Explanation and Prediction in Evolutionary Theory.'

[38]  Cf. Hempel, 'Deductive-Nomological vs. Statistical Explanation,' p. 109.

[39]  For a good discussion of geometrical optics, cf. E. Mach, *The Principles of Physical Optics*. Most contemporary optics texts, if they present geometrical optics at all seriously, still do so only in a way that inextricably intertwines it with electromagnetic theory, and thereby obscures its rather simple and elegant, yet powerful, basis. For a detailed discussion of the example, cf. Wilson, *Explanation, Causation and Deduction*.

[40]  Cf. J. S. Mill, *System of Logic*, Bk. III, Ch. xxii; C. Hempel, 'Aspects of Scientific Explanation,' p. 352.

[41]  G. Bergmann, *Philosophy of Science*, p. 102.

[42]  Scriven, 'Explanation and Prediction in Evolutionary Theory,' p. 480.

[43]  For details, see Wilson, *Explanation, Causation and Deduction*.

[44]  Cf. Grünbaum, *Philosophical Problems of Space and Time*, First Edition, p. 311.

[45]  Cf. Bergmann, *Philosophy of Science*; and Wilson, *Explanation, Causation and Deduction*.

[46]  Salmon, 'Statistical Explanation,' p. 54; B. C. van Fraassen, *The Scientific Image*, p. 28, p. 31; J. Greeno, 'Explanation and Information.'

[47]  Cf. Collins, 'The Use of Statistics in Explanation.'

[48]  Salmon, 'Statistical Explanation,' p. 53ff.

[49]  The following patterns of inference are treated in detail in Wilson, 'Goudge's Contribution to Philosophy of Science'; and *Explanation, Causation and Deduction*.

[50]  D. Hume, *Treatise of Hume Nature*, Parts III and IV of Book I. Cf. Wilson, 'Hume's Defence of Causal Inference.'

[51]  Cf. Mappes and Beauchamp, 'Is Hume Really a Sceptic about Induction?'; and Beauchamp and Rosenberg, *Hume and the Problem of Causation*.

[52]  In this sense, Hume is a falliblist; see Chapter II, below.

[53]  G. E. Moore makes a similar distinction for similar resons in his *Ethics*, p. 118ff.

### Section II

[1]  Cf. Bergmann, *Philosophy of Science*, Ch. 2; Wilson, *Explanation, Causation and Deduction*, Ch. I and II, and 'Explanation in Aristotle, Newton and Toulmin.'

[2]  For greater detail, see Bergmann, *Philosophy of Science*, Ch. 2.

[3]  Cf. Bergmann, *Philosophy of Science*, Ch. 2; Wilson, *Explanation, Causation and Deduction*; Brodbeck, 'Explanation, Prediction and "Imperfect" Knowledge.'

[4]  Cf. Helmer and Rescher, 'On the Epistemology of the Inexact Sciences.'

[5]  Cf. G. H. von Wright, *The Logical Problem of Induction*; and Wilson, *Explanation, Causation and Deduction*, Ch. I.

[6]  Cf. Wilson, 'Explanation in Aristotle, Newton and Toulmin.'

[7]  On senses of "acceptability", see Lakatos, 'Changes in the Problem of Inductive Logic,' p. 380ff.

[7a]  Cf. Wilson, *Explanation, Causation and Deduction*, p. 25ff.

[8]  Cf. Wilson, *Explanation, Causation and Deduction*, Ch. I.

[9]  Cf. Wilson, 'Explanation in Aristotle, Newton and Toulmin.'

[10]  *Ibid.*

[10a]  Cf. H. Hertz, *The Principles of Mechanics*, p. 11.

[11]  Cf. Wilson, *Explanation, Causation and Deduction*, Ch. I, Sec. 3.

[12]  Cf. Wilson, 'Is Operationism Unjust to Temperature?'

[13]  We use this example in Chapter III, below.

[14]  Cf. Wilson, *Explanation, Causation and Deduction*, Ch. I.

[15]  The importance of this point for issues in the logic of explanation is developed at length in Wilson, *Explanation, Causation and Deduction*.

[16]  Cf. *ibid.*; and also Wilson, 'Goudge's Contribution to Philosophy of Science.'
   The remarks here are, of course, of a piece with our discussion in Sec. I, above, of the use of such *ex post facto* explanation in connection with statistical laws.

[17]  Cf. Wilson, *Explanation, Causation and Deduction*.

[18]  Cf. *ibid.*; and also Wilson, 'Goudge's Contribution to Philosophy of Science.'

[19]  Cf. Wilson, 'Kuhn and Goodman: Revolutionary vs. Conservative Science'; and 'A Note on Hempel on the Logic of Reduction.'

[20]  For example, it fits very well the hypothesizing of the existence of, and subsequent identification of the planet Neptune.

[21]  This is Kuhn's idea of a paradigm; cf. *The Structure of Scientific Revolutions*; p. 25ff, p. 35ff. Also Wilson, 'Kuhn and Goodman: Revolutionary vs. Conservative Science.'

[22]  Cf. Wilson, 'Is There a Prussian Hume?'

[23]  Cf. *ibid.*

[23a]  See p. 71.

[24]  Suppe, 'The Search for Philosophic Understanding of Scientific Theories,' p. 162.

[25]  Hanson, 'Is There a Logic of Scientific Discovery?'

[26]  *Ibid.*, p. 621.

[27]  *Ibid.*, pp. 621—2, pp. 626—7.

[28]  *Ibid.*, p. 622.

[29]  *Ibid.*, p. 624.

[30]  *Ibid.*, p. 625.

[31]  This is the difficulty that lies in presenting a case in terms of a single example. Kuhn avoids it only by providing us with a wealth of examples.

[32]  'Is There a Logic of Scientific Discovery?' p. 624.

[33]  *Ibid.*, p. 625.

[34]  *Ibid.*

[35]  Kuhn, *The Structure of Scientific Revolutions*, pp. 17—18.

[36]  *Ibid.*, p. 23.

[37]  *Ibid.*, p. 25ff.

[38]  *Ibid.*, p. 35ff, p. 147.

[39]  *Ibid.*, p. 76, p. 158.

[40]  *Ibid.*, pp. 35—6, p. 147.

[41]  *Ibid.*, p. 76.

42  *Ibid.*, p. 156, p. 158.

43  *Ibid.*, Sec. IX.

44  This is J. S. Mill's way of putting it; see his *System of Logic*, p. 325. Cf. F. Wilson, 'Mill on the Operation of Discovering and Proving General Propositions'; also 'Is There a Prussian Hume?'

45  I. Lakatos, 'Falsifiability and the Methodology of Scientific Research Programmes,' insists that non-falsifiability is due to the alleged fact that the observer can always, by fiat, reject an observation as non-veridical (p. 107f), while insisting that all theories are of the naive falsificationist *logical form* "All A are B" (p. 110). Kuhn rejects this account of the logical form of theories, insisting that greater logical complexity is what prevents falsification (cf. T. Kuhn, 'Logic of Discovery or Psychology of Research,' pp. 14—15; *Structure of Scientific Revolutions*, p. 147).

46  *Structure of Scientific Revolutions*, p. 42.

47  *Ibid.*, p. 36.

48  It is only slightly more complex if we do not assume that the hypotheses are contrary.

49  Cf. Wilson, *Explanation, Causation and Deduction*, Ch. I.

50  Lakatos, 'Falsifiability and the Methodology of Scientific Research Programmes,' p. 118; Kuhn, *Structure of Scientific Revolutions*, p. 154.

51  *Structure of Scientific Revolutions*, Sec. VI.

52  *Ibid.*, Sec. VII.

53  *Ibid.*, p. 76.

54  *Ibid.*, p. 83, p. 158, p. 169.

55  *Ibid.*, Secs. IX, XII.

56  *Ibid.*, p. 155—6.

57  *Ibid.*, pp. 155—6.

58  Neither Lakatos, 'Falsification and the Methodology of Scientific Research Programmes,' nor Laudan, *Progress and Its Problems*, ever suggest criteria for when it would be rational to take up and use, put to the test, a theory never hitherto worked on. This is a serious deficiency in their positions. Kuhn in contrast, we see, has no such gap in his account of scientific practice.

59  *Structure of Scientific Revolutions*, p. 156.

60  Hanson, 'Is There a Logic of Scientific Discovery?' p. 623, p. 624.

61  *Ibid.*, p. 624.

62  Cf. Lakatos, 'Falsification and the Methodology of Scientific Research Programmes,' pp. 177—80.

63  Kuhn, 'Comment on the Relations of Science and Art,' p. 342.

64  *Structure of Scientific Revolutions*, p. 169.

65  *Ibid.*, p. 159.

66  L. Laudan, *Progress and Its Problems*, charges that Kuhn's position is inadequate because he does not give, once for all, a definite degree of inefficiency at which one and all should start search for a new paradigm. We see that such a demand is unreasonable. Laudan's criticism is plainly mistaken.

67  *Structure of Scientific Revolutions*, p. 83.

68  *Ibid.*, pp. 155—6.

69  *Ibid.*, p. 156, p. 158.

70  See p. 60.

[71] This feature of Kuhn's position closes a gap present in the methodologies of Lakatos and Laudan; see Note 58, above.

[72] The methodologies of Lakatos and Laudan have a gap at this point, too; see Note 58, above.

## CHAPTER II

*Section I*

[1] Cf. W. E. Johnson, *Logic*, pp. 42—3.

These considerations, by the way, fully answer criticisms of formal logic based on the supposed "oddity" of material conditionals. (For these criticisms, see, e.g., P. F. Strawson, *Introduction to Logical Theory*, p. 82ff.)

[2] The parallel between factual and predictive conditionals, on the one hand, and counterfactual conditionals, on the other, has been noted by Mackie, 'Counterfactuals and Causal Laws.'

[3] Mackie, 'Counterfactuals and Causal Laws' argues that it is *plausible* to construe the former as sustained by the latter, but he does not go on to make the crucial stronger case that they *ought* to be so construed. It is only this latter sort of argument that can successfully reply to those who propose to analyse counterfactuals in terms of possible worlds.

[4] Hume, *Treatise*, p. 77.

[4a] Cf. Wilson, 'Review of N. Swartz's *The Concept of Physical Law*.'

[5] Cf. Wilson, 'The Lockean Revolution in the Theory of Science.'

[6] Cf. *ibid.*

[7] Locke, *Essay concerning Human Understanding*, p. 45.

[8] *Ibid.*, p. 43.

[9] *Ibid.*, p. 44.

[10] Aaron, *John Locke*, p. 81.

[11] Cf. Wilson, 'The Lockean Revolution in the Theory of Science.'

[12] Cf. Wilson, 'Dispositions Defined: Harré and Madden on Defining Disposition Concepts.' Some of the ideas that generate the illusion of objective necessities are discussed in detail in connection with some views of Scriven in Wilson, *Explanation, Causation and Deduction*, Ch. III. See also Addis, 'Ryle's Ontology of Mind.'

[13] Locke, *Essay concerning Human Understanding*, pp. 452—5.

[14] *Ibid.*, p. 305.

[15] *Ibid.*, pp. 438—71.

[16] *Ibid.*, p. 556ff.

[17] *Ibid.*, pp. 558—9; cf. also pp. 584—5.

[18] Cf. Wilson, 'Acquaintance, Ontology and Knowledge.'

[19] Cf. *ibid.*

[20] Cf. Wilson, 'Weinberg's Refutation of Nominalism,' and 'Review of M. Mandelbaum's *The Anatomy of Historical Knowledge*.'

[21] Cf. Wilson, 'Review of J. Yolton's *Metaphysical Analysis*,' and 'Review of Y. Ganthier's *Méthodes et concepts de la logique formelle* and R. Hébert's *Mobiles du discours philosophique*.'

[22] Cf. F. Wilson, 'Explanation in Aristotle, Newton, and Toulmin.'

23 Hume, *Treatise*, p. 157ff.
24 *Ibid.*, p. 158.
25 N. Malebranche, *De la recherche de la vérité*, Vol. III, p. 205.
26 P. Jones, *Hume's Sentiments*, p. 24.
27 Malebranche, *De la recherche de la vérité*, Vol. III, pp. 314—16.
28 Hume, *Treatise*, p. 159.
29 *Ibid.*, p. 160.
30 Cf. Wilson, 'Acquaintance, Ontology and Knowledge.'
31 Hume, *Treatise*, p. 159.
32 *Ibid.*, p. 89.
33 *Ibid.*, pp. 79—80.
34 *Ibid.*, p. 163.
35 *Ibid.*, p. 103.
36 *Ibid.*, p. 162.
37 Hume, *Enquiries*, p. 29.
38 For details, see Wilson, 'Acquaintance, Ontology and Knowledge'; and Bergmann, 'The Revolt against Logical Atomism.'
39 Hume, *Treatise*, p. 155.
40 *Ibid.*, p. 156.
41 *Ibid.*, p. 172.
42 Chisholm, 'Law Statements and Counterfactual Inference,' p. 230.
43 Cf. Bergmann, 'Comments on Professor Hempel's "Concept of Cognitive Significance"'; and his 'Intentionality' where he discusses the whole issue of *context*.
43a N. Swartz, *The Concept of Physical Law*, p. 195ff, objects to this subjectivist account of necessity, as does Armstrong, *What Is a Law of Nature*, p. 61ff. A similar objection is raised by Lewis, *Counterfactuals*, p. 74, and is discussed in Chapter III, Section IV, below. See also Wilson, 'Review of N. Swartz's *The Concept of Physical Law*.'
44 Cf. Wilson, 'Hume's Defence of Science.'
45 Hume, *Enquiries*, p. 160. See also J. Lenz, 'Hume's Defence of Causal Inference'; A. J. Ayer, *The Problem of Knowledge*, p. 75; and Wilson, 'Hume's Defence of Causal Inference' and 'Hume's Cognitive Stoicism.'
46 Hume, *Treatise*, p. 112.
47 *Ibid.*, p. 116. This discussion of the role of education cannot be fully understood until Hume explains, in Book II, the role of the principle of sympathy in the understanding of human nature. See in particular *ibid.*, pp. 316—17. See also P. Ardal, *Passion and Value in Hume's Treatise*, pp. 46ff. This illustrates, what is too often not realized, that Book I of the *Treatise* cannot be fully understood if it is read independently of Books II and III.
48 *Treatise*, p. 123.
49 *Ibid.*, pp. 132ff.
50 *Ibid.*, pp. 173ff.
51 *Ibid.*
52 *Ibid.*, p. 139.
53 I.e., without gathering more evidence, or enlarging the sample.
54 This is a bit too simple. Contrary generalities can be confirmed if we pick our samples correctly; clearly, we are presupposing a principle of total evidence. But the

point remains, since even the total evidence is but a sample relative to total population — the population of the universe. For more detail on Hume on the logic of research, see Wilson, 'Is There a Prussian Hume?'

[55] Compare the discussion of subjective and objective justification in G. E. Moore, *Ethics*, pp. 118—21, where it arises in the context of a discussion of utilitarianism.

[56] See T. Beauchamp and T. A. Mappes, 'Is Hume Really a Sceptic about Induction?'; and T. Beauchamp and A. Rosenberg, *Hume and the Problem of Causation*.

[57] Hume, *Treatise*, pp. 131—5, p. 154.

[58] *Ibid.*, p. 153. See Wilson, 'Hume's Defence of Causal Inference.'

[59] Hume, *Treatise*, pp. 152—3.

[60] *Ibid.*, Bk. I, Part III, Sec. xiii. See Wilson, 'Hume's Defence of Causal Inference.'

[61] Hume, *Treatise*, p. 175.

[62] I.e., upon the mind of the observer, the one thinking about the process. See Hume, *Treatise*, p. 408: "The necessity of any action, whether of matter or of the mind, is not properly a quality in the agent, but in any thinking or intelligent being, who may consider the action, and consists in the determination of his thought to infer its existence from some preceding objects . . . ."

[63] *Ibid.*, pp. 406—7.

[64] *Ibid.*, p. 156.

[65] *Ibid.*, p. 153.

[66] *Ibid.*, p. 409.

[67] *Ibid.*, p. 117.

[68] *Ibid.*, p. 175.

[69] *Ibid.*

[70] Butler, Preface to *sermons, Works*, Vol. 2, pp. xxi—xxii.

[71] *Ibid.*, pp. xxvi—xxvii.

[72] Butler, 'Dissertation II,' *Works*, Vol. 1., p. 330.

[73] Hume, *Treatise*, p. 620.

[74] For details, see Wilson, 'Hume's Defence of Causal Inference.'

[75] Cf. Wilson, 'Is There a Prussian Hume?'

[76] Kneale, 'Natural Laws and Contrary-to-Fact Conditionals.'

[77] Popper, 'A Note on Natural Laws and So-Called "Contrary-to-Fact" Conditionals.'

[78] Cf. G. Molnar, "Kneale's Argument Revisited."

[79] The example used by Molnar, *ibid.*, is that there is no river of Coca Cola; this example is beloved also by Swartz, *The Concept of Physical Law*.

[80] Swartz, in contrast, in his *The Concept of Physical Law*, p. 57, gleefully accepts that *all* unrestricted negative existential claims like Kneale's example or Molnar's (Note 79) are laws. But this leaves him open, as the Humean is not, to objections by such lovers of necessity as Armstrong, *What Is a Law of Nature?*, p. 13, p. 18.

[81] Armstrong, *What Is a Law of Nature?*, p. 13ff, directs our attention to "single case uniformities" as possible counterexamples to the Humean. Consider all the properties of a particular; in many cases, as subset of these will constitute an individuating conjunction. Let A be such an individuating conjunction, and B any other property of the particular that conjunction individuates. Then "All A are B" will be a true generality. But it is not law, and therefore so much the worse for the Humean, Armstrong concludes. Now, it is true that such single case generalities will, objectively, be as capable of prediction and therefore of explanation as, say, Newton's Laws. But,

equally clearly, almost all such generalities will be *very* imperfect, and will, therefore, provide only *very* weak explanations.

Moreover, the evidence for such single case generalities will almost always not be scientific, and so they will not be reckoned by the Humean to be justifiably law-assertible.

These two points suffice, it seems to me, to defend the Humean against Armstrong. Swartz, *The Concept of Physical Law*, oddly avails himself of neither of these plays; cf. Wilson, 'Review of N. Swartz's *The Concept of Physical Law*.'

[82] N. Swartz, *The Concept of Physical Law*, p. 68, avoids the problem by simply denying — without giving reasons — that any objectively true generality can sustain a counterfactual conditional. See Note 1b of Section II, below.

*Section II*

[1] Kneale, *Probability and Induction*, p. 75.

[1a] Swartz, *The Concept of Physical Law*, in contrast to the Humean, accepts that all (objectively true) generalities are (physical) laws — though he does point out (p. 3ff) that for only some (physical) laws is the evidence for their truth scientific, and so only some (physical) laws are to be reckoned as "scientific laws."

[1b] Swartz, *The Concept of Physical Law*, p. 68ff, denies that all true generalities can, objectively speaking, sustain the assertion of counterfactuals; rather, he says, only those that are *pervasive* can sustain counterfactuals. Unfortunately, he neither explains what it is to be *pervasive*, nor why only generalities of this sort and not others can support counterfactuals: why should the fact that a law pervades the actual world justify the claim that it pervades possibilities as well?

[2] Chisholm, 'The Contrary-to-Fact Conditional.'

[2a] See p. 15, above.

[3] As we argued in Chapter I, Sec. I, this is *also* explanatory — though only imperfectly so, since the laws of geometrical optics on which the prediction is based are cross-section laws rather than process laws. (For a detailed discussion of this example, see Wilson, *Explanation, Causation, and Deduction*.) The point is that predictions based on such laws, in contrast to predictions based on statistical laws, give, as we saw in Chapter I, Sec. I, reasons for *being*, and not *merely* reasons for expecting. This renders them explanatory where statistical laws are not.

[3a] Perhaps Swartz (see Note 1b, above) takes *pervasiveness* (whatever that is) to be a mark of explanatory power; and if he then confuses a weak explanation with a non-explanation (cf. Chapter I, above), then he might think of non-pervasive generalities as non-explanatory, and for that (bad) reason incapable of supporting counterfactuals.

[4] For a similar point, see Mackie, 'Counterfactuals and Causal Laws,' p. 72f.

[5] See Note 1, above.

[6] Kneale, 'Natural Laws and Contrary-to-Fact Conditionals,' p. 123.

[7] *Ibid.*

[8] For details on the structure of Darwin's theory, cf. Wilson, *Empiricism and Darwin's Science*.

[9] As does Armstrong, *What Is a Law of Nature?* (But see also Note 43a of Section I, above.)

*Section III*

[1] M. Tooley, 'The Nature of Laws.'

[2] D. Armstrong, *Universals and Scientific Realism*, Vol. II, pp. 149—153.

[3] Tooley, pp. 678—83, and Armstrong, p. 153f, both allow laws to have a more complex logical structure than this, but we need not pursue these complexities.

[3a] Another problem is this. Since, besides (1) we also have laws like

$$(x)\,[(Fx \lor Hx) \supset Gx]$$

we would have for $\mathcal{N}$ not only (2)

$$\mathcal{N}(F,\,G)$$

but also

$$\mathcal{N}(F,\,H,\,G)$$

and so on for many other cases. This means that $\mathcal{N}$, unlike other relations, has variable polyadicity, a very strange property.

Armstrong, *What Is a Law of Nature?*, p. 150ff, struggles with this ontological problem.

See also H. Hochberg, 'Natural Necessity and Laws of Nature.'

[4] Bergmann made this point many years ago; see his essay 'On Nonperceptual Intuition.' See also H. Hochberg, 'Possibilities and Essences in Wittgenstein's *Tractatus*.'

[5] Goodman, *Fact, Fiction and Forecast*, specifies one set of conditions, Popper, *Conjectures and Refutations*, specifies a somewhat different, but not incompatible set. For what we are about, such details may safely be ignored.

[5a] Armstrong, *What Is a Law of Nature?* subsequently struggles with the issue and for a variety of reasons ends up embracing (one is not surprised) Aristotelianism. This ties in with his earlier insistence on a physical basis for all dispositions; cf. Wilson, 'Dispositions Defined: Harré and Madden on Defining Disposition Concepts.'

Some of the relevant ontological problems are discussed in Wilson, *Explanation, Causation and Deduction*, Section (3.6).

[6] Tooley, 'The Nature of Laws,' p. 668, offers a second argument; this argument is effective against a view which treats laws as mere regularities but not against the Humean view which insists that any assertion of a law involves an assertion of necessity.

[7] Tooley, 'The Nature of Laws,' p. 669.

[8] *Ibid.*

[9] *Ibid.* One can find the suggestion made in, e.g., Hempel, 'Studies in the Logic of Confirmation.'

[10] For a discussion of measurement, see Bergmann, 'The Logic of Measurement.'

[11] See Note 8, above.

[12] Cf. Wilson, *Explanation, Causation and Deduction.*

[13] Tooley, 'The Nature of Laws,' pp. 690—3.

[14] *Ibid.*, p. 692.

[15] *Ibid.*

[16] R. Carnap, *The Logical Foundations of Probability*, Sec. 110. See also his 'Replies and Systematic Expositions,' p. 977.

<sup>17</sup> Cf. G. H. von Wright, *The Logical Problem of Induction*, p. 80.
<sup>18</sup> Cf. *ibid.*, pp. 76—79.
<sup>19</sup> Carnap, *The Logical Foundations of Probability*.
<sup>20</sup> Carnap, *The Logical Foundations of Probability*, Sec. 110 G.
<sup>21</sup> Cf. E. Nagel, 'Carnap's Theory of Induction'; I. Lakatos, 'Changes in the Problem of Inductive Logic.'
<sup>22</sup> Otherwise the apparatus of Q-predicates (cf. Carnap, *The Logical Foundations of Probability*, pp. 124—30) does not work.
<sup>23</sup> See Section I, above.
<sup>24</sup> Cf. Wilson, 'Acquaintance, Ontology and Knowledge.'
<sup>25</sup> Cf. E. B. Allaire, 'Existence, Independence, and Universals'; G. Bergmann, 'Stenius on the *Tractatus*.'
<sup>26</sup> Cf. R. Grossmann, 'Conceptualism,' and 'Sensory Intuition and the Dogma of Localization.'
<sup>27</sup> Tooley, 'The Nature of Laws,' p. 674. He refers to D. Lewis, 'How to Define Theoretical Terms.' See also Wilson, *Reasons and Revolutions*.
<sup>27a</sup> Cf. Wilson, Addison Analyzing Disposition Concepts.'
<sup>28</sup> Cf. Wilson, *Empiricism and Darwin's Science*.
<sup>29</sup> Tooley, 'The Nature of Laws,' p. 673.
<sup>30</sup> Cf. Wilson, 'Acquaintance, Ontology, and Knowledge'; G. Bergmann, 'The Revolt against Logical Atomism.' We must, of course, distinguish nominalism from atomism; cf. Wilson, 'Weinberg's Refutation of Nominalism' and 'Review of M. Mandlebaum's *The Anatomy of Historical Knowledge*.'
<sup>31</sup> Cf. Wilson, *Reasons and Revolutions*; Lakatos, 'Falsification and the Methodology of Scientific Research Programmes'; Kuhn, *The Structure of Scientific Revolutions*; Laudan, *Progress and Its Problems*. That the last three of these works make substantially this point is one of the main themes of the first of these works.
<sup>32</sup> Cf. Hempel, 'Studies in the Logic of Confirmation,' and a vast subsequent literature.
<sup>33</sup> Cf. Wilson, *Reasons and Revolutions*.
<sup>34</sup> Cf. von Wright, *The Logical Problem of Induction*; G. Bergmann, 'Some Comments on Carnap's Logic of Induction'; and W. Hay, 'Professor Carnap and Probability' and 'Review of Carnap's *Continuum of Inductive Methods*.'
<sup>35</sup> See Chapter I, above. Cf. Wilson, 'Mill on the Operation of Discovering and Proving General Propositions.'
<sup>36</sup> The point is the simple one that *ought implies can*. For the use of this principle, and its converse that *must implies ought*, see F. Wilson, 'Mill's Proof that Happiness Is the Criterion of Morality', 'Hume's Defence of Causal Inference,' and 'Hume's Cognitive Stoicism.'
<sup>37</sup> Cf. Wilson, *Explanation, Causation and Deduction*, and *Empiricism and Darwin's Science*. Also the discussion in Chapter III, Sec. V, below, of some views of A. Rosenberg.

## CHAPTER III

*Section I*

<sup>1</sup> For a discussion of a number of attempts to construe (1) as not involving the assertion of a law, see Wilson, *Explanation, Causation, and Deduction*, Chapter III.

[2]   See D. Lewis, 'Causation.'
[3]   See D. Lewis, *Counterfactuals.*
[4]   Lewis, 'Causation,' p. 185ff.
[5]   Lewis, *Counterfactuals*, p. 73.


*Section II*

[1]   Lewis, *Counterfactuals*, p. 1.
[2]   For the notion of the "quantifier-free development of a sentence," placed in particular in the context of confirmation theory, see C. G. Hempel, 'Studies in the Logic of Confirmation.' See also N. Swartz, *The Concept of Physical Law*, p. 79ff.
[3]   Lewis, *Counterfactuals*, p. 7ff.
[4]   *Ibid.*, p. 7.
[5]   *Ibid.*, p. 9.
[6]   *Ibid.*
[7]   *Ibid.*, pp. 9—10.
[8]   *Ibid.*, p. 11.
[9]   *Ibid.*, p. 14.
[10]  *Ibid.*, p. 19ff.
[11]  *Ibid.*, p. 20.
[12]  *Ibid.*, p. 19.
[13]  When it is not important to do so, we will ignore the subtlety of Lewis' concept being a *variably* strict condition, and treat it simply as a strict conditional, speaking of the truth-conditions of A → B in terms of *the* A-world nearest to ours.
[14]  Our criticisms of Lewis', and other possible-world accounts of counterfactuals, will be largely philosophical, rather than technical, directed at the details of formalisms. For the latter sort of criticism, which, given the intentions of Lewis *et al.*, are not to be ignored, see K. Fine, 'Review of D. Lewis' *Counterfactuals*' and D. Nute, 'David Lewis and the Analysis of Counterfactuals.'


*Section III*

[1]   Lewis, *Counterfactuals*, p. 8, p. 65; Stalnaker, 'A Theory of Conditionals.'
[2]   As Mackie has put it, "If we interpret counterfactuals . . . as arguments, we cannot say that they are true or false or that they are implied by other statements." (See his 'Counterfactuals and Causal Laws,' p. 97.) Of course, one can give reasonable *non-strict* senses of 'true', 'false', 'valid', etc., that can be applied to counterfactuals. (Cf. Lewis, *Counterfactuals*, p. 65ff.) We shall so speak when nothing turns on the matter.
[3]   Cf. Wilson, *Explanation, Causation, and Deduction*, Ch. III.
[4]   W. V. O. Quine, 'On What There Is,' p. 4.
[5]   Cf. Wilson, 'Logical Necessity in Carnap's Later Philosophy,' Ch. I.
[6]   D. Lewis, 'Counterpart Theory and Quantified Modal Logic,' pp. 110—1.
[7]   See also R. Stalnaker, 'Possible Worlds' and A. Plantinga, 'Actualism and Possible Worlds,' for others who similarly abandon the philosophical task, and take an unexplicated modal notion as primitive.
[8]   Cf. D. Lewis, 'Anselm and Actuality.'
[9]   Cf. N. Rescher, 'The Ontology of the Possible.'

[10] On the notion of "implicit definition", see Wilson, 'Implicit Definition Once Again', 'Logical Necessity in Carnap's Later Philosophy,' Ch. III, and 'Barker on Geometry as *A Priori.*'

[11] Lewis, 'Counterpart Theory and Quantified Modal Logic,' p. 111; and *Counterfactuals*, p. 88ff.

[12] Lewis, *Counterfactuals*, p. 87. How come, by the way, that properties can be identical across worlds (otherwise inter-world similarity can't be defined) but individuals have only counterparts? Surely the problems about individuals recur at the level of properties!

[13] Lewis, 'Counterpart Theory and Quantified Modal Logic,' p. 111.

[14] *Ibid.*, p. 122.

[14a] For an excellent discussion of the notion of "resemblance," see Butchvarov, *Resemblance and Identity*. See also Wilson, 'Resemblance, Identity and Universals: Comments on March on Sorting out Sovites.'

[15] Lewis, 'Counterpart Theory and Quantified Modal Logic', p. 112.

[16] Indeed, Lewis himself slips into this way of speaking: "Among my common opinions that philosophy must respect (if it is to deserve credence) are not only my naive belief in tables and chairs, but also my naive belief that these tables and chairs might have been arranged otherwise." (*Counterfactuals*, p. 88) Such discourse can, of course, be *interpreted* in terms of counterpart theory, but, then, as has been remarked *vis à vis* Humean paraphrases of counterfactuals, "Any analysis can be saved by paraphrasing the counterexamples." (Stalnaker, 'A Theory of Conditionals,' p. 174) This is not really true, but it is, I think, an important warning, and the point is that Lewis doesn't heed it. Cf. D. Nute, 'David Lewis and the Analysis of Counterfactuals,' pp. 359—60.

[17] Cf. R. M. Adams, 'Theories of Actuality,' p. 199.

[18] We return to this point below; see Note 27 of Section VII.

[19] But then, how can an *actual* perceiving of a non-actual possibility bridge the gap between two possible worlds?

[20] Cf. R. M. Adams, 'Theories of Actuality,' p. 201; G. Bergmann, 'Sketch of an Ontological Inventory,' and 'Notes on Ontology.'

[21] Bergmann, 'Diversity,' pp. 26—7.

[22] G. Frege, 'On the Foundations of Geometry,' p. 5.

[23] Cf. Wilson, 'Barker on Geometry as *A Priori.*'

[24] We can, in the present context, safely ignore *defined terms.*

[25] Bergmann, 'Sketch of an Ontological Inventory.'

[26] Cf. Bergmann, 'Diversity'; also Wilson, 'Effability, Ontology and Method.'

[26a] Recall Note 13 of Section II, above.

[27] Cf. W. Lycan, 'The Trouble with Possible Worlds,' p. 295; also J. Hattiangadi, 'Meaning, Reference and Subjunctive Conditionals.'

[28] Cf. Bergmann, 'Sketch of an Ontological Inventory,' and 'Notes on Ontology'; also Stalnaker, 'Possible Worlds,' p. 230.

[29] For more detail, cf. Wilson, 'Logical Necessity in Carnap's Later Philosophy,' Ch. I.

[30] Essentially, it is the apparatus of normal forms. It is related to the notion of the quantifier-free development of a formula; see Note 2 to Chapter II, Sec. II, above. See also N. Swartz, *The Concept of Physical Law*, Chapter 7. It relates, too, to Hintikka's notion of a "model set"; see his 'The Modes of Modality.'

[31] This needs to be adjusted to fit the case of an infinite universe; see Wilson, 'Logical Necessity in Carnap's Later Philosophy,' Ch. I.

[32] Cf. Bergmann, 'Intentionality,' and 'Analyticity'; and M. J. Cresswell, 'The World Is Everything that Is the Case.'
[33] Cf. Wilson, 'Logical Necessity in Carnap's Later Philosophy,' Ch. I.
[33a] Compare J. Hintikka, 'The Modes of Modality.'
[34] Lewis, *Counterfactuals*, p. 84.
[35] *Ibid.*, p. 88.
[36] This is, of course, simply Moore's old distinction between commonsense and its analysis; cf. G. Bergmann, 'Logical Positivism, Language, and the Reconstruction of Metaphysics.'
[37] Lewis, *Counterfactuals*, p. 85.
[38] Cf. Wilson, 'Logical Necessity in Carnap's Later Philosophy.'
[39] Cf. H. Hochberg, '"Possible" and Logical Absolutism,' and 'Professor Storer on Empiricism.'

*Section IV*

[1] Lewis, *Counterfactuals*, p. 73.
[2] *Ibid.*, p. 74ff.
[3] *Ibid.*, p. 74. See Note 43a to Chapter II, Section I, above.
[4] Lewis, 'Causation,' p. 185.
[5] *Ibid.*, pp. 185—6.
[6] *Ibid.*, p. 185.
[7] *Ibid.*, p. 187.
[8] Lewis, *Counterfactuals*, pp. 75—6.
[9] *Ibid.*, p. 75.
[10] This also replies to Swartz and Armstrong; see Note 43a to Chapter II, Section I, above.

*Section V*

[1] A. Rosenberg, 'Causation and Counterfactuals: Lewis' Treatment Reconsidered.'
[2] Cf. C. G. Hempel and P. Oppenheim, 'Studies in the Logic of Explanation,' and Hempel, 'Studies in the Logic of Confirmation.'
[3] Cf. R. Carnap, 'Testability and Meaning,' and *The Logical Foundations of Probability*.
[4] See Chapter I, Section II, above; and Wilson, *Empiricism and Darwin's Science*, and 'Mill on the Operation of Discovering and Proving General Propositions.'
[5] Cf. Wilson, 'Explanation in Aristotle, Newton and Toulmin.'
[6] For more on generic laws, see Wilson, *Reasons and Revolutions*.
[7] Lewis, 'Causation,' p. 190.
[8] Rosenberg, 'Causation and Counterfactuals,' pp. 216—7.
[9] *Ibid.*, p. 217.
[10] Cf. Bergmann, 'Stenius on the *Tractatus*.'
[11] See Wilson, *Explanation, Causation and Deduction*, Ch. II.
[12] Cf. Carnap, *The Logical Foundations of Probability*.
[13] Rosenberg, 'Causation and Counterfactuals,' p. 211.
[14] *Ibid.*, p. 212.
[15] *Ibid.*

16  *Ibid.*, p. 214.
17  *Ibid.*
18  Lewis, 'Causation,' p. 184.
19  Rosenberg, 'Causation and Counterfactuals,' p. 212.
20  *Ibid.*, p. 210.
20a  See Note 14a to Section III, above.
21  A. Kaminsky, 'On Literary Realism.'
22  R. Stalnaker, 'A Theory of Conditionals.'
23  Stalnaker has some minor differences with Lewis concerning the nature of possible worlds; cf. Stalnaker, 'Possible Worlds.'
24  Stalnaker, 'A Theory of Conditionals,' pp. 169—72.
25  Stalnaker, *ibid.*, asserts A → B is true if B holds in *the* A-world which is most similar to the actual world. This ignores the possibility, raised by Lewis, that two A-worlds might tie as closest to the actual world, with only one of them being a B-world.

Stalnaker's system validates the principle

$$[(A \to B) \lor (A \to \sim B)]$$

for all A and B that are accessible from the actual world. But this principle fails, on Lewis' analysis (*Counterfactuals*, p. 79ff), if (A & B)-worlds and (A & ~B)-worlds tie for similarity to the actual world. Lewis shows (*ibid.*, Sec. 3—4) that Stalnaker's system is equivalent to a special case of Lewis', namely, the case where there are no ties-for-most-similar and no asymptotic-approaches-to-most-similar.

26  Stalnaker, 'A Theory of Conditionals,' pp. 171—2.
27  *Ibid.*, p. 172, p. 176.
28  *Ibid.*, p. 486.

## Section VI

1  Cf. Note 1 of Sec. IV, above.
2  Such views are discussed in Wilson, *Explanation, Causation, and Deduction.*
3  See Chapter I, Sec. II, above; also Wilson, *Empiricism and Darwin's Science*, and *Reasons and Revolutions.*
4  Cf. R. Chisholm, 'Law Statements and Counterfactual Inference.'
5  N. R. Campbell, *Foundations of Science*, pp. 122—32. Cf. his remark (p. 129), "Any fool can invent a logically satisfactory theory to explain any law." Note that this is *not* (what is usually said to be) Duhem's point, that when a law is falsified one can always find an auxiliary hypothesis that will save it from refutation. Rather, Campbell's point is one about any given set of acceptable laws, that we can always find a deductive organization of these known generalizations that will have the required balance of simplicity and scope.
6  Campbell, *ibid.*, p. 129ff.
7  *Ibid.*, p. 132.
8  I. Lakatos, 'Falsification and the Methodology of Scientific Research Programmes'; Wilson, *Reasons and Revolutions*; and also the final Section of Ch. I, above.
9  J. S. Mill, *System of Logic*, Bk. III, Ch. IV, Sec. 2.
10  Cf. Note 2 of Section IV, above.

[11] Cf. Note 3 of Section IV, above.
[12] Lewis, *Counterfactuals*, p. 74.
[13] Cf. Chapter II, above.
[14] Cf. Wilson, 'Hume's Defence of Causal Inference.'
[15] See Note 3 of Section IV, above.

*Section VII*

[1] Lewis, *Counterfactuals*, p. 26.
[2] Cf. J. Bennett, 'Counterfactuals and Possible Worlds,' p. 384.
[3] Or, the conjunction of L and auxiliary premises about individual facts is false. But this makes no difference for the present point.
[4] Lewis, *Counterfactuals*, p. 28.
[5] Cf. W. E. Johnson, *Logic*, Part I, pp. 40—2.
[6] Goodman, *Fact, Fiction, and Forecast*, pp. 4—5.
[7] Cf. P. F. Strawson, *Introduction to Logical Theory*, p. 82ff.
[8] Cf. Wilson, *Explanation, Causation, and Deduction*, Ch. II and Ch. III, esp. Sec. 5.
[9] Lewis, *Counterfactuals*, p. 28.
[10] *Ibid.*
[11] *Ibid.*
[12] *Ibid.*, pp. 27—8.
[13] *Ibid.*, p. 27.
[14] This example is adapted from J. Bennett, 'Counterfactuals and Possible Worlds,' pp. 387—8.
[15] *Ibid.*, p. 388.
[16] Stalnaker, 'A Theory of Conditionals,' pp. 175—6, notes this ambiguity of counter-factual conditionals. For him, however, the ambiguity is said to be "pragmatic" rather than semantic, as it is for the Humean. We discuss this *apparent* difference between the Humean analysis and the possible-world analysis in subsection (i) of Section VIII below.
[17] Lewis, 'Causation,' p. 187.
[18] Compare the remarks on "standing conditions" by W. Sellars, 'Counterfactuals,' pp. 145—6.
[19] On this point, with which some have taken issue, see also Wilson, *Explanation, Causation, and Deduction*, Ch. II and Ch. III.
[20] Of course, if we are correct then Lewis' analysis does not *quite* fit any of ordinary usage; for, if we are correct, then, contrary to Lewis, A → B is *not* justifiably assertible when A and B are both true.
[21] Cf. Chisholm, 'The Contrary-to-Fact Conditional,' p. 483.
[22] See also Wilson, *Explanation, Causation, and Deduction*, for more detail.
[23] Cf. Goodman, *Fact, Fiction, and Forecast*, pp. 4—5.
[24] The symmetry of explanation and prediction is discussed in detail in Wilson, *Explanation, Causation and Deduction*.
[25] We discuss this further, below; see Section VIII.
[26] Cf. Note 5 of Sec. IV, above.
[27] Cf. Note 18 of Sec. III, above.
[28] We discuss this point on semifactuals below, in the next Section.

[29] For Lewis on counterfactual independence, cf. Note 6 of Sec. IV, above.

[30] The issue of whether we can know causal relations obtain when we do not know any relevant laws is discussed in Wilson, *Explanation, Causation and Deduction*, Sections (2.5) and (2.6).

The case we have made for laws being included as part of counterfactual assertions is elaborated for the similar case of explanations, *ibid.*, Ch. III.

[31] Cf. Goodman, *Fact, Fiction and Forecast*, pp. 13—15.

[32] Stalnaker, 'A Theory of Conditionals,' p. 170, introduces, as a technical device, the "absurd world" in which all contradictions hold.

[33] We discuss these issues further; see below, subsection (iv) of Section VIII.

*Section VIII*

[1] In his *Fact, Fiction and Forecast.*

[2] Cf. Chisholm, 'The Contrary-to-Fact Conditional,' p. 493ff.

[3] Cf. Wilson, 'Logical Necessity in Carnap's Later Philosophy,' Ch. I.

[4] Stalnaker, 'A Theory of Conditionals,' p. 176.

[5] *Ibid.*, p. 170.

[6] Cf. Note 27 to Section V, above.

[7] Stalnaker, 'A Theory of Conditionals,' p. 176.

[8] *Ibid.*

[9] Cf. Strawson, *Introduction to Logical Theory*, p. 211ff.

[10] Stalnaker, 'A Theory of Conditionals,' p. 174.

[11] *Ibid.*, p. 176.

[12] *Ibid.*, p. 173.

[13] Cf. Goodman, *Fact, Fiction and Forecast*, pp. 5—6; Chisholm, 'The Contrary-to-Fact Conditional,' pp. 486—7.

[14] Note that the generalizations are *law*-asserted; they are *not* asserted *merely* because they are vacuously true (if they are vacuous). So the point made holds even where the relevant laws in fact have no instances. Cf. Chisholm, 'The Contrary-to-Fact Conditional,' pp. 486—7.

[15] The Stalnaker-Lewis position acquires the *appearance* of neatness through nothing more then an artful and judicious employment of symbols, which give a veneer of order and simplicity to what closer inspection reveals to be quite messy. (One might compare here the now-fashionable "Intelim" rules used in systems of natural deduction in logic: there is a superficial neatness about two rules for each connective, one for introducing it and one for eliminating it, but this neatness disappears as soon as one recognizes the very great differences among the various rules.)

Far better, it seems to me, to let the messiness hang out!

So in the present essay one will find no "general theory" of explanation which is just a symbolic simplification of complex realities. We avoid both the quasi-mathematical symbols now so much in fashion, and such quasi-mathematical jargon as "selection function." Nor do we indulge the current taste for introducing the apparatus of set-theory. Nor, finally, do we use any of the equally fashionable symbolizations of Chomskyan structuralist semantics.

[16] Cf. Goodman, *Fact, Fiction, and Forecast*, p. 5.

[17] Stalnaker, 'A Theory of Conditionals,' p. 174.

[18] Cf. Goodman, *Fact, Fiction, and Forecast*, p. 6; Chisholm, 'The Contrary-to-Fact Conditional,' p. 492.

[19] Stalnaker, 'A Theory of Conditionals,' p. 174.

[20] *Ibid.*

[21] Lewis, 'Causation,' p. 190.

[22] We return to this point, below.

[23] Cf. Chisholm, 'The Contrary-to-Fact Conditional,' p. 487.

[24] Goodman, *Fact, Fiction, and Forecast*, p. 6.

[25] Chisholm, 'The Contrary-to-Fact Conditional,' p. 492.

[26] Stalnaker, 'A Theory of Conditionals,' p. 174.

[27] *Ibid.*, p. 167.

[28] *Ibid.*, p. 168.

[29] We discuss this qualification — that $\mathscr{F}$ be cotenable with A — in greater detail below.

[30] See Note 43, below.

[31] Cf. Note 4 and Note 10 to Section VII, above.

[32] Cf. Note 24, above.

[33] The law (u') rides "piggyback" on (t'). For more on this matter, see Chapter III, Sec. (3.6), of Wilson, *Explanation, Causation and Deduction*.

[34] For further discussion of this sort of point, see the material cited in the previous note.

[35] Lewis, *Counterfactuals*, p. 31; Stalnaker, 'A Theory of Conditionals,' p. 173.

[36] Stalnaker, 'A Theory of Conditionals,' p. 173.

[37] Cotenability is discussed in detail, below.

[38] Bennett, 'Counterfactuals and Possible Worlds,' p. 384.

[39] *Ibid.*, pp. 384—5.

[40] Ice that has partially thawed and re-frozen supports less weight than fresh-frozen ice of the same thickness.

[41] See Wilson, *Explanation, Causation and Deduction*, in particular the discussion of Scriven in Ch. II and of Davidson in Ch. III.

[42] Cf. Wilson, *Explanation, Causation and Deduction*.

[43] See Note 30, above.

[44] Lewis, *Counterfactuals*, p. 32; Bennett, 'Counterfactuals and Possible Worlds,' p. 385; Stalnaker, 'A Theory of Conditionals,' p. 173.

[45] Bennett, 'Counterfactuals and Possible Worlds,' p. 385.

[46] *Ibid.*

[47] Lewis, *Counterfactuals*, p. 32.

[48] *Ibid.*, p. 33.

[49] *Ibid.*, p. 35; Bennett, 'Counterfactuals and Possible Worlds,' p. 385; Stalnaker, 'A Theory of Conditionals,' p. 173.

[50] Lewis, *Counterfactuals*, p. 35.

[51] *Ibid.*

[52] Stalnaker, 'A Theory of Conditionals,' p. 174.

[53] See Wilson, *Explanation, Causation and Deduction*, Ch. II and III.

[53a] Lewis, *Counterfactuals*, p. 16.

[54] Lewis, 'Causation,' p. 181, p. 189.

[55] Cf. *ibid.*, p. 183; and Anscombe, 'Causality and Determination,' p. 67.

[56] Cf. Anscombe, 'Causality and Determination,' p. 67. Also Ducasse, 'Causality: A Critique of the Humean Analysis'; I have discussed Ducasse in *Explanation, Causation,*

*and Deduction*, Section (1.2). Also Davidson, 'Causal Relations,' p. 92; Davidson's position is discussed in detail in Wilson, *Explanation, Causation and Deduction*, Ch. III.
[56a] See Wilson, *Explanation, Causation and Deduction*, Ch. II and III.
[57] Cf. Notes 4 and 5 of Sec. IV, above.
[58] Lewis, 'Causation,' pp. 186—7.
[59] *Ibid.*
[60] Cf. Note 7 of Sec. IV, above.
[61] Lewis, 'Causation,' p. 188.
[62] Consider the simple case of two $C_i$, to wit, $C_1$ and $C_2$, and two $A_i$, to wit, $A_1$ and $A_2$. The $C_i$ are not compossible; that means that

$$\sim C_1 \vee \sim C_2$$

holds. We must show that no two $(A \& C_i)$ are compossible, i.e., that

$$\sim (A \& C_1) \vee \sim (A \& C_2)$$

But the latter is logically equivalent to

$$(\sim A \vee \sim C_1) \vee (\sim A \vee \sim C_2)$$

which clearly follows by addition and association from the non-compossibility of the $C_i$.
[63] Again limit ourselves to the case of two $A_i$ and two $C_i$. Since $\mathscr{L}$ and $\mathscr{F}$ entail $A_i \supset C_i$ for all $i$, we have as premises

$$A_1 \supset C_1$$
$$A_2 \supset C_2$$

and also the non-compossibility of the $A_i$ and the $C_i$:

$$\sim A_1 \vee \sim A_2$$
$$\sim C_1 \vee \sim C_2$$

We must show that $(A \& C_1) \supset A_1$. The proof for $(A \& C_2) \supset A_2$ will be exactly parallel.

Proceed by conditional proof. Assume $(A \& C_1)$, or, what is the same, given the definition of 'A', assume

$$(A_1 \vee A_2) \& C_1$$

which is logically equivalent to

$$(A_1 \& C_1) \vee (A_2 \& C_1)$$

From the first disjunct, $A_1$ follows directly. In the second disjunct, $A_2$ enables us to deduce $C_2$ while $C_1$ enables us to deduce $\sim C_2$, which is a contradiction, so that $A_1$ follows from this disjunct also. Since $A_1$ follows from both disjuncts, it follows from this disjunct also. Since $A_1$ follows from both disjuncts, it follows from the disjunction as a whole, which was the assumption of our conditional proof. Q.E.D.
[64] Again working with but two $A_i$ and two $C_i$, we must show that

$$\sim A_1 \vee \sim A_2$$
$$\sim (A \& C_1) \vee \sim (A \& C_2)$$
$$A \& C_1$$
$$\sim A_1$$

form a consistent set. To do this, assign truth-values to the $A_i$ and $C_i$ so that $\sim A_1$, $C_1$, $A_2$, $\sim C_2$ are all true. Then each of the original set is true, and that set is therefore consistent.

[65] Lewis, 'Causation,' p. 188.

[66] *Ibid.*, pp. 188—9.

[67] Cf. *ibid.*, p. 190.

[68] Lewis has a bad conscience, and in fact more or less admits this point in *ibid.*, Note 10, p. 189.

[69] See Note 66, above.

[70] For a discussion in detail on how the Humean draws this distinction, see Wilson, *Explanation, Causation and Deduction*, Ch. II.

[71] Lewis, 'Causation,' pp. 189—91; see Note 7 of Sec. V, above.

[72] Lewis, 'Causation,' p. 190.

[73] *Ibid.*, p. 191.

[74] Lewis, *Counterfactuals*, pp. 75—7.

[75] Cf. Kim, 'Causes and Counterfactuals,' p. 193.

[76] See Note 25 of Sec. VII, above.

[76a] See Wilson, *Explanation, Causation and Deduction*, Ch. II.

[77] See Note 67, above.

[78] Lewis, 'Causation,' p. 191.

[79] *Ibid.*, p. 187.

[80] *Ibid.*

[81] *Ibid.*

[82] See Note 7 of Sec. IV, above.

[83] Lewis, 'Causation,' p. 191.

[84] *Ibid.*

[85] See Note 49, above.

[86] See Note 16, above.

[87] See Note 22, above.

[88] See Note 32, above.

[89] See Note 33, above.

[90] See Note 31, above.

[91] See Note 98, below.

[92] See, e.g., Note 33 of Sec. VII, above.

[93] Goodman, *Fact, Fiction, and Forecast*, p. 8.

[94] *Ibid.*, pp. 9—11.

[95] *Ibid.*, p. 13.

[96] See Note 24, above.

[97] See Note 34, above.

[98] These issues are discussed further in Wilson, *Explanation, Causation and Deduction*, Ch. III.

[99] Goodman, *Fact, Fiction, and Forecast*, p. 13.

[100] *Ibid.*, p. 11.

[101] *Ibid.*, p. 13.

[102] For further discussion, see Wilson, *Explanation, Causation and Deduction*, Ch. III.

[103] See Wilson, *Explanation, Causation and Deduction*, Ch. II and Ch. III.

[104] Goodman, *Fact, Fiction and Forecast*, p. 14.

[105]  This law is, clearly, imperfect. But this is irrelevant. It is its law sttaus that counts — and it *is* a law, its imperfection notwithstanding. For more detail about such things, see Wilson, *Explanation, Causation and Deduction*, Ch. II and Ch. III.

[106]  W. Sellars, 'Counterfactuals,' p. 136.

[107]  Cf. *ibid.*, p. 146.

[108]  Goodman, *Fact, Fiction and Forecast*, p. 15.

[109]  Lewis, *Counterfactuals*, p. 57 and p. 70, suggests other definitions of cotenability. However, they have little to recommend them. Certainly, they are not the sense of 'cotenable' that Goodman introduced. For criticism of Lewis' notions, see J. Bennett, 'Counterfactuals and Possible Worlds,' pp. 389—90. In any case, we shall give a rationale for the notion of "cotenability" that we shall use.

[110]  Goodman, *Fact, Fiction, and Forecast*, pp. 16—7.

[111]  Cf. Wilson, 'Logical Necessity in Carnap's Later Philosophy,' Chapters I and II.

[111a]  As we mentioned above (Note 31 to Chapter III, Sec. III) the apparatus of state descriptions needs to be modified to handle infinite universes.

At this point we avoid a number of problems with the restriction imposed by assumption (*) which limits properties to a set of causally relevant variables, i.e., which takes "possible worlds" to be possible states of a system. In particular we take generic laws, including functional laws like $y = f(x)$, for granted. Thus, all "possible worlds" turn out to be similar in these respects, although they may differ in specific laws (see Section V, above); and interworld similarity, insofar as it is law-determined, is determined by specific laws. This is no great restriction since Lewis, like Rosenberg (see Section V, above) tends to ignore generic laws; and (as we shall argue below) where the context of discourse is defined by the purpose of interference, this restriction is fully justified. The restriction permits us to ignore technical difficulties with such things as second-order properties in state descriptions. On the other hand, the restriction must be relaxed, and generic laws must be taken into account, in other contexts (see Notes 142a, 147a, below), especially that of research.

For an important discussion of the notion of the "state" of a "system", see Bergmann, 'The Logic of Quanta.'

[112]  From Goodman, 'Reply to Cooley,' p. 532.

[113]  Cf. J. Bennett, 'Counterfactuals and Possible Worlds,' p. 400; J. Mondadori and A. Morton, 'Modal Realism: The Poisoned Pawn,' p. 247; G. H. von Wright, 'On the Logic and Epistemology of the Causal Relation,' p. 98. But von Wright, unlike the others, places his account firmly in the context of a discussion of causation and action rather than of counterfactuals. Von Wright's position is discussed in Wilson, *Explanation, Causation and Deduction*, Ch. II, Section (2.4).

[114]  Cf. N. Rescher, *Hypothetical Reasoning*; and 'Belief-Contravening Suppositions and the Problem of Contrary-to-Fact Conditionals.'

[115]  Rescher, *Hypothetical Reasoning*, p. 18.

[116]  *Ibid.*, p. 34.

[117]  *Ibid.*, p. 30.

[118]  The rules of *importation*

$$A \rightarrow (A \,\&\, C) \therefore A \rightarrow C$$
$$A \rightarrow C \therefore A \rightarrow (A \,\&\, C)$$

clearly hold for counterfactual implication " $\rightarrow$ ".

[119] Rescher, *Hypothetical Reasoning*, p. 30ff.

[120] *Ibid.*, p. 35ff.

[121] *Ibid.*, p. 38.

[122] *Ibid.*

[123] *Ibid.*, p. 32f.

[123a] See above, p. 249.

[124] Rescher, *Hypothetical Reasoning*, p. 33.

[125] *Ibid.*, pp. 33, 35.

[126] Goodman has made this point; see Rescher, *Hypothetical Reasoning*, Note 1 to p. 34.

[127] See the remarks of Goodman cited *ibid.*

[128] Rescher, *Hypothetical Reasoning*, Note 1 to p. 34, pp. 79—80.

[129] Goodman, *Fact, Fiction, and Forecast.*

[130] Cf. J. W. N. Watkins, 'The Paradoxes and Confirmation.'

[131] Rescher, *Hypothetical Reasoning*, p. 85.

[132] If reasoned prediction and explanation can be based not only on deduction from laws, but also on inductive or, more accurately, statistical relations, that is, on statistical relations unmediated by any *law-deduction* of the predicted (explained) events, *then, perhaps,* Rescher can sustain his case that inductive relations can distinguish between logically equivalent hypotheses. It is not at all clear how the case would go, however. But Rescher does hold that there are explanations which are statistical and not law-deductive. See his *Scientific Explanation.* On the other hand, it is not at all clear that Rescher even has a case that there are scientific explanations which are purely statistical, as our discussion in Chapter I, Sec. I attempted to establish.

[133] Goodman, *Fact, Fiction and Forecast*, Note 8, p. 14.

[134] See Note 112, above.

[135] Rescher, *Hypothetical Reasoning*, pp. 34—5.

[136] Lewis, 'Causation,' pp. 188—9.

[137] Lewis, *Counterfactuals*, p. 52; italics added.

[138] *Ibid.*, p. 5; italics added.

[139] Mondadori and Morton, 'Modal Realism: The Poisoned Pawn,' p. 247; italics added.

[140] Stalnaker, 'A Theory of Conditionals,' p. 171; italics added.

[141] See Note 158, below.

[141a] A somewhat similar suggestion is made by M. Leman, 'On the Logic and Criteriology of Causality.' But Leman makes the idea of intervention and production not just context-determining, but part of the very analysis of counterfactuals. In fact, he takes activity ("production") to be an unanalyzable operator which appears as a primitive feature of counterfactual conditionals, and then proceeds to analyze concepts like "cause" in terms of this and still further primitives like *force*. But in fact, the notions of activity, production and force presuppose rather than account for the notion of lawfulness. I have discussed the general issue whether the notion of cause is to be analyzed in terms of a concept like that of *production* in Wilson, *Explanation, Causation, and Deduction*, Sec. (2.4). See also Note 147b, below.

[141b] See Note 141a, above, and 147b, below.

[142] Whether one is *strict* or *liberal* in one's ascriptions of responsibility, and if liberal then which is the alternative bearer of responsibility, will, of course, be a *moral* decision.

[142a]  See Note 111a, above.

[143]  Cf. Chisholm, 'The Contrary-to-Fact Conditional.'

[144]  Popper, 'Kant's Critique and Cosmology'; Kuhn, *The Structure of Scientific Revolutions*. Nor, of course, need the positivist disagree with this; see Wilson, *Reasons and Revolutions*.

[145]  Cf. Francis Bacon, *The Great Instauration*, p. 16ff,

[146]  This distinction between acceptance$_1$ and acceptance$_2$ is defended in more detail in Wilson, *Reasons and Revolutions*.

[147]  Kuhn, *The Structure of Scientific Revolutions*; Wilson, *Empiricism and Darwin's Science*.

[147a]  See Note 111a, above.

[147b]  This notion of human effort should not be thought of as unanalyzable in terms of the notions of law and of causal law. The contrary is supposed by, for example, von Wright, 'On the Logic and Epistemology of the Causal Relation'; and is critically analyzed in Wilson, *Explanation, Causation and Deduction*, Sec. (2.4). See Note 141a, above.

[148]  Cf. Note 1 of Section V, above.

[149]  Cf. J. Mackie, 'Counterfactuals and Causal Laws.'

[150]  For some attempts along these lines, see Wilson, *Reasons and Revolutions*.

[151]  See, however, Wilson, 'Hume's Defence of Causal Inference,' for Hume's justification of scientific inference; and also Wilson, 'Hume's Defence of Science.'

[152]  Compare a similar point in Wilson, 'Dispositions: Defined or Reduced?'

[153]  Lewis, *Counterfactuals*, pp. 71—2.

[154]  Cf. J. Bennett, 'Counterfactuals and Possible Worlds,' p. 395.

[155]  Lewis, *Counterfactuals*, p. 76.

[156]  *Ibid.*

[157]  *Ibid.*

[158]  See Note 141, above.

[159]  Bennett, 'Counterfactuals and Possible Worlds,' p. 395, is surely wrong when he disagrees with Lewis on this point.

[160]  Bennett arrives at a somewhat similar conclusion without, however, supplying any clear rationale other than that of tinkering with Lewis' position so as to render it immune to certain counterexamples; see his 'Counterfactuals and Possible Worlds,' pp. 399—400.

[161]  Lewis, *Counterfactuals*, p. 67.

[162]  See Note 153, above.

[163]  It may, however fit Rescher's position (see his *Hypothetical Reasoning*), whom Lewis cites (see his *Counterfactuals*, Note 2, p. 70), since Rescher tends to ignore the role of laws which the Humean holds to be of the essence of legitimate counterfactual assertion (see Notes 120, 121, above).

*Section IX*

[1]  See Wilson, *Explanation, Causation and Deduction*, Ch. I and Ch. II; and *Empiricism and Darwin's Science*, Part I, Sec. I.

[2]  See Wilson, *Explanation, Causation and Deduction*, Ch. II; and 'Explanation in Aristotle, Newton, and Toulmin.'

[3]  For more on this topic, see Wilson, *Explanation, Causation and Deduction*, Ch. III.

[4]  Cf. *ibid.*

[5]  E. Adams, *The Logic of Conditionals.* Cf. A. Appiah, 'Generalising the Logic of Conditionals.' The idea has been critically discussed in D. Lewis, 'Probabilities of Conditionals and Conditional Probabilities.'

[6]  See Chapter I, Section I for this distinction.

[7]  Compare John Aaron's remarked quoted above; see Note 10 to Chapter II, Section I.

# BIBLIOGRAPHY

Aaron, J., *John Locke*. Oxford, 1955.

Adams, E., *The Logic of Conditionals*. Dordrecht, Holland, 1975.

Adams, R. M., 'Theories of Actuality'. In M. J. Loux, *The Possible and the Actual.*

Addis, L., *The Logic of Society*. Minneapolis, 1975.

Addis, L., 'Ryle's Ontology of Mind'. In Addis and Lewis, *Moore and Ryle: Two Ontologists.*

Addis, L. and Lewis, D., *Moore and Ryle: Two Ontologists.* The Hague, 1965.

Allaire, E. B., 'Existence, Independence, and Universals'. In E. B. Allaire *et al., Essays in Ontology.*

Allaire, E. B., '*Tractatus* 6.3751'. *Analysis* **19** (1959).

Allaire, E. B., *et al., Essays in Ontology*. The Hague, 1963.

Anscombe, G. E. M., 'Causality and Determinism'. In E. Sosa, *Causation and Conditionals.*

Appiah, A., 'Generalising the Probabilistic Semantics of Conditionals'. *Journal of Philosophical Logic* **13** (1984).

Árdal, P., *Passion and Value in Hume's Treatise*. Edinburgh, 1966.

Armstrong, D., *Universals and Scientific Realism*. Cambridge, 1978.

Armstrong, D., *What Is a Law of Nature?* Cambridge, 1983.

Austin, J. L., *Sense and Sensibilia*. London, 1962.

Ayer, A. J., *The Problem of Knowledge*. Harmondsworth, Middlesex, 1956.

Bacon, F., *The Great Instauration*. In E. A. Burtt, ed., *The English Philosophers from Bacon to Mill*, New York, 1939.

Bain, A., *Mind and Body*. London, 1872.

Bayle, P., *Historical and Critical Dictionary*. Selections, trans. and ed. R. H. Popkin, New York, 1965.

Beauchamp, T. and Rosenberg, A., *Hume and the Problem of Causation*. Oxford, 1981.

Bennett, J., 'Counterfactuals and Possible Worlds'. *Canadian Journal of Philosophy* **4** (1974).

Bergmann, G., 'Comments on Professor Hempel's "The Concept of Cognitive Significance"'. In his *The Metaphysics of Logical Positivism.*

Bergmann, G., 'Dispositional Properties and Dispositions'. *Philosophical Studies* **6** (1955).

Bergmann, G., 'Analyticity'. In his *Meaning and Existence.*

Bergmann, G., 'Diversity'. *Proceedings and Addresses of the American Philosophical Association*, 1967—68.

Bergmann, G., 'Ideology'. In Brodbeck, *Readings in the Philosophy of the Social Sciences.*

Bergmann, G., 'Intentionality'. In his *Meaning and Existence.*

Bergmann, G., *Logic and Reality*. Madison, Wisc., 1964.

Bergmann, G., 'The Logic of Measurement'. In *Proceedings of the Sixth Hydraulics Conference* (State University of Iowa Studies in Engineering). Iowa City, Iowa, 1956.

Bergmann, G., 'The Logic of Quanta'. In Feigl and Brodbeck, *Readings in the Philosophy of Science*.

Bergmann, G., 'Logical Positivism, Language, and the Reconstruction of Metaphysics'. In his *The Metaphysics of Logical Positivism*.

Bergmann, G., *Meaning and Existence*. Madison, Wisc., 1959.

Bergmann, G., *The Metaphysics of Logical Positivism*. New York, 1954.

Bergmann, G., 'Notes on Ontology'. *Noûs* **15** (1981).

Bergmann, G., 'On Non-Perceptual Intuition'. In his *The Metaphysics of Logical Positivism*.

Bergmann, G., *Philosophy of Science*. Madison, Wisc., 1957.

Bergmann, G., 'The Problem of Relations in Classical Psychology'. In his *The Metaphysics of Logical Positivism*.

Bergmann, G., 'Realistic Postscript'. In his *Logic and Reality*.

Bergmann, G., 'The Revolt against Logical Atomism'. In his *Meaning and Existence*.

Bergmann, G., 'Sketch of an Ontological Inventory'. *Journal of the British Society for Phenomenology* **10** (1979).

Bergmann, G., 'Some Comments on Carnap's Logic of Induction'. *Philosophy of Science* **13** (1946).

Bergmann, G., 'Stenius on the *Tractatus*'. In his *Logic and Reality*.

Black, M. (ed.) *Philosophical Analysis*. Ithaca, N.Y., 1950.

Brodbeck, M., 'Explanation, Prediction, and "Imperfect" Knowledge'. In Feigl and Maxwell, *Minnesota Studies in the Philosophy of Science*, Vol. III.

Brodbeck, M., 'The Philosophy of John Dewey'. In E. B. Allaire *et al., Essays in Ontology*.

Brodbeck, M. (ed.) *Readings in the Philosophy of the Social Sciences*. New York, 1968.

Brody, B. (ed.) *Readings in the Philosophy of Science*. Englewood Cliffs, N.J., 1970.

Butler, R. J. (ed.) *Analytical Philosophy, First Series*. London, 1962.

Butler, S., *Works*. Ed. S. Halifax. London, 1874.

Campbell, N. R., *Foundations of Science*. New York, 1957.

Capaldi, N., *David Hume*. Boston, 1975.

Carnap, R., *The Logical Foundations of Probability*. Chicago, 1950.

Carnap, R., 'Testability and Meaning'. In Feigl and Brodbeck, *Readings in the Philosophy of Science*.

Chappell, V. C. (ed.) *Hume: A Collection of Critical Essays*. Garden City, N.Y., 1966.

Chisholm, R., 'The Contrary-to-Fact Conditional'. *Mind*, NS **55** (1946). In Feigl and Sellars, *Readings in Philosophical Analysis*.

Chisholm, R., 'Law Statements and Counterfactual Inference'. In E. H. Madden, *The Structure of Scientific Thought*.

Coburn, N. H., 'Logic and Professor Ryle'. *Philosophy of Science* **21** (1954).

Collins, A., 'The Use of Statistics in Explanation'. *British Journal for the Philosophy of Science* **17** (1966).

Cresswell, M. J., 'The World Is Everything that Is the Case'. In M. Loux, *The Possible and the Actual*.

Davidson, D., 'Causal Relations'. In E. Sosa, *Causation and Conditionals*.

Feigl, H. and Brodbeck, M. (eds.) *Readings in the Philosophy of Science*. New York, 1953.

Feigl, H. and Maxwell, G. (eds.) *Current Issues in Philosophy of Science*. New York, 1961.

Feigl, H. and Maxwell, G. (eds.) *Minnesota Studies in the Philosophy of Science*, Vol. III. Minneapolis, 1956.

Feigl, H. and Sellars, W. (eds.) *Readings in Philosophical Analysis*. New York, 1949.

Fine, K. 'Review of Lewis' *Counterfactuals*'. *Mind* **84** (1975).

Fraasen, B. C. van. *The Scientific Image*. Oxford, 1980.

Frank, P., *The Validation of Scientific Theories*. New York, 1961.

Frank, P., 'The Variety of Reasons for the Acceptance of Scientific Theories'. In his *The Validation of Scientific Theories*.

Frege, G., 'On the Foundations of Geometry'. *Philosophical Review* **69** (1960).

Galileo Galilei, *Dialogues concerning Two New Sciences*. Trans. H. Crew and A. de Salvio, 1914. Reprinted New York, n.d.

Gardner, P. (ed.) *Theories of History*. Glencoe, Ill., 1959.

Goodman, N., *Fact, Fiction and Forecast*, Third Edition. Indianapolis, 1979.

Goodman, N., 'The Problem of Counterfactual Conditionals'. *Journal of Philosophy* **44** (1947). Reprinted as Ch. I of his *Fact, Fiction and Forecast*.

Greeno, J., 'Explanation and Information'. In Salmon, *Statistical Explanation and Statistical Relevance*.

Grossmann, R., *Ontological Reduction*. Bloomington, Indiana, 1973.

Hanson, N. R., 'Is There a Logic of Scientific Discovery?' In Brody, *Readings in the Philosophy of Science*.

Hattiangadi, J., 'Meaning, Reference, and Subjunctive Conditionals'. *American Philosophical Quarterly* **16** (1979).

Hausman, A., 'Hume's Theory of Relations'. *Noûs* **1** (1967).

Hay, W., 'Professor Carnap and Probability'. *Philosophy of Science* **19** (1952).

Hay, W., 'Review of Carnap's *Continuum of Inductive Methods*'. *Philosophical Review* **62** (1953).

Helmer, O. and Rescher, N., 'On the Logic of the Inexact Sciences'. *Management Studies* **6** (1959).

Hempel, C. G., *Aspects of Scientific Explanation and Other Essays*. New York, 1965.

Hempel, C. G., 'Aspects of Scientific Explanation'. In his *Aspects of Scientific Explanation and Other Essays*.

Hempel, C. G., 'Deductive-Nomological vs. Statistical Explanation'. In Feigl and Maxwell, *Minnesota Studies in the Philosophy of Science*, Vol. III.

Hempel, C. G., 'Studies in the Logic of Confirmation'. In his *Aspects of Scientific Explanation and Other Essays*.

Hempel, C. G. and Oppenheim, P., 'Studies in the Logic of Explanation'. In Brody, *Readings in the Philosophy of Science*.

Hertz, H., *The Principles of Mechanics*, trans. D. Jones and J. Walley, Introduction by R. S. Cohen. New York: Dover, 1956.

Hintikka, J., 'The Modes of Modality'. *Acta Philosophica Fennica* **16** (1963).

Hochberg, H., 'Natural Necessity and Laws of Nature'. *Philosophy of Science* **48** (1981).

Hochberg, H., 'Possibilities and Essences in Wittgenstein's *Tractatus*'. In E. D. Klemke (ed.) *Essays on Wittgenstein*. Urbana, Ill., 1971.

Hochberg, H., '"Possible" and Logical Absolutism'. *Philosophical Studies* **6** (1955)

Hochberg, H., 'Professor Storer on Empiricism'. *Philosophical Studies* **5** (1954).

Hook, S. (ed.) *Determinism and Freedom*. New York, 1961.

318 BIBLIOGRAPHY

Hooker, C. A., 'Philosophy and Meta-Philosophy of Science: Empiricism, Popperianism and Realism'. *Synthese* **32** (1975).

Hume, D., *Enquiries concerning Human Understanding and the Principles of Morals*. Ed. L. A. Selby-Bigge. Second Edition. London, 1902.

Hume, D., 'Of the Standard of Taste'. In Vol. 3 of his *Philosophical Works*, ed. T. H. Green and T. H. Grose, 4 vols. London, 1886.

Hume, D., *Treatise of Human Nature*. Ed. L. A. Selby-Bigge. Oxford, 1888.

Johnson, W. E., *Logic*. In three volumes. London, 1921, 1922, 1924.

Jones, P., *Hume's Sentiments*. Edinburgh, 1982.

Kaminsky, A., 'On Literary Realism'. In J. Halperin, ed., *Theory of the Novel*, London, 1974.

Kaufmann, F., *The Methodology of the Social Sciences*. London, 1944.

Kim, J., 'Causes and Counterfactuals'. In E. Sosa, *Causation and Conditionals*.

Kneale, W., *Probability and Induction*. London, 1949.

Kneale, W., 'Natural Laws and Contrary-to-Fact Conditionals'. *Analysis* **10** (1950).

Kuhn, T., 'Comment on the Relations of Science and Art'. In his *The Essential Tension*.

Kuhn, T., *The Essential Tension*. Chicago, 1977.

Kuhn, T., 'Logic of Discovery or Psychology of Research'. In Lakatos and Musgrave, *Criticism and the Growth of Knowledge*.

Kuhn, T., 'Mathematical vs. Experimental Traditions in the Development of Physical Science'. In his *The Essential Tension*.

Kuhn, T., *The Structure of Scientific Revolutions*, Second Edition. Chicago, 1970.

Lakatos, I., 'Changes in the Problems of Inductive Logic'. In Lakatos, *The Problem of Inductive Logic*.

Lakatos, I., 'Falsifiability and the Methodology of Scientific Research Programmes'. In Lakatos and Musgrave, *Criticism and the Growth of Knowledge*.

Lakatos, I. (ed.) *The Problem of Inductive Logic*. Amsterdam, 1968.

Lakatos, I. and Musgrave A. (eds.) *Criticism and the Growth of Knowledge*. London, 1970.

Laudan, L., *Progress and Its Problems*. Berkeley, 1977.

Leman, M., 'On the Logic and Criteriology of Causality'. *Logique et Analyse* **27** (1984).

Lenz, J., 'Hume's Defence of Causal Inference'. In V. Chappell, *Hume*.

Lewis, D., 'Anselm and Actuality'. *Noûs* **4** (1970).

Lewis, D., 'Causation'. In E. Sosa, *Causation and Conditionals*.

Lewis, D., *Counterfactuals*. Cambridge, Mass., 1973.

Lewis, D., 'Counterpart Theory and Quantified Modal Logic'. In M. Loux, *The Possible and the Actual*.

Lewis, D., 'Probabilities of Conditionals and Conditional Probabilities'. *Philosophical Review* **85** (1976).

Locke, J., *An Essay concerning Human Understanding*. Ed. P. H. Nidditch. London, 1975.

Loux, M. J. (ed.) *The Possible and the Actual*. Ithaca, 1979.

Lycan, W., 'The Trouble with Possible Worlds'. In M. Loux, *The Possible and the Actual*.

Mach, E., *The Principles of Physical Optics*. Trans. J. Anderson and A. Young. New York, 1926.

Mackie, J. L., 'Counterfactuals and Causal Laws'. In Butler, *Analytical Philosophy*, First Series.

Madden, E. H. (ed.) *The Structure of Scientific Thought.* Boston, 1960.

Malebranche, N., *De la recherche de la vérité.* Ed. G. Rodin-Lewis. Paris, 1962.

Mandelbaum, M., *The Anatomy of Historical Knowledge.* Baltimore, 1977.

Mappes, J. and Beauchamp, T., 'Is Hume Really a Sceptic about Induction?'. *American Philosophical Quarterly* **12** (1975).

Mill, J. S., *System of Logic.* Eighth Edition. London, 1872.

Mises, R. von., *Positivism.* New York, 1956.

Mondadori, J. and Morton, A., 'Modal Realism: The Poisoned Pawn'. In M. Loux, *The Possible and the Actual.*

Moore, G. E., *Ethics.* London, 1912.

Nagel, E., 'Carnap's Theory of Induction'. In Schilpp, *The Philosophy of Rudolf Carnap.*

Neurath, O., *Empiricism and Sociology.* Ed. M. Neurath and R. S. Cohen. Dordrecht, Holland, 1973.

Neurath, O., 'Empirical Sociology'. In his *Empiricism and Sociology.*

Norton, D. F., Capaldi, N., and Robison, W. (eds). *McGill Hume Studies.* San Diego, 1976.

Nute, D., 'David Lewis and the Analysis of Counterfactuals'. *Noûs* **10** (1976).

Plantinga, A., 'Actualism and Possible Worlds'. In M. Loux, *The Possible and the Actual.*

Popper, K., *Conjectures and Refutations.* Third Edition. London, 1969.

Popper, K., 'Kant's Critique and Cosmology'. In his *Conjectures and Refutations.*

Popper, K., 'A Note on Laws and So-Called "Contrary-to-Fact" Conditionals'. *Mind*, NS **58** (1949).

Rescher, N., *Hypothetical Reasoning.* Amsterdam, 1964.

Rescher, N., *Scientific Explanation.* New York, 1970.

Rescher, N., 'The Ontology of the Possible'. In M. Loux, *The Possible and the Actual.*

Rosenberg, A., 'Causation and Counterfactuals: Lewis' Treatment Reconsidered'. *Dialogue* **18** (1979).

Russell, B., 'Mysticism and Logic'. In his *Mysticism and Logic.*

Russell, B., *Mysticism and Logic.* London, 1917.

Russell, B., 'The Place of Science in a Liberal Education'. In his *Mysticism and Logic.*

Ryle, G., '"If", "So" and "Because"'. In M. Black, *Philosophical Analysis.*

Salmon, W., 'Statistical Explanation'. In his *Statistical Explanation and Statistical Relevance.*

Salmon, W., *Statistical Explanation and Statistical Relevance.* Pittsburgh, 1971.

Schilpp, P. A. (ed.) *The Philosophy of Rudolf Carnap.* La Salle, Ill., 1963.

Scriven, M., 'Explanation and Prediction in Evolutionary Theory'. *Science* **130** (1959).

Scriven, M., 'Explanations, Predictions and Laws'. In Feigl and Maxwell, *Minnesota Studies in Philosophy of Science*, Vol. III.

Scriven, M., 'Truisms as the Grounds for Historical Explanations'. In Gardner, *Theories of History.*

Sellars, W., 'Counterfactuals'. In E. Sosa, *Causation and Conditionals.*

Sellars, W., 'Is There a Synthetic *A Priori*?' In his *Science, Perception and Reality.*

Sellars, W., *Science, Perception and Reality.* London, 1963.

Sosa, E. (ed.) *Causation and Conditionals.* London, 1975.

Spinoza, B., *Ethics.* In *The Chief Works,* trans. R. Elwes, New York, 1951.

Stalnaker, R., 'Possible Worlds'. In M. Loux, *The Possible and the Actual.*

Stalnaker, R., 'A Theory of Conditionals'. In E. Sosa, *Causation and Conditionals.*

Strawson, P. F., *Introduction to Logical Theory.* London, 1952.

Sumner, L. W., Slater, J. G., and Wilson, F. (eds.) *Pragmatism and Purpose.* Toronto, 1980.

Suppe, F., 'Afterword — 1977', to his *The Structure of Scientific Theories.*

Suppe, F., 'The Search for Philosophic Understanding of Scientific Theories'. In his *The Structure of Scientific Theories.*

Suppe, F. (ed.) *The Structure of Scientific Theories,* Second Edition. Urbana, Ill., 1977.

Swartz, N., *The Concept of Physical Law.* Cambridge, 1985.

Taylor, T. (trans.) *The Philosophical and Mathematical Commentaries of Proclus.* Two Volumes. London, 1788—89.

Tooley, M., 'The Nature of Laws'. *Canadian Journal of Philosophy* **7** (1977).

Veblen, T., 'The Place of Science in Modern Civilization'. In his *The Place of Science in Modern Civilization and Other Essays.* New York, 1930.

Watkins, J. W. N., 'The Paradoxes of Confirmation'. In B. Brody, *Readings in the Philosophy of Science.*

Weinberg, J., *Abstraction, Relation and Induction.* Madison, Wisc., 1965.

Weinberg, J., 'Induction'. In his *Abstraction, Relation and Induction.*

Wilson, F., 'Acquaintance, Ontology and Knowledge'. *New Scholasticism* **54** (1970).

Wilson, F., 'Addison Analysing Disposition Concepts', *Inquiry* **28** (1985).

Wilson, F., 'Barker on Geometry as *A Priori*'. *Philosophical Studies* **20** (1969).

Wilson, F., 'Dispositions: Defined or Reduced?' *Australasian Journal of Philosophy* **47** (1969).

Wilson, F., 'Effability, Ontology and Method'. *Philosophy Research Archives* **9** (1983).

Wilson, F., *Empiricism and Darwin's Science.* In preparation.

Wilson, F., *Explanation, Causation and Deduction.* Dordrecht, Holland, 1985.

Wilson, F., 'Explanation in Aristotle, Newton and Toulmin'. *Philosophy of Science* **36** (1969).

Wilson, F., 'Goudge's Contribution to Philosophy of Science'. In Sumner, Slater, and Wilson, *Pragmatism and Purpose.*

Wilson, F., 'Dispositions Defined: Harré and Madden on Defining Disposition Concepts', *Philosophy of Science* **52** (1985).

Wilson, F., 'Hume and Ducasse on Causal Inference from a Single Experiment'. *Philosophical Studies* **35** (1979).

Wilson, F., 'Hume's Cognitive Stoicism'. *Hume Studies,* 1985 Supplement.

Wilson, F., 'Hume's Defence of Causal Inference'. *Dialogue* **22** (1983).

Wilson, F., 'Hume's Defence of Science', *Dialogue,* forthcoming.

Wilson, F., 'Hume's Theory of Mental Activity'. In D. F. Norton *et al., McGill Hume Studies.*

Wilson, F., 'Implicit Definition Once Again'. *Journal of Philosophy* **62** (1965).

Wilson, F., 'Is There a Prussian Hume?' *Hume Studies* **8** (1982).

Wilson, F., 'Kuhn and Goodman: Revolutionary vs. Conservative Science'. *Philosophical Studies* **44** (1983).

Wilson, F., 'The Lockean Revolution in the Theory of Science'. In *Early Modern Philosophy: Epistemology, Metaphysics and Politics*, ed. S. Tweyman and G. Moyal. Delmar, N.Y., 1985.

Wilson, F., 'Mill on the Operation of Discovering and Proving General Propositions'. *Mill News Letter* **17** (1982).

Wilson, F., 'Mill's Proof that Happiness Is the Criterion of Morality'. *Journal of Business Ethics* **1** (1982).

Wilson, F., 'A Note on Hempel on the Logic of Reduction'. *International Logic Review* **13** (1982).

Wilson, F., 'The Notion of Logical Necessity in Carnap's Later Philosophy'. In A. Hausman and F. Wilson, *Carnap and Goodman: Two Formalists.* The Hague, 1967.

Wilson, F., *Reasons and Revolutions.* In preparation.

Wilson, F., 'Resemblance, Identity and Universals: Comments on March on Sorting Out Sorites'. *Canadian Journal of Philosophy*, forthcoming.

Wilson, F., 'Critical Notice of I. Hacking's *The Emergence of Probability*'. *Canadian Journal of Philosophy* **8** (1978).

Wilson, F., 'Review of N. Rescher's *Scientific Explanation*'. *Dialogue* **11** (1972).

Wilson, F., 'Review of J. Yolton's *Philosophical Analysis*'. *Philosophy of Science* **37** (1970).

Wilson, F., 'Review of M. Mandelbaum's *The Anatomy of Historical Knowledge*'. *Philosophical Review* **88** (1979).

Wilson, F., 'Review of N-Swartz's *The Concept of Physical Law*'. *Philosophy of Science*, forthcoming.

Wilson, F., 'Review of Y. Gauthier's *Méthodes et concepts de la logique formelle* and R. Hébert's *Mobiles du discours philosophique*'. *University of Toronto Quarterly* **48** (1979).

Wilson, F., 'The Role of a Principle of Acquaintance in Ontology'. *The Modern Schoolman* **47** (1969).

Wilson, F., 'Weinberg's Refutation of Nominalism'. *Dialogue* **8** (1969).

Wright, G. H. von, *The Logical Problem of Induction*, Second Edition. London, 1965.

Wright, G. H. von, 'On the Logic and Epistemology of the Causal Relation'. In E. Sosa, *Causation and Conditionals.*

# INDEX OF NAMES

Aaron, R., 83
Adams, E., 289
Aristotle, ix, 34, 82, 83
Armstrong, D. M., 111–114

Bacon, F., 12, 277
Bain, A., 3
Bennett, J., 209–211, 213, 214
Bergmann, G., 145–147
Blake, W., 2
Bode, J. E., 151
Bradley, F. H., ix
Bromberger, S., 18
Butler, R., 93

Campbell, G., 3
Campbell, N. R., 170
Capaldi, N., xv
Carnap, R., 123, 124, 128–131, 148, 155, 156, 159, 165, 254
Chisholm R., ix, 88, 100, 202, 203, 206
Cooley, J., 256, 257
Coulomb, C. A. de, 35, 52, 53, 130

Darwin, C., 106
Descartes, R., 82, 83

Einstein, A., 70, 279
Euclid, 1

Fraassen, B. C. van, 23
Frege, G., 145

Galilei, Galileo, 5, 34
Goodman, N., ix, x, xiv, 169, 195, 202, 203, 206, 207, 244–246, 248, 250–252, 254, 256, 259, 262, 264, 266, 287
Greeno, J., 23

Habermas, N., 4
Hanson, N. R., 60–65, 70

Hempel, C. G., 12–17, 24, 28, 155, 156, 165
Hertz, H., 40
Hooke, R., 35, 41, 156
Hooker, C. A., 8, 9

Kneale, W., 94, 95, 99, 100, 105, 111
Kuhn, T., x, 42, 50, 60, 61, 65–67, 69, 70, 277, 279

Lakatos, I., 69, 116, 170
Lewis, David, xiii–xv, 133–288 (= Ch. III).
Locke, J., xi, 1, 2, 82–86

Malbranche, N., 86
Maxwell, J. C., 70
Mill, J. S., ix, 12, 131
Molière, 84
Mondadori, J., 273
Morton, A., 273

Neurath, O., 4, 5
Newton, I., 3, 9, 29, 35, 39, 53, 54, 130, 159, 211

Oppenheim, P., 12—17, 24, 28

Peacock, T. L., 1
Plato, 83
Popper, K., 27, 49, 65, 94, 95, 266, 277
Proclus, 1

Quine, W. V. O., 141–144, 149

Rescher, N., 261–272, 275, 276
Retz, C. de, 91
Rosenberg, A., 155–163, 165, 279
Russel, B., 2, 4, 5, 45

Salmon, W., 23, 24, 26
Scriven, M., 17–22, 102
Sellars, W., 249
Spinoza, B. de, 2
Stalnaker, R., 168, 197–199, 201–208, 217, 218, 263, 273, 278, 280
Suppe, F., 60, 61

Taylor, T., 1–3, 11, 83

Tooley, M., 111–114, 117, 118, 120, 123–132

Veblen, T., 4

Waals, J. D. van der, 42, 117, 130, 166, 167
Weber, M., 31
Wittgenstein, L., 148

# INDEX OF SUBJECTS

acceptance of generalities
  as hypotheses for research purposes (=
      acceptance₁) 36, 56–7, 61–6,
      69, 277
  as means to knowledge 37
  pragmatic interest in 37
  use in explanation, prediction and as-
      serting counterfactuals (= ac-
      ceptance₂) 36, 50, 56–7, 61–6,
      69, 277
accidental generalities xi–xii, 81, 99–111
  and belief-contravening assump-
      tions 98–9, 105, 107
  cannot be used to sustain assertion of
      counterfactual conditionals xii,
      81
  objectively considered
      not different from laws xi, 81, 87, 96
      have explanatory and predictive po-
          wer of laws xii, 97
  *See also*: laws and accidental generali-
      ties
acquaintance, principle of xi, 125–8, 146,
      147
  Hume's appeal to 87
  Locke's appeal to 85–6
actuality
  contains all possible worlds 133, 143,
      145–7
  as unanalyzable property of possible
      worlds 142, 145

barometer, as counterexample
  to analysis of counterfactual conditions
      in terms of laws 102, 230, 272
  to symmetry of explanation and
      prediction 20–2
belief-contravening assumptions xiii, xiv,
      98–9, 105, 107, 111, 243–80
  rule that laws are to be retained 262
      often justified 268
      sometimes unjustified 271–3, 276–80

causation
  asymmetries due to interest in
      manipulation 18–23, 228, 232,
      284, 288
  and correlation 21, 234, 268, 269, 288
  as deduction from laws 133
  as not involving laws 133, 152
  in terms of counterfactual dependen-
      ce 152, 225–43
common cause, principle of 23–7
conditionals: use in inferences 74, 174
conditionals, counterfactual
  in analysis of causal dependence 152
  analysis involving laws xiv, 79, 133,
      192, 242
      ambiguous on analysis 196, 243,
          285–8
      and belief-contravening assump-
          tions 109–11, 243–80
      as condensed arguments 79, 139,
          223, 247
      connective with principle of predicta-
          bility 248, 263
      lack truth conditionals 139
      normative defence of, in terms of
          cognitive interests 140, 191–5,
          274–80
      proposed counterfactuals xiv, 182,
          199, 203–43, 280–8
      results close to analysis in terms of
          possible worlds 260, 268
      squares best with ordinary
          usage 219–24, 278, 284
  analysis in terms of possible worlds xiv,
      80, 133–9, 172–95, 289
      abiguous upon analysis 197, 279
      evidence against 184, 193, 219–23,
          240, 278, 280
      evidence for 181, 199, 203–15, 225–43
      proponents do not take into account
          accepted cognitive interests
          274–80

rejected by reference to relevant cognitive interests 173, 191–5, 274–80
results close to analysis involving laws 260, 268
truth conditions 139
as variably strict conditionals 136–9, 221
assertion of
in context of research 277
requires rules for belief-contravening assumptions xii, 109–11, 243–80
sustained by causal laws vs. sustained by non-causal laws 268
sustained by laws xii, 10, 81, 87
not sustained by accidental generalities xii, 81, 87
denials of 192, 202, 223–4, 245
and factual conditionals 200
and material conditionals 72–3, 176–9
and predictive conditionals 76, 188
problem of
actual connection presupposed 73, 80
apparent reference to unrealized possibilities 72, 73, 141–50
approach of modal logic and set theory ix, 289
- philosophically problematic ix
empiricist approach to ix, 72, 289
- related to deductive-nomological model of explanation x, 77, 190, 288
- resources considered too meagre ix
part of philosophy of language ix, 289
part of philosophy of science ix, 289
conditionals, factual 77
conditionals, material 72, 208
assertion odd in absence of connection 73, 75, 177
conditions for assertion 73–81, 176–95
rhetorical use 76, 178, 262–6
conditionals, predictive 76, 179, 188
and counterfactual conditionals 76, 179, 188, 289
conditionals, semifactual 199, 206, 241, 243

and counterfactual conditionals 200, 206
control, idea of
and conditionals 77, 140, 191–5, 274–80, 284, 288
and explanation 6–8, 12, 77
cotenability xiv, 195, 209, 213, 222, 244–55
analysis of 252–5
ties in with notion of process law 253
condition imposed by purposes of counterfactual discourse 209, 213, 244
and principle of predictability 246, 248

descriptions, definite
appealed to in defence of objective necessary connections 126–32
role in explanations 46, 56, 187
role in theories 45, 56
Determinism, Principle of 35, 123, 129
and imperfect law 35
not justified a priori 126–7
role in theories 35, 43–6, 67–71, 130
supported a posteriori 130–2
discovery, logic of: see research

electrostatics: example of a theory 52, 130
explanation
basic idea is control 6–9, 12, 77, 80
better and worse 22, 47, 288 (see also explanation, ideal)
contrasted with nonexplanation 16, 23
deductive-nomological model x, 13–16, 47, 48, 56, 133
connected to use of statistics 24
defined by principle of predictability 14, 16
Humean limits 27
law premiss essential for 15, 16, 48
purported counter-example 17–27
related to counterfactual conditionals x, 10, 80
related to factual conditionals 77
ex post facto 25, 47, 57
and deductive-nomological model 25, 48

often involves definite descriptions 45, 56
only apparently uninformative 48
and use of statistics 24–8
ideal 29–34
    constituted by process knowledge 29
    determined by cognitive interests 30
    - and imperfect: see laws, imperfect
    - and prediction 14–28, 46
    purported asymmetrical cases 18, 20, 46
    symmetrical 14
    see also predictability, principle of; reason for being; reason for expecting
statistical, so called 12, 23, 26

generalities: see accidental generalities; laws; laws and accidental generalities

Humean, the ix–xiv, 27, 60, 65, 68, 69, 72, 81, 82, 88, 89, 92–101, 105, 110–14, 117, 118, 129–34, 139–41, 150–2, 155, 171–6, 179–87, 190, 192–224, 232–7, 240–3, 259, 260, 269, 280–9

interests, cognitive
idle curiosity and pragmatic 4, 30, 33
in laws that permit control 6–8, 12, 77, 80
in regularities as timeless patterns 2, 9, 11
in truth 11
none in mystical necessary connections 2, 3
not taken into account in proposed analysis of counterfactuals of possible world 274
used to justify analysis of counterfactuals involving laws 140, 173, 191–5, 245–51, 260, 263–4, 274–80
used to justify norms for scientific explanation x, 8, 30, 92
used to justify norms for scientific method 92

used to reject possible world's analysis of counterfactuals 173, 191–5, 274–8
intuitive "know-how": role in science 32

language of science
empiricist's 56, 101, 102, 146, 147, 150, 151
excludes objective necessary connections 112
has logical framework of Principia Mathematica 16, 112, 146, 147, 255
laws
acceptance of 36–40, 92, 93
analysis of counterfactual conditionals inseparable from ix, xiv, 80, 192, 194, 195
analysis of counterfactual conditionals which does not involve xiv, 80, 133–9
analysis of, as regularities that fit into axiomatic system 134, 151–5, 168–72
    inadequacy of 169–72
cognitive interest in 9
do not have simplistic logical form 50, 65, 156, 165, 240
and factual conditionals 77
generalities are worthy of acceptance as when evidence is scientific xi, 88–9, 91–3, 99–111
Humean limits of 27–8, 68, 89–99, 244, 254
are hypotheses 37, 38 (see also Humean limits of)
are a matter-of-fact regularities xi, 2, 7
may relate physical and non-physical factors 20
and predictive conditionals 76, 188
provide knowledge of means to and reasons for being 11, 18, 20
provide reasons for expecting 11
and semifactual conditionals 200, 206
sustain assertion of counterfactual conditionals xii, 10, 79–83, 97
are timeless patterns 2, 9, 11
see also: laws and accidental generalities